Advances in Prevention of Foodborne Pathogens of Public Health Concern during Manufacturing

Advances in Prevention of Foodborne Pathogens of Public Health Concern during Manufacturing

Special Issue Editors

Aliyar Cyrus Fouladkhah
Bledar Bisha

MDPI • Basel • Beijing • Wuhan • Barcelona • Belgrade

MDPI

Special Issue Editors

Aliyar Cyrus Fouladkhah Bledar Bisha
Tennessee State University University of Wyoming
USA USA

Editorial Office
MDPI
St. Alban-Anlage 66
4052 Basel, Switzerland

This is a reprint of articles from the Special Issue published online in the open access journal *Microorganisms* (ISSN 2076-2607) from 2018 to 2019 (available at: https://www.mdpi.com/journal/microorganisms/special_issues/foodborne_pathogens_manufacturing).

For citation purposes, cite each article independently as indicated on the article page online and as indicated below:

LastName, A.A.; LastName, B.B.; LastName, C.C. Article Title. *Journal Name* **Year**, *Article Number*, Page Range.

ISBN 978-3-03921-932-2 (Pbk)
ISBN 978-3-03921-933-9 (PDF)

Contents

About the Special Issue Editors

Aliyar Cyrus Fouladkhah is a graduate of the Food Science and Human Nutrition Master's and Food Microbiology Doctoral programs of Colorado State University (CSU). He holds a Graduate Certificate in Applied Statistics and Data Analysis from CSU's Statistics Department and completed his training in the Public Health Department of Yale University in the Advanced Professional MPH program (AP MPH), tracked in Applied Biostatistics and Epidemiology, a Graduate Certification in Food and Drug Regulatory Affairs, and a Certificate in Climate Change and Human Health.

He has been a member of an authors' team of over 100 peer-reviewed publications, conference proceedings, newsletters, and popular-press articles. He is currently serving as the faculty director of Public Health Microbiology laboratory in Nashville, where his program has been extramurally funded by federal and private industry agencies and provided education opportunities for undergraduate, MS, Ph.D., and post-doctoral students/fellows. He has served as president of Rocky Mountain (Colorado) and Volunteer (Tennessee) sections of IFT and as exam item writer and reviewer for both certified food scientists and certified in public health national examinations and as instructor of public health workshops in Guatemala, Dominican Republic, and South Africa. He is currently serving as the Chair of Health and Medical Sciences division of Tennessee Academy of Science.

Bledar Bisha is an Associate Professor of food microbiology and interim department head in the Department of Animal Science at the University of Wyoming. His expertise is in the control and detection of foodborne pathogens. He received his veterinary degree from the Agricultural University of Tirana, Albania, and his M.Sc. and Ph.D. degrees in Food Science and Technology (Food Microbiology) from Iowa State University, Ames, IA. Dr. Bisha later pursued his postdoctoral studies at Colorado State University, Ft. Collins, CO.

His current research focus at the University of Wyoming is on the detection and control of foodborne pathogens and sample preparation for rapid microbial detection, primarily involving the utilization of microfluidics (paper-based analytical devices), bacteriophage, and mass spectrometry-based methods. Additionally, Dr. Bisha conducts research on the ecology and epidemiology of foodborne pathogens (including antimicrobial resistance) and on novel control methods for these pathogens in food, including physical, biological, and chemical methods. He has authored over 34 peer-reviewed journal articles and book chapters, eight conference and popular articles, and over 70 presentations. Dr. Bisha serves on the editorial boards of two scientific journals and serves as an ad hoc reviewer for multiple other peer reviewed journals.

microorganisms

MDPI

Editorial

Safety of Food and Water Supplies in the Landscape of Changing Climate

Aliyar Cyrus Fouladkhah [1,*], Brian Thompson [2] and Janey Smith Camp [3]

[1] Public Health Microbiology Laboratory, Tennessee State University, Nashville, TN 37209, USA
[2] School of Public Health, Yale University, 60 College St, New Haven, CT 06510, USA;
   brian.thompson@yale.edu
[3] Department of Civil and Environmental Engineering, Vanderbilt University, Nashville, TN 37235, USA;
   janey.camp@vanderbilt.edu
* Correspondence: afouladk@tnstate.edu or aliyar.fouladkhah@aya.yale.edu; Tel.: +1-970-690-7392

Received: 15 September 2019; Accepted: 16 October 2019; Published: 18 October 2019

In response to evolving environmental, production, and processing conditions, microbial communities have tremendous abilities to move toward increased diversity and fitness by various pathways such as vertical and horizontal gene transfer mechanisms, biofilm formation, and quorum sensing [1,2]. As such, assuring the safety of water and food supplies from various natural and anthropogenic microbial pathogens is a daunting task and a moving target. Recent outbreaks of *Listeria monocytogenes* in South Africa associated with a ready-to-eat product (affecting close to 1000 individuals) and the 2018 outbreak of Shiga toxin-producing *Escherichia coli* O26 associated with ground meat in the United States (leading to the recall of more than 132,000 pounds of products) are bitter reminders of the devastating influences of foodborne diseases on the public health and food manufacturing [3,4].

Recent epidemiological studies of world populations indicate that 420,000 people lose their lives every year due to foodborne diseases, with around one-third of those being 5 years of age or younger. It is further estimated that every year, 1 in 10 individuals experience foodborne diseases around the globe, leading to an annual loss of 33 million healthy life years [5]. These episodes of food and water illnesses, hospitalizations, and deaths are concerns for both developing economies and developed nations. In the United States, as an example, epidemiological data derived from active surveillance data of the Centers for Disease Control and Prevention reveals that every year 31 main foodborne pathogens cause 9.4 million episodes of illnesses and about 56,000 cases of hospitalizations, leading to at least 1351 deaths of American adults and children [6].

In addition to these public health challenges, foodborne diseases are a major cause of consumer insecurity and economic burden to private industry, healthcare facilities, and government agencies due to costs associated with medical treatments and secondary costs related to food recalls and outbreak investigations [1]. Foodborne nontyphoidal *Salmonella enterica* serovars, as an example, cause an estimated 1,027,561 illnesses annually in the United States, with 27.2% and 0.5% hospitalization and death rates, respectively [6], leading to annual public health burden of 32,900 disability-adjusted life years [7]. Similarly, from 1998 to 2018, the bacterium had been the causal agent of >2500 single or multi-state outbreaks in the United States [8]. Overall, the cost of foodborne diseases is estimated to be $77.7 billion annually in the United States [9].

In addition to economic losses, consumers' insecurity, and hospitalization, illness, and death episodes, victims of foodborne diseases may suffer prolonged and potentially life-long health complications after exposures to foodborne pathogens. Some of these main sequelae are Guillain–Barré syndrome, reactive arthritis, post-infectious irritable bowel syndrome, hemolytic uremic syndrome, and end-stage renal disease that could occur after infections with foodborne pathogens such as *Campylobacter* spp., *Salmonella enterica* serovars, and various serogroups of Shiga toxin-producing

Escherichia coli. These additional public health burdens are calculated using epidemiological metrics such as the above-mentioned disability-adjusted life year [7].

Changes in the climate will unequivocally have pronounced effects on the proliferation of microbial pathogens and consequently the prevalence of foodborne diseases. As an example, it has been reported that only a 1 °C increase (above 5 °C) in temperature of an environment could lead to 5% to 10% increase in cases of salmonellosis [10]. In the United States alone, a 5% increase in illness episodes could translate to >50,000 additional cases of illnesses of nontyphoidal *Salmonella* serovars every year.

Similarly, the safety of water supplies is also interconnected with the changing climate. The World Health Organization estimates that approximately 2 million deaths each year are attributed to waterborne diarrheal diseases, with the vast majority of these deaths occurring in children [11]. This is largely attributed to the fact that 785 million people lack basic drinking-water service, with 144 million of these people reliant upon surface water [12]. Climatic conditions such as flooding and drought can influence the fate and transport of pathogenic microorganisms, as well as their fate and proliferation rates in the environment. The potential impacts of climate change on water supplies are primarily centered on anticipated changes in precipitation and increasing temperatures.

Increased precipitation can lead to runoff and flooding. Increased nutrient loading of surface waters due to runoff in both urban and rural areas coupled with warm temperatures can contribute to increased multiplication of cyanobacterial blooms and their harmful counterparts [13–15]. Flooding is attributed to increased risk of gastrointestinal illness when ground and surface sources for drinking water are impacted and not treated sufficiently. This presents potential concerns for citizens worldwide that do not have access to treated drinking water and may also present challenges in conventional treatment processes. During flooding events, surface and ground waters can become contaminated by sewer flooding and overflows that can result in higher risk of exposure to enteric pathogens [16]. In fact, there is a significant historic correlation between extreme rainfall events and outbreaks of waterborne diseases [17]. Conversely, drought can affect river flows, flushing rates, and eutrophication processes, which can lead to increased concentrations of Cyanobacteria and pathogens attributed to diarrheal diseases [11,18].

Surface water temperatures in streams have been shown to directly correlate to ambient air temperatures [19,20]. Therefore, one can anticipate that increasing ambient temperatures caused by climate change will in turn increase temperatures of surface waters, which serve as sources for drinking water, agricultural irrigation, and other domestic purposes that impact human health, especially in developing nations where drinking water treatment might not be as ubiquitous.

Climate change is one of the most significant challenges facing the public health and the safety and security of our food and water supplies. Without a major overhaul of our current energy production, political, and transportation systems, there will continue to be massive greenhouse gasses (GHGs) emissions into the atmosphere, further driving the changes in the climate. Beyond that, inertia in the climate systems will force continued climate change irrespective of GHG emission abatements [21]. Therefore, it is imperative that we better understand the risks to the safety of our food and water supply posed by climate change for the conduct of vulnerability assessments and the development of climate mitigation, adaption, and resilience programs.

Human-emitted GHGs are driving climate change [22] and altering many planetary systems in potentially irrevocable ways (e.g., the melting of the cryosphere, the warming of the oceans, changing rainfall patterns, etc.) [23]. Given that the climate will continue to warm throughout at least the first half of the 21st century [21], it is crucial to project the effects of future climate change. The Intergovernmental Panel on Climate Change (IPCC) has projections on the climatic effects of climate change across a range of different GHGs emissions scenarios (i.e., RCP2.6, RCP4.5, RCP6.0, RCP8.5) [23]. More GHGs emissions will result in an increased average surface temperature, greater precipitation, and higher sea levels. These consequences of future climate change can work individually or synergistically to threaten the safety of our food and water supply by impacting the fate and proliferation of foodborne and waterborne pathogens.

While the direct link between climate change and infectious diseases is inherently not characterized [24,25], we can infer their relationship by assimilating the impact of climatic factors and these diseases [26]. Many foodborne and waterborne diseases show strong cyclical periodicity based on precipitation and temperature—factors that are impacted by climate change [26,27]. The large rainfall events that will become commonplace due to climate change will challenge the safety of our water supply by causing sanitary and combined sewer overflows [28–30]. Further, these large rainstorms spread etiological agents of viral, parasitic, and bacterial infections [20,26,31]. To highlight one challenge, climate change is increasing sea surface temperature and causing sea level rise, both of which could fuel cholera outbreaks [32]. The increases in sea surface temperature promote greater *Vibrio* multiplication, as an example, and the rises in sea level could facilitate *Vibrio* infiltration into local water sources.

The public health and our food production and processing infrastructures in the 21st century will undoubtedly face paramount challenges due to global warming and subsequent changes in environmental conditions. Emerging and re-emerging zoonotic infectious diseases and subsequently increases in pesticides and veterinary drugs use and residues; increases in the prevalence of drug-resistant microorganisms in the food chain and healthcare facilities; the enhanced proliferation and prevalence of waterborne and foodborne bacteria, viruses, and parasitic agents in various regions and commodities; increases in the prevalence of toxigenic fungi and mycotoxins in the production environment and the food and feed chain; and increases in harmful algal blooms affecting fishery products will undoubtedly represent crucial challenges to our water and food safety and security in the 21st century. These will almost certainly affect the vulnerable populations from developing countries the most—those who have contributed the least to the current changes in the climate. Susceptible and at-risk populations, including the very young, elderly, pregnant women, and the immunocompromised, will also be most severely affected by this main public health challenge of our time.

Without intervention at the population level, the availability, access, utilization, and stability of an array of food and agricultural crops and water resources could almost certainly be jeopardized in the landscape of changing climate [33]. Although solutions to these challenges are inherently a moving target, the genetic wealth of plant, animal, and aquatic species could be a great resource for the development of climate resilience, adaption, and mitigation programs [34]. Developing evidence-based food and agricultural systems for climate change mitigation, expanding adaption programs tailored for small and emerging entrepreneurs, strengthening regional and international cooperation, and financing climate-smart food and agricultural systems are some of the current proposed interventions [35].

The current special issue provides a collection of research and review articles that discuss mitigating and prevention strategies associated with some of the most important foodborne and waterborne pathogens in the United States and around the globe. The public health burden of these pathogens will continue to gain further importance and momentum in future years in the landscape of the changing climate.

Author Contributions: A.C.F., B.T., and J.S.C. co-wrote, revised, and edited the manuscript.

Funding: Financial support in part from the National Institute of Food and Agriculture of the United States Department of Agriculture (Projects 2017-07534; 2017-07975; 2017-06088) and information from Climate Reality Leadership Corps is acknowledged gratefully by the corresponding author.

Acknowledgments: Technical contributions and administrative support of the members of the Public Health Microbiology Laboratory is sincerely appreciated by the authors. The authors also appreciate the administrative support of the editorial team of *Microorganisms*.

Conflicts of Interest: The authors declare no conflict of interest. The content of the current publication does not necessarily reflect the views of the funding agencies.

References

1. Fouladkhah, A. The Need for Evidence-Based Outreach in the Current Food Safety Regulatory Landscape Commentary Section. *J. Ext.* **2017**, *55*, 2COM1.

2. Fouladkhah, A. Meat safety: Past, present, and future outlook. In *The Marketplace: Strategies for Today and Tomorrow, Proactive Strategies to Deal with Changes*; White Paper Provided to 2011 International Livestock Congress; Woerner, W.C., Ed.; ILC-USA: Denver, CO, USA, 2011.

3. Centers for Disease Control and Prevention. Outbreak of *E. coli* Infections Linked to Ground Beef. September 2018. Available online: https://www.cdc.gov/media/releases/2018/s0920-recalled-ground-beef.html (accessed on 17 October 2019).

4. World Health Organization Disease Outbreak News. Listeriosis—South Africa. 2018. Available online: https://www.who.int/csr/don/28-march-2018-listeriosis-south-africa/en/ (accessed on 17 October 2019).

5. World Health Organization. Global Burden of Food Safety. 2015. Available online: http://www.who.int/foodsafety/areas_work/foodborne-diseases/ferg/en/ (accessed on 17 October 2019).

6. Scallan, E.; Hoekstra, R.M.; Angulo, F.J.; Tauxe, R.V.; Widdowson, M.A.; Roy, S.L.; Jones, J.L.; Griffin, P.M. Foodborne illness acquired in the United States—Major pathogens. *Emerg. Infect. Dis.* **2011**, *17*, 7–15. [CrossRef] [PubMed]

7. Scallan, E.; Hoekstra, R.M.; Mahon, B.E.; Jones, T.F.; Griffin, P.M. An assessment of the human health impact of seven leading foodborne pathogens in the United States using disability adjusted life years. *Epidemiol. Infect.* **2015**, *143*, 2795–2804. [CrossRef] [PubMed]

8. Allison, A.; Daniels, E.; Chowdhury, S.; Fouladkhah, A. Effects of elevated hydrostatic pressure against mesophilic background microflora and habituated *Salmonella* serovars in orange juice. *Microorganisms* **2018**, *6*, 23. [CrossRef]

9. Scharff, R.L. Economic burden from health losses due to food-borne illness in the United States. *J. Food Prot.* **2012**, *75*, 123–131. [CrossRef] [PubMed]

10. World Health Organization. Food Safety—Climate Change and the Role of WHO. 2019. Available online: https://www.who.int/foodsafety/publications/all/Climate_Change_Document.pdf?ua=1 (accessed on 17 October 2019).

11. World Health Organization. Health and Sustainable Development—Waterborne Disease Related to Unsafe Water and Sanitation. 2019. Available online: https://www.who.int/sustainable-development/housing/health-risks/waterborne-disease/en/ (accessed on 17 October 2019).

12. World Health Organization. Fact Sheet: Drinking Water. 2019. Available online: https://www.who.int/news-room/fact-sheets/detail/drinking-water (accessed on 17 October 2019).

13. Paerl, H.W.; Huisman, J. Blooms like it hot. *Science* **2008**, *320*, 57–58. [CrossRef]

14. Paerl, H.W.; Hall, N.S.; Calandrino, E.S. Controlling harmful cyanobacterial blooms in a world experiencing anthropogenic and climatic-induced change. *Sci. Total Environ.* **2011**, *409*, 1739–1745. [CrossRef]

15. Hunter, P.R. Climate change and waterborne and vector-borne disease. *J. Appl. Microbial.* **2003**, *94*, 37–46. [CrossRef]

16. Ten Veldhuis, J.A.E.; Clemens, F.H.L.R.; Sterk, G.; Berends, B.R. Microbial risks associated with exposure to pathogens in contaminated urban flood water. *Water Res.* **2010**, *44*, 2910–2918. [CrossRef]

17. Curriero, F.C.; Patz, J.A.; Rose, J.B.; Lele, S. The association between extreme precipitation and waterborne disease outbreaks in the United States, 1948–1994. *Am. J. Public Health* **2011**, *91*, 1194–1199. [CrossRef]

18. Newcombe, G.; Chorus, I.; Falconer, I.; Lin, T.F. Cyanobacteria: Impacts of climate change on occurrence, toxicity and water quality management. *Water Res.* **2012**, *46*, 1347. [CrossRef] [PubMed]

19. Morrill, J.C.; Bales, R.C.; Conklin, M.H. Estimating stream temperature from air temperature: Implications for future water quality. *J. Environ. Eng.* **2005**, *131*, 139–146. [CrossRef]

20. Cann, K.F.; Thomas, D.R.; Salmon, R.L.; Wyn-Jones, A.P.; Kay, D. Extreme water-related weather events and waterborne disease. *Epidemiol. Infect.* **2013**, *141*, 671–686. [CrossRef] [PubMed]

21. Watson, R.T.; Albritton, D.L.; Barker, T.; Bashmakov, I.; Canziani, O.F.; Christ, R.; Cubasch, U.; Davidson, O.R.; Gitay, H.; Griggs, D.J.; et al. *Climate Change 2001: Synthesis Report*; A Contribution of Working Groups I, II and III to the Third Assessment Report of the Intergovernmental Panel on Climate Change; IPCC: Geneva, Switzerland, 2011.

22. Pachauri, R.K.; Allen, M.R.; Barros, V.R.; Broome, J.; Cramer, W.; Christ, R.; Church, J.A.; Clarke, L.; Dahe, Q.; Dasgupta, P.; et al. *Climate Change 2014: Synthesis Report*; Contribution of Working Groups I, II and III to the Fifth Assessment Report of the Intergovernmental Panel on Climate Change; IPCC: Geneva, Switzerland, 2014; p. 151.

23. Stocker, T.F.; Qin, D.; Plattner, G.K.; Tignor, M.; Allen, S.K.; Boschung, J.; Nauels, A.; Xia, Y.; Bex, V.; Midgley, P.M. *Climate Change 2013: The Physical Science Basis*; IPCC: Geneva, Switzerland, 2013.

24. Liang, L.; Gong, P. Climate change and human infectious diseases: A synthesis of research findings from global and spatio-temporal perspectives. *Environ. Int.* **2017**, *103*, 99–108. [CrossRef]

25. Kolstad, E.W.; Johansson, K.A. Uncertainties associated with quantifying climate change impacts on human health: A case study for diarrhea. *Environ. Health Perspect.* **2010**, *119*, 299–305. [CrossRef]

26. Semenza, J.C.; Herbst, S.; Rechenburg, A.; Suk, J.E.; Höser, C.; Schreiber, C.; Kistemann, T. Climate change impact assessment of food-and waterborne diseases. *Crit. Rev. Env. Sci. Tech.* **2012**, *42*, 857–890. [CrossRef]

27. Hashizume, M.; Faruque, A.S.; Wagatsuma, Y.; Hayashi, T.; Armstrong, B. Cholera in Bangladesh: Climatic Components of Seasonal Variation. *Epidemiology* **2010**, *21*, 706–710. [CrossRef]

28. Jagai, J.S.; DeFlorio-Barker, S.; Lin, C.J.; Hilborn, E.D.; Wade, T.J. Sanitary sewer overflows and emergency room visits for gastrointestinal illness: Analysis of Massachusetts data, 2006–2007. *Env. Health Perspect.* **2017**, *25*, 117007. [CrossRef]

29. Donovan, E.; Unice, K.; Roberts, J.D.; Harris, M.; Finley, B. Risk of gastrointestinal disease associated with exposure to pathogens in the water of the Lower Passaic River. *Appl. Environ. Microbiol.* **2008**, *74*, 994–1003. [CrossRef]

30. U.S. Environmental Protection Agency Report to Congress on Impacts and Control of Combined Sewer Overflows and Sanitary Sewer Overflows. 2004. Available online: https://www.epa.gov/sites/production/files/2015-10/documents/csossortc2004_full.pdf (accessed on 17 October 2019).

31. Schwartz, B.S.; Harris, J.B.; Khan, A.I.; Larocque, R.C.; Sack, D.A.; Malek, M.A.; Faruque, A.S.; Qadri, F.; Calderwood, S.B.; Luby, S.P.; et al. Diarrheal epidemics in Dhaka, Bangladesh, during three consecutive floods: 1988, 1998, and 2004. *Am. J. Trop. Med. Hyg.* **2006**, *74*, 1067–1073. [CrossRef]

32. Lobitz, B.; Beck, L.; Huq, A.; Wood, B.; Fuchs, G.; Faruque, A.S.G.; Colwell, R. Climate and infectious disease: Use of remote sensing for detection of Vibrio cholerae by indirect measurement. *Proc. Natl. Acad. Sci. USA* **2000**, *97*, 1438–1443. [CrossRef] [PubMed]

33. Food and Agriculture Organization of the United Nations. Climate Change and Food Security: Risks and Responses. 2016. Available online: http://www.fao.org/3/a-i5188e.pdf (accessed on 17 October 2019).

34. Food and Agriculture Organization of the United Nations. Climate Change, Coping with the Roles of Genetic Resources for Food and Agriculture. 2015. Available online: http://www.fao.org/3/a-i3866e.pdf (accessed on 17 October 2019).

35. Food and Agriculture Organization of the United Nations. The State of Food and Agriculture-Climate Change, Agriculture, and Food Security. 2016. Available online: http://www.fao.org/3/a-i6030e.pdf (accessed on 17 October 2019).

![microorganisms logo] *microorganisms*

MDPI

Review

Outbreak History, Biofilm Formation, and Preventive Measures for Control of *Cronobacter sakazakii* in Infant Formula and Infant Care Settings

Monica Henry [1] **and Aliyar Fouladkhah** [1,2,*]

[1] Public Health Microbiology Laboratory, Tennessee State University, Nashville, TN 37209, USA;
    mhenry3@my.tnstate.edu
[2] Cooperative Extension Program, Tennessee State University, Nashville, TN 37209, USA
*   Correspondence: afouladk@tnstate.edu or aliyar.fouladkhah@aya.yale.edu; Tel.: +1-970-690-7392

Received: 18 January 2019; Accepted: 9 March 2019; Published: 12 March 2019

Abstract: Previously known as *Enterobacter sakazakii* from 1980 to 2007, *Cronobacter sakazakii* is an opportunistic bacterium that survives and persists in dry and low-moisture environments, such as powdered infant formula. Although *C. sakazakii* causes disease in all age groups, infections caused by this pathogen are particularly fatal in infants born premature and those younger than two months. The pathogen has been isolated from various environments such as powdered infant formula manufacturing facilities, healthcare settings, and domestic environments, increasing the chance of infection through cross-contamination. The current study discusses the outbreak history of *C. sakazakii* and the ability of the microorganism to produce biofilms on biotic and abiotic surfaces. The study further discusses the fate of the pathogen in low-moisture environments, articulates preventive measures for healthcare providers and nursing parents, and delineates interventions that could be utilized in infant formula manufacturing to minimize the risk of contamination with *Cronobacter sakazakii*.

Keywords: *Cronobacter sakazakii*; powdered infant formula; *Cronobacter* outbreaks; preventive measures; infant care setting

1. Introduction

Cronobacter sakazakii is a recently classified and an emerging and opportunistic pathogen, found in a number of low-moisture foods including in powdered infant formula. Capable of causing morbidity in all age groups, this pathogen affects neonates and infants leading to life-threatening health complications, such as neonatal meningitis, urinary tract infection, sepsis, and seizures [1,2]. Historically, the pathogen was known as yellow-pigmented *Enterobacter cloacae* [3] until it was reclassified in 1980 as *Enterobacter sakazakii* by Farmer et al. [4,5]. With advancements in identification methods such as partial 16S ribosomal DNA, as well as hsp60 sequencing and polyphasic analyses, the genus undergone further reclassification in recent years [1,4]. Two proposals for defining the new novel genus *Cronobacter* were posited in 2007 and 2008 [4,6], that were further defined in 2012 [7]. The genus *Cronobacter* was derived from the Greek term "Cronos," a Titan of ancient mythology who swallowed his infants when they were born, in fear of being replaced by them [5]. The species epithet *sakazakii*, was proposed by Farmer et al. in 1980, in honor of the Japanese microbiologist, Riichi Sakazaki (1920–2002), a bacterial taxonomist also involved in nomenclature development of this pathogen [5,8].

C. sakazakii is a peritrichously flagellated, rod-shaped, and non-spore-forming pathogen. It is recognized as facultative anaerobic where its preferable growth is without oxygen presence but can grow with a small amount of oxygen. The growth temperature range is 6–45 °C with an optimum multiplication temperature of 37–43 °C. It can also survive low-moisture environments, such as infant formula, with a water activity of 0.30 to 0.83 for up to 12 months [9]. Ranging from 9–44%, *C. sakazakii* can be found in environmental samples from domestic and manufacturing facilities [10]. Thermal resistance can play a major part in the survival rate of *C. sakazakii*. Lukewarm water with temperature ranging from 52–58 °C has been confirmed to reduce the pathogen in reconstituted powdered infant formula [11]. *C. sakazakii* can be cultured on tryptic soy agar showing a distinct morphology of yellow-pigmented colonies. The vast majority of reported cases worldwide are from the United States, France, UK, Belgium, Philippines, Brazil, Israel, Spain, Hungary, Japan, Mexico, China, and Switzerland [1]. It is noteworthy that recent studies indicate that non-*sakazakii* species of *Cronobacter* including *malonaticus*, *turicensis*, *universalis*, *dublinensis*, *muytjensii*, and *condimenti*, could potentially cause morbidity and life-threatening complications in infants and adults [12,13]. Except for *C. condimenti* that has not been involved in any documented clinical episode, the other six species have clinical significance, with *C. sakazakii* and *C. malonaticus* as the major pathogenic species of public health concern followed by *C. turicensis*, *C. universalis*, *C. muytjensii*, and *C. dublinensis* [14–16]. As further delineated in Section 3.1, unlike vast majority of foodborne pathogens of public health concern, pathogenic species of *Cronobacter* are currently not part of the notifiable disease surveillance systems in nearly all public health infrastructures of North America and European Union, thus, the true epidemiological picture of these pathogens will continue to be unknown. Currently, identification of *Cronobacter* species can only be made through the use of species-specific PCR analyses or by whole genome sequencing.

2. Outbreak and Sporadic Episodes

C. sakazakii is an emerging pathogen in neonates and infants that was first known internationally before being recognized in the United States. In the late 1920s, there had been a report of "yellow-pigment coliforms" by Pangalos as the first published information on *Cronobacter* species from a case of septicemia [17]. Fast forwarding to the 1950s, strains from potable and/or river water samples from Metropolitan Water Board in London, submitted to England's National Collection of Type Cultures, seem to have similar physiological traits of possibly *Cronobacter* species. Nevertheless, between 1958 and 2016, there are approximately eight countries which reported cases of *C. sakazakii* suggesting that its reemergence and increased prevalence is reflective of its increased public health concern [5]. Here are listings of outbreaks in chronological order from the first documented case of *C. sakazakii* outbreak to the current. A summary of these outbreaks was also uploaded and is available in a public repository that can be accessed at https://doi.org/10.7910/DVN/TZ5PV9 (accessed on 21 February 2019).

As delineated earlier, our understanding of pathogenic *Cronobacter* species have been subject of redefinitions in recent years [14–16]. Until the correct identification of clinical isolates of *Cronobacter* species is achieved, the epidemiology of infections caused by these pathogens will always be lacking.

2.1. 1958: England

In 1958, there were two reported cases of neonatal meningitis at the Osterhills Hospital in England who died within two days apart. Patient 1 was a male, born on May 29, 1958. He had an average birth weight of 3034 g and was born after 38 weeks of gestation. After 10 days of life, the infant was discharged from the hospital but was quickly readmitted the next day after signs of grunting, jaundice, and loss of appetite. Samples were taken from the brain, cerebrospinal fluid, bronchus, urine, and blood where *Enterobacter cloacae* (reclassified in later years to *Enterobacter sakazakii* [18]) was isolated and a diagnosis of meningitis was confirmed. Intramuscular injection of oxytetracycline was given to the patient. However, within 48 h, the patient died.

Patient 2 was a female, born on June 5th, 1958, with her twin brother. After an emergency cesarean section, the newborns were premature after 32 weeks of gestation with the patient and her brother weighing 2013 g and 1191 g, respectively. The brother began to show signs of good progression over five days, however, the patient did not. On day 5 of life, the patient had immediate signs of collapsing cerebral, jaundice and an urticarial rash. Samples from bronchus, liver, marrow, and spleen were taken, but hours later the patient died.

Both of the neonates had similar findings in the necropsy report, one including abnormalities in the brain which may lead to the presence of meningitis. Pertaining to the respiratory tract, patient 1 had no evidence of inflammation unlike patient 2 who contained scanty yellow fluids with consolidated lungs. The strains that were isolated from both neonates were identical and reported as yellow-pigmented Coliform. Being abnormal that patient 2 twin brother was not affected with the pathogen given they were nursed in the same environment, had the same treatments, and shared the same nurse with expectations of using different incubators [19]. Due to lack of microbiological, epidemiological, and bioinformatic evidence, the true source of infection is undetermined in this historic outbreak.

2.2. 1965: Denmark

A case is described of neonatal meningitis complicated by brain abscess and hydrocephalus. The etiological agent was an uncommon *Enterobacter* morphologically similar to a strain isolated from the spinal fluid in two cases of neonatal meningitis in 1961 at St. Albans, England [20].

2.3. 1979: Macon, Georgia

The first documented case for *C. sakazakii* in the United States of America was at The Medical Center of Georgia, Macon, Georgia in 1979. A male infant who was born healthy was fed on nursery routine formula feedings and was only 30 g lesser than normal birth weight when he was discharged after four days of life. On day 6 of life, the patient became irritable, eating less, and was coming down with a fever. After the temperature was taken at 38.9 °C, the patient was hospitalized for further testing. The patient had a normal urinalysis, platelet count, and umbilicus had no signs of infections. What seemed to be abnormal was the high heart rate of 192 bpm, low leukocyte count, and maintained elevated axillary temperature. Diagnosed with possible sepsis, the patient's blood samples were then collected and tested positive for *C. sakazakii*. A combination of injections of ampicillin (75 mg every 12 h) and gentamycin (7.5 mg every 12 h) were given to the patient and blood samples were taken again. After seeing that the blood sample isolate was susceptible to ampicillin, gentamycin was discontinued. After six days of being in the hospital and two weeks of age, the dosage of ampicillin increased to 100 mg every 8 h. Blood samples were taken after a week of the new treatment and no *C. sakazakii* could be detected. The patient seemed to be doing well and after ten more days of therapy, ampicillin was discontinued and he was observed for another day. The patient was discharged from the hospital and later came back for a two-month check-up with reported normal development, weight of 5120 g, and no signs of a *C. sakazakii* infection [21].

2.4. 1977–1981: Netherlands

Eight infants were infected with *C. sakazakii* over the timeline of four years from 1977 to 1981. This epidemic was the largest in the Netherlands to be reported. Five of the patients (1–5) were admitted into the same hospital (A). Two of the eight patients (6 and 7) were at different hospitals (B and C) at birth and transferred to the same initial hospital when developing symptoms of illness (D). One patient (8) was at another hospital not related to the rest of the patients (E). Patients 6–8 were in the same area in another part of the country from the general hospital of patients 1–5. Out of the eight cases, only two patients, patient 1 and 8, survived [22].

2.4.1. Hospital A

Patient 1 was a male that was in good condition until he started exhibiting complications on day 5 of life in September of 1977. He was born prematurely with a weight of 2830 g after 36 weeks of gestation. Leading up to day 5 of life, the patient's temperature began to rise to 38.2 °C which is considered as fever in infants. Along with the raising of the temperature, leukocyte counts were 5500/mm^3 and protein concentration was low from samples taken from the cerebrospinal fluid (CSF).

The patient began treatment with ampicillin and kanamycin for 48 h. After a second CSF sample, leukocyte counts decreased while protein and glucose concentrations were still low and the patient's temperature still elevated at 39 °C. As a new treatment, gentamicin was given to the patient for an additional 15 days and he recovered with a low leukocyte count and high protein and glucose concentration. However, the patient was diagnosed with a severed neurological sensory development upon recovery [22].

Patient 2 was a female in good condition until she started to show symptoms on day 3 of life in April of 1979. Born with a weight of 2400 g after 39 weeks of gestation. Antibiotics were given to the patient with a combination of ampicillin and kanamycin and no progress was made. Gentamicin was then given and still no progress was made. The patient did not survive the infection [22].

Patient 3 was a female in good condition until day 3 of life. Born with a weight of 1670 g after 32 weeks of gestation, the patient was given ampicillin and gentamicin. Patient 4 was a male that was in good condition until day 4 of life. Born with a weight of 1900 g after 32 weeks of gestation, this patient was given ampicillin and gentamicin. Patient 5 was a female that was in good condition until she started to exhibit complications on day 5 of life. Born with a weight of 2690 g after 38 (Full term) weeks of gestation, the patient also received ampicillin and gentamicin [22].

2.4.2. Hospital D

Patient 6 was a male that was in good condition until the day 5 of life in February of 1978. On day 5, the patient was transferred to this hospital, he was born with the weight of 2085 g after 38 weeks of gestation. Chloramphenicol and gentamicin were given to the patient. Patient 7 was a female that was in good condition until she started exhibiting complications on day 5 of life in September of 1979. Patient 7 was also transferred to hospital D. She was born prematurely with a weight of 1370 g. Two antibiotic regimens were administered for the patient, ampicillin and gentamicin, as well as chloramphenicol and gentamicin [22].

2.4.3. Hospital E

Patient 8 was a female that was in good condition until exhibiting symptoms on day 9 of life in April of 1979. Born weighing 850 g after 30 weeks of gestation. A combination of ampicillin and gentamicin were given to the patient, she later survived and was diagnosed with severed neurological sensory development [22].

2.5. 1980: Indianapolis, Indiana

Unlike previously documented episodes that the affected patients were only few days old, this case is the first documented case where an infant was older than one month of life and did not develop symptoms associated with *C. sakazakii* infection during the hospitalization or shortly after. A female after five weeks of age develop a fever and seizure episodes was admitted to a hospital in Indianapolis, Indiana. A cerebrospinal fluid sample (CSF) exhibited high leukocyte count (15,600 per mm^3), a low protein concentration (295 mg/dl), and very-low glucose concentration (4 mg/dl). A combination of ampicillin (400 mg/kg every 24h) and chloramphenicol (100 mg/kg every 24 h) had begun. *C. sakazakii* was isolated from the CSF samples and treatments were continued with only ampicillin. After six days of therapy, *C. sakazakii* was continuing to proliferate in the serosanguinous fluid and gentamicin (7.5 mg/kg every 24 h) was added into the treatment. On day 15, *C. sakazakii* continued to persist and computed tomography (CT) showed massive ventricular dilation. The patient was transferred to a hospital for more advanced observation and assistance, the combination of ampicillin and gentamycin were continued. After the last positive sample of the pathogen from the patient's ventricular fluid, ampicillin and gentamicin were continued to be administered for 21 days. Ventricular fluid was then tested negative 24 h after discontinuation of the antibiotics. After two months, the patient was discharged, however, the circumference of the head continued to increase and developing skills were severely delayed [23].

2.6. 1981: Oklahoma City, Oklahoma

A male born from a healthy delivery was admitted into the hospital at five weeks of age from symptoms of fever, grunting, and fatigue. The temperature was taken with a high reading of 39.2 °C and heart rate of 180 bpm. Neurological symptoms included the absence of rooting and sucking reflexes and incomplete Moro reflex. The CSF sample exhibited leukocyte count 2871/mm^3, glucose concentration of 46 mg/dl and protein concentration of 168 mg/dl. Along with CSF, blood and urine samples were taken and sub-cultured to sheep blood agar and *C. sakazakii* was detected. Gentamicin and ampicillin were given to the patient until the test showed the pathogen was susceptible to ampicillin then gentamicin was discontinued. After 14 days of treatment with ampicillin, the patient was discharged in good condition [24].

2.7. 1982: Greece

C. sakazakii was first isolated in Greece in 1982 from the fecal samples of two thalassaemic children. Limited pieces of information were available in the reviewed citation about the cases [25].

2.8. 1984: Greece

A neonatal intensive care unit in Greece had 11 neonate-associated *C. sakazakii* infections reported from September 10 to October 17, 1984. The neonates had swabs from the throat and rectum on first or second day after admission and again after three to four days, and follow-up weekly sampling thereafter. After strains were plated on blood agar, MacConkey's, Chapman's, and Sabouraud's agar; twenty-eight strains were identified as *C. sakazakii*. Along with the neonates being tested, environmental surfaces, medical fluids in the unit, and 77 fingertips of the staff were also tested for the presence of *C. sakazakii*. Isolates of *C. sakazakii* was not found on abiotic environmental surfaces, medical fluids, nor the staff. Out of the 11 patients, seven survived [25].

2.9. 1986–1987: Reykjavik, Iceland

Three cases were reported in Reykjavik, Iceland of neonates contracted with *C. sakazakii* in 1986–1987 [26].

Case 1: A male born on March 18, 1986, after 36 weeks of gestation. He had a birth weight of 3144 g and appeared to be healthy with feeding of breast milk and powdered infant formula. On day 5 of life, the patient's health began to deteriorate and his spinal fluid was taken for microbiological analyses. *C. sakazakii* was isolated from the cultured spinal fluid sample and blood. Treatments of ampicillin and gentamicin began immediately along with cefuroxime within 12 h. After two weeks, cultures of *C. sakazakii* were still positive from ventricular fluids and chloramphenicol was added to the treatments for two months. The patient was discharged from the hospital at three months of age with his mental and physical development considered "markedly impaired". At the age of two years, the patient was diagnosed with severed neurological sensory development and quadriplegic [26].

Case 2: A male born December 14, 1986, with a weight of 2508 g had Down's syndrome. He was orally fed reconstituted powdered infant formula hours after his anoplasty surgery and exhibited no health complications until day 5 of life when he started to eat poorly. Patient's health began to deteriorate quickly and electrocardiograms and ultrasonograms were taken. The *C. sakazakii* infection was confirmed in the spinal fluid. Treatments of ampicillin and cefotaxime were not successful and the patient did not survive. Meningitis was confirmed from the autopsy [26].

Case 3: A male of a twin was born after 38 weeks of gestation on January 6, 1987, with a weight of 3308 g, reportedly healthy and was feeding on breast milk and reconstituted powdered infant formula until day 5. On day 6, he had a fever and his health was deteriorating quickly. Cerebrospinal fluid (CSF) samples were taken with a high leukocyte count and *C. sakazakii* was isolated but the blood was negative. Ampicillin and cefotaxime were started and health improvements were shown. The second testing of CSF was negative for *C. sakazakii* and antibiotics discontinued after three weeks. He was discharged after one month but CT scans did show cystic cavity in the left frontal lobe. He exhibited seizure disorder and delays in developmental areas [26].

2.10. 1988: USA

For the reported two cases, limited pieces of information were provided about the patients' progress in the cited literature [27].

2.11. 1988: Memphis, Tennessee, USA

In March of 1988 at a neonatal intensive care unit in Memphis, Tennessee four infants, two with bacteremia, another with a urinary tract infection, and one with bloody diarrhea had isolates of *C. sakazakii*. It was found that all four infants were fed from the same infant formula batch used in a blender. The staff's cleaning procedure was cleaning the blender with tap water and handwashing agents, but after being cultured a heavy growth of *C. sakazakii* was found on the preparation equipment. The blender was discontinued for use until it was sterilized, after being sterilized no cultures of *C. sakazakii* was detected [28].

2.12. 1989: Porto, Portugal

At a hospital in Porto, Portugal, there were 187 cases of meningitis where 15 patients were neonatal, 79 infants, and 93 between the ages of 1–14 years. Among these cases, 15 patients died and two out of the fifteen were infected with *C. sakazakii*. They both were neonates with infant formula consumption [29].

2.13. 1990: Maryland, USA

One case was reported in the literature with limited pieces of information about the patient's prognosis and survival [27].

2.14. 1990: Ohio

At a children's hospital in Cincinnati, Ohio, a 2520 g male was born of 35 weeks, exhibiting symptoms of poor feeding, apnea, and bradycardia after day 2 of life. Blood samples were taken from the patient and *C. sakazakii* was found. Ampicillin and cefotaxime were administered on day 4, CT scans were taken and the patient was shown to have increased tension on the left side of the brain. After three weeks of treatment, the patient showed no sign of inflammation and was discharged on day 28. About two weeks after discharge, the patient, with reported consumption of infant formula, was readmitted into the hospital with symptoms of poor feeding and fever. He was diagnosed with meningitis and the antibiotics of ampicillin and cefotaxime were started again. After the treatment cerebrospinal fluid results came back negative but after another CT scan was taken, an abscess was found in the brain. The cyst was drained (did not tested positive for *C. sakazakii*), and the patient had a resolving cerebral infraction [30].

2.15. 1993–1998: Jerusalem, Israel

At a hospital in Jerusalem from 1993–1998, four cases of neonates were infected with *C. sakazakii*, the bacterium was additionally isolated from a blender used to mix and prepare infant formula [31].
Case 1: In 1993, a neonatal born in a full term and fed infant formula in the hospital tested positive for *C. sakazakii* [31].
Case 2: In 1995, a healthy female born by caesarian section after 36 weeks of gestation developed conjunctivitis, she was fed infant formula and tested positive for *C. sakazakii* [31].
Case 3: In 1997, *C. sakazakii* was found in a 6-year old boy from bone marrow transplantation for lymphoblastic leukemia [31].

Case 4: In 1998, a vaginally-delivered, full term female infant was admitted with a diagnosis of meningitis which *C. sakazakii* was cultured from the infant's CSF [31].

2.16. 1994: France

Thirteen cases were reported in the cited literature with limited pieces of information about the patients' prognosis and survival [32].

2.17. 1998: Brussels, Belgium

Between June and July of 1998, 12 cases of neonates in neonatal intensive care unit of a Hospital were being contracted with *C. sakazakii*. This is the largest documented case in the history of this pathogen infecting neonates. All 12 patients had a birth weight of <2000 g and were orally fed a powdered infant formula before the development of neonatal necrotizing enterocolitis. Only two patients, male twins, died from this outbreak [33].

2.18. 1999–2000: Jerusalem, Israel

In the same hospital as the previous case in Jerusalem, 2 patients contracted *C. sakazakii* in 1999 and 2000. Patient 1 was a female born at 27 weeks of 620 g in December of 1999. On the ninth day of life, the infant was diagnosed with *C. sakazakii* infection after being fed with infant formula. The patient responded well to cefotaxime and survived. Patient 2 was a female of 36 weeks gestation with a weight of 2155 g. She was delivered by caesarian section because of fetal distress. Delivered 3 weeks after patient 1 in January 2000, *C. sakazakii* was cultured from CSF on day 4 of life. Treatments of cefotaxime and gentamicin were given but severe damage occurred in the brain. The patient survived and after three months was discharged with neurological problems. The infant was fed infant formula before being infected with *C. sakazakii* [31].

2.19. 2000: North Carolina, USA

One case was reported in the cited literature with limited information about patient's survival and prognosis [27].

2.20. 2001: Knoxville, Tennessee, USA

In 2001, a neonate was born of 1276 g through a caesarean section at 33.5 weeks. Patient being underweight at birth, intensive care was needed for the infant. Along with the low weight, the patient had a fever, tachycardia, decreased vascular perfusion, and neurologic abnormalities at 11 days. After another nine days, the patient passed away with a trace of *C. sakazakii*. This patient and 49 others were microbiologically screened in addition to obtaining environmental samples from the infant formula and preparation area. By the end of the screening, it was determined that the infant that died was infected through the powdered infant formula feed and no other patient was infected [34]. It is noteworthy that this outbreak is the first documented incidence of a manufactured lot of powdered infant formula being intrinsically contaminated. Thus, this outbreak epidemiologically linked powdered infant formula with *Cronobacter*.

2.21. 2002: Wisconsin, USA

One case was reported in the cited literature with limited pieces of information about patient's survival and prognosis [27].

2.22. 2002: Chandigarh, India

A female born at 34 weeks of gestation, weighing 1400 g, was re-admitted to the hospital on July 2002. The neonate was put on oral rehydration due to respiratory problems. After day 5 of life, the infant was put on a ventilator and developed sepsis with meningitis after development of grunting, episodic apnoea, chest retraction, and tachypnoea. Cerebrospinal fluid samples exhibited a high protein concentration, low glucose concentration, and elevated leukocyte count. The infant started an antibiotics chemotherapy of ciprofloxacin and netilmicin after a positive blood culture for *C. sakazakii*. The isolate showed resistance to ciprofloxacin, cefotaxime, and ceftazidime and sensitivity to gentamicin, amikacin, netilmicin, and co-trimoxazole. The infant did not survive the infection [35].

2.23. 2003: USA

Six neonatal cases with infant formula consumption were reported in the literature with limited information about the diagnosis and survival of the cases [27].

2.24. 2004.: France

On October 25th, 2004, and December 7, 2004, from two different hospitals, two neonatal cases were documented with *C. sakazakii* infection. Another two cases were identified at two separate locations by a regional public health surveillance system in early December. In all episodes, there were four cases and four hospitals involving neonates being contracted with *C. sakazakii*. Two of the four patients died [36].

2.25. 2004: USA

Two cases were reported in the cited literature with limited pieces of information about the patients' diagnosis and prognosis [27].

2.26. 2005: USA

Two cases were reported in the referenced study with limited information about the cases' medical history, prognosis, and survival [27].

2.27. 2006: Chandigarh, India

In the same hospital as the last case in India, another case of *C. sakazakii* occurred four years later. A two-month female infant was on breastfeed and was admitted to the hospital in July 2006 with a cough and respiratory distress. Prior to the patient's admission, the mother had hypertension and diabetes that required insulin; this caused the infant to suffer from jaundice on day 3 of life. After three days of being in the hospital, the infant developed sepsis and was transferred to pediatric intensive care. Blood cultured positive for *C. sakazakii* and was resistant to many of the antibiotics while exhibited sensitivity to the ceftriaxone-sulbactam combination. Treatment with ceftriaxone-sulbactam started with vancomycin, initially for five days. After availability of susceptibility data, only ceftriaxone-sulbactam administration was continued for additional two weeks and stopped when the blood culture was negative for *C. sakazakii*. The infant was discharged and reported afebrile [35].

2.28. 2007: Bilbao, Spain

A healthy male born 31 weeks and weighed 1715 g until day 3 of life when exhibited poor feeding. On day 5 of life, the patient was diagnosed with sepsis and was on a combination of ceftazidime and vancomycin. After the tenth day of treatment, an improvement was observed in clinical analytics. The patient was fed breast milk for the duration of the hospital and after 24 days, physical examination, serial brain scans, and psychomotor development were normal [37].

2.29. 2010: Queretaro, Mexico

In 2010, two infants were infected by *C. sakazakii* in a hospital in Queretaro, Mexico. The infants who were fed infant formula developed bloody diarrhea. Antibiotics (cefotaxime and vancomycin for case 1 and clindamycin and amikacin for case 2) were given and the two patients recovered [38].

2.30. 2011: Missouri, Florida, Oklahoma, and Illinois, USA

In 2011, four states in the United States (Missouri, Florida, Oklahoma, and Illinois) had cases of *C. sakazakii* infections. In Missouri, a 10-day old infant died from *C. sakazakii*, the bacterium later found in the infant formula, bottle of nursery water, and the serving container. Immediately, a major retailer recalled that brand of powdered infant formula from its stores nationwide on December 22, 2011. A leading regulatory agency of the country tested factory-sealed containers of the formula and nursery water of the same batch and no cultures of *C. sakazakii* were found. In Florida, an infant died of *C. sakazakii* infection, however, the strain of that case could not be obtained, nor in the case in Oklahoma. The strains from the Missouri and Illinois cases were gathered but it was not genetically related to that which was isolated from the reconstituted powdered infant formula and nursery water [39].

2.31. 2015: Sydney, Australia

In 2015, a male infant was born prematurely after 27 weeks of gestation without any signs of health complications. However, patient's health suddenly deteriorated on day 10 of life and blood cultures tested positive for *C. sakazakii*. After an unsuccessful antibiotic treatment with meropenem, patient was redirected to palliation after discussion with parents and died at 11 days after birth. *C. sakazakii* was isolated from breast milk expressed by a handheld breast pump that had not been properly sterilized before use [40]. The isolates from patient's blood and the expressed milk were identical based on bioinformatic evidence derived from whole genome sequencing of the patient and breast milk isolates [40].

2.32. 2016: Pennsylvania, USA

In April 2016, a female born at 26 weeks of gestation and weight of 1405 g was healthy until 21 days of life when she was diagnosed with sepsis. Samples taken from the cerebrospinal fluid and blood showed *C. sakazakii* presence. Treatments of ampicillin and cefepime were given however, seizures developed and the brain had liquefaction necrosis. The infant did not receive any powdered infant formula, however, pasteurized donor human milk and expressed maternal milk were given during the first week after birth. *C. sakazakii* was isolated from the breast pump kit and the kitchen sink drain from the mother's home [41].

It is noteworthy that in addition to the above-referenced episodes of infant morbidity and mortality associated with pathogenic *Cronobacter*, a study of 2012 [42], have summarized 68 cases from 1958 to 2003 and 30 cases belonging to 2004 to 2010. The study had accumulated the information based on personal communications, health records from the U.S. Centers for Disease Control and Prevention, the U.S. Food and Drug Administration and the World Health Organization, and published records. Since patients' specific prognosis and condition were not provided, those studies are not included in the current list of outbreaks. The 2012 study concludes that *Cronobacter* can infect both healthy term and hospitalized preterm neonates, and further recommends use of ready-to-feed formula for infants <2 months [42]. Differences among various types of infant formula are presented in Section 3.2 of the current study.

3. Recommendations for Parents and Caregivers

3.1. Vulnerable Population

C. sakazakii is a pathogen found primarily in dry and dehydrated food vehicles with low water activity, such as herbal teas, starches, and most concerning in powdered infant formula. The bacterium usually causes no health complications in adults, while could lead to sepsis, severe meningitis, and possible deaths in infants that are less than 12 months old [2]. Centers for Disease Control and Prevention estimates that four to six infants are infected with *C. sakazakii* each year in the United States. Although Minnesota Department of Public Health requires reporting of *C. sakazakii* in infants under one year of age within one business day of positive test results [43], in other states, *C. sakazakii* is not a reportable disease in infants nor adults, therefore almost certainly this is an underreported infection in the United States [2]. In neonates and infants, the symptoms start with fever and poor feeding, crying, and very low energy. In some severe cases, seizures, brain abscess, and high leukocyte counts occur, that could lead to long-lasting brain problems such as severed neurological sensory development [2]. The disease is more prevalent in premature infants that have low birth weights and possibly diagnosed with malnutrition such as low iron. Though very rare, *C. sakazakii* could also infect people of all ages, it is typically more severe for the elderly. Immunocompromised individuals including cancer patients, those having HIV and organ transplants are the adults who are more susceptible to *C. sakazakii* infection. Diagnosis of *C. sakazakii* infection is through blood sampling and usually followed by testing the cerebrospinal fluid for leukocyte count, glucose and protein concentration, and other special testing with the brain [2].

3.2. Infant Formula Manufacturing and Common Exposure Routes of C. sakazakii

The transmission of *C. sakazakii* is widely associated with reconstituted powdered infant formula. Micronutrients used in the formulation of powdered infant formula are heat-labile, therefore, it must be added after the pasteurization/heat treatment to keep the nutritional value in compliance with regulatory standards [10]. Infant formula is designed as a substitution for breastfeeding to mimic the nutritional properties of breast milk. Despite assumptions of many new parents, due to considerable compositional differences, cow's milk could not be utilized for feeding newborns, making the infant formula the only practical alternative to breastfeeding. There are three commercially available sources of infant formula: (1) Powdered formula, the least expensive and most popular, and must be mixed with water; (2) liquid concentrate, which must be mixed with an equal amount of water; and (3) ready-to-feed products that are the most expensive products and require no mixing with additional liquids [44].

The manufacturing processing of each of these infant formulas is different. For powdered infant formula, it undergoes two processes, dry-blend, and wet-blend spray drying. Dry blending to produce powdered infant formula is practiced by many firms in at least 40–50 processing plants worldwide [45]. The process begins with ingredients that have been tested for microbiological contamination and blended in large batches until nutrients/ingredients are distributed uniformly in a batch. It is then passed through sifters to remove oversized particles and other extraneous materials. The sifted product is then transferred to bags, totes, or lined fiberboard drums for storage. For canning the powdered infant formula prior to release to the market, it is flushed with inert gas, sealed, labeled, coded, and packaged into cartons. The packed product then undergoes a final check for microbiological contaminants. The ingredients that are used for this processing method are in dehydrated powdered form tested by the supplier(s). Since this process does not require extensive thermal processing, it is very critical that microbiological testing is conducted as well as working with reputable suppliers with validated food safety management plans in place [45]. Microbiological contamination might be present in low amounts, distributed heterogeneously and, thus, may be difficult to detect in random lot testing alone.

Wet blend-spray drying is another method to produce powdered infant formula. This process also begins with ingredients from suppliers that have been tested then it goes through pasteurization where the destruction of microbial cells will occur due to a thermal treatment in a relatively short amount of time. Next is homogenization, where the size of fat and oil particles are being reduced to have a uniform mixture, some companies may do this step before pasteurization [45]. Since powdered infant formula is designed to mimic the nutritional properties of breast milk, heat-sensitive micronutrients, such as vitamins, amino acids, and fatty acids, are then added after pasteurization where they will otherwise become inactivated/denatured by intensive heat. The mixture may now pass through an evaporator that is heated up to 62–77 °C and transferred through a high-pressure pump to spray dryer nozzles or cooled for storage, then reheated, and pumped directly to the spray dryer. As the mixture passes through the nozzles of the spray dryer, the water is evaporated and dry powder is created at the bottom of the spray dryer ranging in temperature typically from 73–79 °C. It is then cooled by a stream of chilled filtered air and passed through a sifter for packaging. It is also checked a final time for microbiological contaminants through random sampling. One disadvantage with wet blending followed spray drying is that it contains water in its processing and has a higher chance of the proliferation of pathogenic or spoilage bacteria. Liquid concentrate and ready-to-feed infant formula are similarly processed, then pasteurized using ultra-high temperatures (UHT) [46]. For these products, in short, first, the ingredients from suppliers are mixed together then skim milk is added at 60 °C, then fats, oils, and emulsifiers. During the formulation of this mixture, minerals, vitamins, and stabilizing gums are then added at various points due to sensitivity to heat. Next is pasteurization through heat exchange plates at a high temperature, typically from 85–94 °C for a short time of 30 seconds [46]. Homogenization is next for a uniform mixture followed by standardization of the correct parameters for pH, fat content, vitamins, and minerals. The last stage is packaging into containers followed by sterilization with heat. The main difference between liquid concentrate formulas and the ready-to-feed formulas is the amount of water required to be added during the preparation of the product. Liquid concentrated and ready-to-feed infant formulas are safer to use due to the existence of high-heat pasteurization and/or sterilization relative to powdered infant formula, however, it is typically more expensive [46].

In summary, addition of heat sensitive ingredients to improve nutritional value of the formula and meeting the strict nutritional regulatory requirement and difficulties in pasteurizing/sterilizing a product in powdered form are main concerns for safe production of infant formula. Other transmission routes of *C. sakazakii* to infants can be associated with the preparation methods at home or care settings. *C. sakazakii* can affect infants by contaminated bottles, nipples, scoops, and other utensils not being properly cleaned as well as hands not being washed properly. Hospitals and other healthcare facilities, such as day cares, are at greater risks due to a high volume of infants and infant formula used. If not cared properly, children who are breastfed and drink from pre-pumped milk could also be infected with *C. sakazakii* if cross-contamination occurs from improperly sanitized breast milk pumps [41].

3.3. Importance of Breastfeeding in Prevention of C. sakazakii Infections

Both Centers for Disease Control and Prevention (CDC) as well as the World Health Organization (WHO) indicate the leading preventive measure for *C. sakazakii* infection in infants is breastfeeding. According to the CDC breastfeeding data, 82.5% of mothers at the beginning of birth breastfeed alone and this percentage decreased to only 24.9% of mothers who continue to breastfeed with no infant formula for their infants after six months of age. Additionally, 55.3% of mothers continue to breastfeed as well as use infant formula after the infant was six months of age [47]. One main reason parents choose infant formula is due to malnutrition in their child. Breastfeeding is a natural way to prevent undernutrition as well as infectious conditions, such as diarrhea and pneumonia in infants [48]. In addition, epidemiological study indicates mothers who breastfeed might have decreased chances of type II diabetes, depression and breast and ovarian cancers later on in life [48]. Ways to stretch breastmilk is to use breast pumps and pre-storing milk in the freezer. This is great for busy mothers and those who have others watching their child. To ensure safety with pre-pumped breastmilk, the breast pump should be washed and sterilized after every use [48]. Getting the correct storage container such as a freezer bag is a great way to store breast milk. Labeling and dating is also good practice to know exactly when the milk was expressed. Good hygiene is also recommended to safely store breastmilk. One of the biggest problems with breastfeeding is the support from family, friends, co-workers, and the hospital. For co-workers and in a workplace setting, paid maternity leaves is required. Designating an area for breastfeeding if the mother decides to bring the child to work is highly recommended [49]. Despite demonstrated health benefits, and although a low rate of breastfeeding adds as high as $2.2 billion a year to medical costs in the United States, according to Centers for Disease Control and Prevention, most US hospital do not fully support breastfeeding [50]. Once the mother starts on durations of infant formula, it is harder for the mother to get back into a habit of breastfeeding exclusively. At the six month mark, infants are ready for solid food and can start eating products such as infant cereal and pureed vegetables [51]. Mothers with HIV who want to breastfeed are recommended to take antiretroviral treatment to reduce the chances of transferring the infections to their child. Another option is obtaining donated or purchasing expressed breastmilk from mothers who are willing to help. There is a process that the donating mothers must go through. First, an application has to be filled out and sent to the company one wish to donate their milk for medical confirmations. Second, a test kit will be sent to the mother and a series of tests with their milk have to be conducted. If screening requirements are passed, then a nurse will be invited into the home for blood testing and other testings that may require a nurse's assistant. If the medical tests are passed, then the mother could begin to label, filling, freezing, and packaging their breast milk. When arrived at the company, it is important to have the breastmilk still frozen after transportation. Further tests are then conducted then the milk is stored in the freezer until use [52].

3.4. Preventive Measures during the Use of Infant Formula for Reducing Risk of C. sakazakii Infections

As stated before, during powdered infant formula manufacturing, companies must add some of the nutrients (vitamins, minerals, amino acids, and fatty acids) after sterilization to avoid denaturation, thus, powdered infant formula is a non-sterile product. This is where pathogenic bacteria, such as *C. sakazakii* can be introduced into the infant formula. Typically microbiological sampling is conducted in a manufacturing facility, however *C. sakazakii* has the chance of presence in small quantities of a large batch of the powdered infant formula with heterogeneous distribution. Thus, sampling alone does not necessarily assure the safety of the product. Since powdered infant formula is the cheapest and most abundant form of infant formula, it is the most bought and used by parents, hospitals, care centers, and other healthcare facilities [45]. For the preparation of this product, hands must be washed before, during, and after preparing powdered infant formula mix [53]. If soap and clean water are not available, then using a sanitizer that is more than at least 60% alcohol can be used as a replacement [53]. The World Health Organization has specific instructions for the preparation of powdered infant formula in care settings and in the home [54]. Since in care settings there are high volumes of infants and many packages of powdered infant formulas might be in use, the recommendations are on a larger scale than those in the home setting. When preparing infant formula, the preparation area would need to be cleaned and disinfected due to the potential presence of microbial pathogens on abiotic surfaces including *C. sakazakii* [54]. Next is to boil a generous amount of water, relevant to the number of infants going to be fed. This eliminates all bacterial microorganisms in planktonic form. Feeding bottles are also not sterile and must be boiled before use. Microwaving should not be used when preparing powdered infant formula. Then the water would need to be cooled slightly but not under 70 °C, then the proper amount of formula could be added to clean and sterilized feeding cups or bottles. If using a larger container, it should not exceed over 1 L [54]. For mixing, bottles can be shaken to fully mix the water and powdered infant formula, feeding cups can be stirred with a pre-sterilized spoon, and large containers can be stored with spoons but has to be distributed to its respective containers immediately to avoid scalds for infants [54]. In a care setting, bottles then would need to be labeled with infant's name, type of formula, or ID, as well as the time and date prepared and the preparer's name to assure traceability is possible in case of contamination occurrence. If intending to prepare powdered infant formula for later use, it is best to prepare new batches of feeding every time and to eliminate leftovers. Leftover feed is suggested to be thrown away and especially not to be used if stored more than two hours at room temperature [54]. However, it can be stored in a sealed container in the refrigerator for up to 24 h. To re-warm stored infant formula, a separate container filled with boiled water could be used to place the bottle or feeding cup inside to evenly warm up the reconstituted product without damaging the micronutrients. Recommendation for the home setting is similar to the care setting facilities on a smaller scale [55]. Hands must be washed properly, equipment being used for preparation must be cleaned with hot soapy water and be scrubbed inside and outside of the bottle, then rinsed thoroughly. Sterilizing equipment is also recommended to eliminate pathogenic microorganisms. For sterilization, a large pan with water could be used to place equipment inside of the pan filled with water, covering the pan, and to bring the content to a boil. Similar to a recommendation for the care setting, the recommended amount could be poured to water with the temperature not under 70 °C. The bottles then could be shaken or swirled gently. The bottle content temperature could be tested on the skin to ensure that it is warm and not too hot for feeding the infant. If it hasn't been used in two hours at room temperature or more, prepared infant formula would need to be discarded [55]. When transporting the prepared milk, the product would need to be kept cold during transportation to slow down or stop the multiplication of potentially pathogenic bacteria. In other cases, where there is no access to boiling water or preparation of powdered infant formula cannot be made at the time, it is recommended to use ready-to-feed infant formula. It is sterile from the processing manufacturing and ready to use without any additional ingredients [45].

4. Fate and Multiplication of *C. sakazakii* on Biotic Surfaces

As previously discussed, *C. sakazakii* is a Gram-negative, facultatively anaerobic, non-sporulating, motile rod-shaped bacterium and is a member of the Enterobacteriaceae family. *C. sakazakii* multiplies between temperatures of 6–45 °C with an optimum temperature of 37–43 °C. It could survive and/or proliferate in water activity levels ranging from 0.30 to 0.83 [9]. The bacterium is widely associated with powdered infant formula from many outbreaks [19,22,26] as previously articulated, thus, powdered infant formula is the essential biotic reservoir of public health concern for *C. sakazakii*. According to the United States Department of Agriculture National Nutrient Database [56], a typical powdered infant formula has a protein content of 1.02g/scoop, total lipid of 2.35 g/scoop, carbohydrates 4.86 g/scoop, seven minerals: calcium, iron, magnesium, phosphorus, potassium, sodium, and zinc; and 13 vitamins: Vitamin C, B-6, B-12, A (RAE), A (IU), E, D (D2 + D3), K, thiamin, riboflavin, niacin, and folate [56]. Since all powdered infant formula has to follow the same guidelines for nutritional value by the Food and Drug Administration, all powdered infant formulas are nearly identical in formulation unless specified for premature or low iron infants [9].

4.1. Fate and Multiplication of C. sakazakii as Affected by Temperature

From the 1980s to the present time, there are several studies on determining the fate and multiplication of *C. sakazakii*. Strains of *C. sakazakii* isolated from clinical, food, and/or environment could be used for various microbiological challenge studies. In 1980, researchers delineated the multiplication rate of the pathogen using strains sent to the CDC from patients and one from an unopened can of dried milk. Of the 57 strains used in their experiment, all grew at 25, 36, and 45 °C and 50 of the strains grew at 47 °C [57]. It was noted that none of the strains grew at 4 or 50 °C. The pathogen grew in presence of D-glucose without added nutrients such as vitamins, minerals, or amino acids. Additionally, the researchers grew the pathogen in aerobic and anaerobic environments. The strains were monitored on tryptic soy agar and all strains grew on the agar at 36 °C after 24 h. The strains produced bright yellow colonies on the agar, known as a typical characteristic of *C. sakazakii* today. It is noteworthy that this pigmentation alone should not be considered as a species criterion [5].

The growth of *C. sakazakii* in broth was monitored in tryptic soy broth and all strains of *C. sakazakii* produced large amounts of sediments. The biochemical reactions of the 57 strains were also conducted [57].

4.2. Survival Rate

A study published in 1997 shows the survival and multiplication of *C. sakazakii* in powdered infant formula and on Brain Heart Infusion (BHI) medium, as well as the incidence of *C. sakazakii* being present in a Canadian supermarket purchase of one of the popular powdered infant formula brands [58]. Ten strains of *C. sakazakii*, five clinical and five food isolates, were used for this experiment. The minimum growth temperature was 5.5 °C for one clinical strain and two food strains on BHI broth, the study also indicated that none of the remaining strains grew under 5.5 °C. In powdered infant formula, the lag time at 10 °C for the formula that contained the food strain (19 h) was less than the formula with the clinical strain (47 h). For generation time at 10 °C, it ranged from 4.18 to 5.52 h with the formula inoculated with the clinical strain exhibiting the longer time. At 23 °C, the mean generation time was 0.67 h. With generation time of 0.67 h at room temperature, it is evident that leaving reconstituted infant formula on countertops or traveling with it without proper refrigeration could potentially increase the risk of *C. sakazakii* infection. After testing 120 powdered infant formulas in a Canadian supermarket, the researchers also observed a 6.7% presence of *C. sakazakii* in the market [58].

In a later study, published in 2006, the objective was to prevent the multiplication of *C. sakazakii* in media and powdered infant formula using bacteriophages [59]. Bacteriophages are viruses that infect bacteria. There are now exploratory studies proposing the use of bacteriophages to control foodborne bacterial pathogens and spoilage bacteria in live animals, meat, dairy products, seafood, and fresh produce [35]. A total of six *C. sakazakii* strains were used, one clinical and five food isolates, and were incubated in powdered infant formula and BHI at 12, 24, and 37 °C. The *C. sakazakii* bacteriophages were prepared from an environmental water sample (centrifuged and sterilized) with an equal amount of BHI, mixed and incubated overnight at 24 °C. Then, against the six *C. sakazakii* strains, five bacteriophages were used. For the powdered infant formula, *C. sakazakii* bacteriophages at 37 °C, the clinical strain showed a decrease in multiplication starting at 2 h and at 24 °C. At 12 °C, it was not significantly different compared to the control. For the food strain in powdered infant formula, the results were similar at 12 °C as for the clinical strain but at 24 and 37 °C they were both reduced by one log immediately at 2 h and one strain continued to be at its detection limit for the duration of 2–10 h. This exploratory study showed with the correct temperature and bacteriophage concentration, the inactivation of *C. sakazakii* could be achieved. The most effective reduction was at the highest bacteriophage concentration of 10^9 at any incubation temperatures of 12, 24, or 37 °C [35].

4.3. Water Activity

Water activity (A_w) is the measurement associated with the availability of water in biological setting and relates to water presence in the food in free form [9]. Water activity could range from 0.1 to 0.99 in foods. In powdered infant formula, the A_w can start at 0.20 depending on the added nutrients and differing for soy or milk based products. *C. sakazakii* could survive in powdered infant formula for two years at low A_w [60,61]. In a study by Joshua B. Gurtler et al., an experiment was conducted to see the survival rate of *C. sakazakii* in soy-based and milk-based powdered infant formulas at various A_w ranging from 0.25 to 0.86 at 4, 21, and 30 °C for 12 months [62]. There were 10 *C. sakazakii* strains used: five clinical, four food, and one environmental isolate; a total of six powdered infant formulas were used (four milk-based and two soy-based) and water activity divided into two categories of low A_w of 0.25–0.50 and high A_w 0.43–0.86. The A_w was adjusted by adding amounts of saturated salt solutions to lower or raise the A_w. In high A_w infant formula, the rate of inactivation of *C. sakazakii* increased as the storage temperature increased. This study notes that augmenting the pathogen inactivation is possible with the increase of A_w and temperature during storage. This study also states that the clinical strains survived longer than the food strains in powdered infant formulas at high A_w [62].

4.4. Thermal Inactivation

Thermal inactivation validation studies are conducted to determine the highest temperature a pathogen can withstand. When heat treatment is applied, the decimal reduction time, D-value, could be calculated. If at a specific temperature the pathogen is being reduced, it could be reported in the context of 1 log reduction or 90% deactivation. This is where one could mathematically assimilate if the pathogen is being reduced over a course of treatment. A study published in 2003 delineates that *C. sakazakii* is not thermotolerant, but resistant to osmotic stress and drying [63]. The temperatures used for heat treatment for the 22 *C. sakazakii* strains were at 53, 54, 56, or 58 °C at different time intervals in a water bath. After heat treatments, the samples were immediately cooled in iced water for 1 min and then enumerated to see the heat treatment results. For the preparation of dry stressed strains, plates of *C. sakazakii* were kept without a lid in a 25 °C incubator for air-drying. This was monitored for 46 days for *C. sakazakii* survival after air-drying. All 22 *C. sakazakii* strains multiplied in 47 °C BHI broth. For heat resistant phenotype, the D-value at 58 °C in phosphate buffer pH of 7 had a mean value of 0.48 min and in reconstituted infant formula it did not have a significant difference. For resistance to dry stress, *C. sakazakii* at 25 °C was decreased by 1–1.5 log unit after 46 days. This study concludes that relative to many other members of the Enterobacteriaceae family, *C. sakazakii* appears to be more resistant to osmotic and dry stress [63].

5. Survival and Biofilm Formation on Abiotic Surfaces

C. sakazakii has not only been reported to survive on various biotic surfaces, such as infant formulas, fruits, vegetables, or human intestines, but also on abiotic surfaces, such as stainless steel and polyester plastic. The pathogen is also capable of forming a sessile community of bacterial biofilms on both biotic and abiotic environments. Specifically, the bacterium is capable of forming a polyanionic extracellular polysaccharide also known as ESP [9]. Study of Kumar et al. delineates the process of biofilm formation into three parts: conditioning of a surface, adhesion of cells, and formation of microcolonies. Within the food industry, surfaces of equipment can be coated with nutrients from the food product enhancing the biofilm formation of the pathogen [64]. Adhesion is in two stages, reversible adhesion followed by an irreversible adhesion. The former starts with weak interactions within the bacterial cells and the substratum. The ability to maintain levels of the van der Walls attraction forces, electrostatic forces and hydrophobic interactions determines the next stage to irreversible adhesion. Irreversible adhesion is a repulsive force that prevents the bacterial cells and the biofilm community from disassociation from the surface. The next stage will be formation of a microcolony where the irreversibly-attached cells are dividing. This formation of microcolony begins the formation of a visual thick layer of organisms formed on abiotic surfaces. ESP is also being produced for a firmer attachment to the abiotic surfaces [64].

One of the main surfaces of concern for *C. sakazakii* infection is the feeding tubes of neonates after being fed reconstituted powdered infant formula. In a study by Hurrell et al., biofilm formation was conducted on enteral feeding tubes [65]. Twelve strains of *C. sakazakii* were used in this study from various patients, infant formulas, enteral feeding tubes, and raw materials. The tubing materials that were selected were polyvinyl chloride (PVC) and polyurethane (PU). The pathogen was inoculated into powdered infant formula and incubated overnight at 37 °C and then diluted after 18 h. This cocktail was then aseptically syringed through the tubes to observe the biofilm formation in two-hour intervals. As a result, 5 strains of *C. sakazakii* produced biofilm mass with a density of 10^8 to 10^{10} CFU/mL. One strain was from a patient, three from powdered infant formula, and one from an enteral feeding tube. The doubling time for the *C. sakazakii* strains isolated from the enteral feeding tubes was between 22–27 min [65].

Another study with biofilm formation on enteral feeding tubes utilized clinical, food, and environmental isolates [66]. Sterile feeding tubes were used and the tubes were inoculated with *C. sakazakii*. For biofilm formation, the tubes were incubated at 4 °C for 24 h and then rinsed and submerged in phosphate-buffered saline. Next, it was divided into two groups monitored for a course of 10 days: 2, 4, 6, 8, and 10 days at 12 °C and 1, 2, 4, 6, 8, and 10 days at 25 °C. The results show attachment and growth of sessile cells were similar among the strains used. The multiplication rates were much higher at 25 °C than 12 °C by 3–4 logs, however, in both temperatures the rate was constant over the course of ten days showing no significant differences. Along with feeding tubes, other plastic surfaces are crucial such as bottles and other equipment [66].

In another study, biofilm formation on plastic surfaces was investigated. Four strains of *C. sakazakii* were used from the American Type Culture Collection (ATCC). For biofilm formation, plastic microtiter plates were used [67]. The plates were inoculated with *C. sakazakii* and incubated for 24 h at 37 °C. The plates were rinsed with distilled water and then submerged with methanol for 15 min. Crystal violet was used to visualize pathogenic growth on the plastic microtiter plates. The results show that out of the four strains used in this experiment, not all produced biofilms in artificial media alone. For biofilm formation on plastic surfaces, 13.9% populated in Brain Heart Infusion and tryptic soy broth mixture and 6.9% populated in nutrient broth [67].

6. Current Decontamination Strategies

6.1. Disinfectants to Eliminate Biofilm Attachment

Various validated antimicrobial agents could be used for inactivation of *C. sakazakii* from biotic surfaces associated with production and preparation of infant formula as well as feeding tubes and equipment in hospitals [9]. Just like decontamination of other microbial pathogens, cleaning alone may lead to modest reductions in removal of a pathogen [68]. To ensure that pathogens are eliminated or reduced to a microbiologically acceptable level, validated sanitation is recommended directly after cleaning. Sanitation could be achieved by physical means such as heated water, UV radiation, or could be achieved using chemical agents such as chlorine-based, iodophors, quaternary ammonium compound sanitizers, and hydrogen peroxide [9].

In general, chlorine-based sanitizers could be very effective against planktonic cells of bacteria, yeast, and molds. Similarly, iodophors are effective against Gram-positive and Gram-negative bacteria, bacterial spores, viruses, and fungi. Quaternary ammonium compounds are also known as an efficacious sanitizer due to their ability to clean and to sanitize surfaces against an array of microorganisms at acidic pH and higher temperatures. Hydrogen peroxide is also a very effective antimicrobial against planktonic bacterial cells, spores, and viruses [9].

In the study of Kim et al., the objective was to see the effectiveness of disinfectants in eliminating *C. sakazakii* [69]. Due to widespread use in infant formula preparation areas, laboratories and hospitals, food services, and child day care settings, quaternary ammonium and phenolic disinfectants were evaluated in their study. Overall, 13 disinfectants were studied in their investigation from various suppliers against biofilm formation of *C. sakazakii* on stainless steel. This experiment was evaluated with two strains of *C. sakazakii* studying biofilm formation on stainless steel at days 6 and 12 as well as the treatment times of 0, 1, 5, and 10 min for submersion in disinfectants. For quaternary ammonium compound-based disinfectant, *C. sakazakii* was reduced to less than 0.30 log CFU/mL within 1 min of submerging into the sanitizer. The quaternary ammonium compound-based sanitizer applied as a spray product were, however, showed only modest reductions. Another effective sanitizer, peroxyacetic acid/hydrogen peroxide, resulted in a log reduction of >2.4 log CFU when applied for 10 min. This study shows that quaternary ammonium compound-based and peroxyacetic acid sanitizers could be very effective for inactivation of the pathogen in planktonic and biofilm stages if used at optimized conditions. Our recent studies, however, indicate that previously validated sanitizers against planktonic cells might not be able to completely eliminate one- and two-week mature biofilms from stainless steel [70,71]. Thus, commercial adoption of a cleaning and sanitizing program requires careful consideration of existing literature and conduct of microbiological validation studies against sessile and planktonic cells for specific intrinsic and extrinsic conditions of a product and processing area.

6.2. Thermal Inactivation

Thermal inactivation could be an efficacious method for decontamination of *C. sakazakii* from biotic and abiotic surfaces. As briefly introduced in Section 6.1, elevated heat could be used as a physical mean for decontamination of significant surfaces associated with the production, manufacturing, and preparation of infant formula. It could be utilized as a processing aid for assuring the safety of the product, through high temperature short time (HTST) pasteurization. Decontamination of *C. sakazakii* could typically be achieved at temperatures 70 °C or higher [72]. In a study, twelve strains were used in reconstituted infant formula prepared based on manufacture's instruction, then 15 mL of infant formula was inoculated with 1.5 mL of *C. sakazakii* and injected into the heating coil apparatus at set temperatures of 58 °C [72]. After being in contact with the controlled heat, the samples were immediately placed on ice to discontinue further heat decontamination. The temperature 58 °C was the set point for z- and D-values to be calculated followed by treatments at temperatures of 56, 60, 65, and 70 °C. The study concluded that to fully inactivate a heat-resistant strain of *C. sakazakii*, temperatures of 70 °C or greater are needed. This elevated temperature could lead to nutrient loss and unwanted changes in organoleptic properties of infant formula [72], thus preservation of heat labile micronutrients is considered as a major curtailment for successful implementation of sterilization or pasteurization of powdered infant formula from *C. sakazakii* in commercial manufacturing. This indicates the need for innovative and emerging technologies [70].

6.3. High-Pressure Processing

High-pressure processing (HPP) or high hydrostatic pressure (HHP), is a non-thermal method involving pressurization of a packaged food in a water-filled closed chamber, for a short duration to inactivate microorganisms [9]. The technology popularity in private industry is gaining momentum in recent years with purchase rates reaching nearly 200 units around the world [73]. Beneficiary aspects of using HPP is the preservation of the color, flavor, freshness and physical properties of foods with minimal damages to nutritional values [74]. The HPP machines can have low to high pressures ranging from <100 to 1000 MPa [75]. According to the National Advisory Committee on Microbiological Criteria for Foods, pasteurization that had been traditionally known as a heat-based intervention is now redefined and HPP is a part of pasteurization definition as a non-thermal pasteurization method [73].

In a study by Arroyo et al., four strains of *C. sakazakii* were used, exposed to elevated hydrostatic pressures ranging from 200 to 600 MPa and for 0 to 10 min [76]. This study also investigated four food vehicles for inoculation: orange juice, chicken soup, vegetable soup, and rehydrated powdered milk. In all food vehicles, the most pressure-resistant strain showed around 3 log reductions at 500 MPa, reaching the study detection limit. The study results indicated utilization of elevated hydrostatic pressure could eliminate the pathogen from biotic surfaces, it also articulates that various isolates of *C. sakazakii* could exhibit considerably different sensitivity to hydrostatic pressure [76]. Our recent studies also exhibit that various phenotypes of *C. sakazakii*, such as rifampicin-resistant variants and pressure-stressed isolates could be inactivated by over 5 logs, using elevated hydrostatic pressure of up to 380 MPa, in rates that are comparable with wild-type isolates [70]. Effects of hydrostatic pressure on retention of heat liable micro and macronutrients of infant formula is currently a knowledge gap of literature and could be considered as the main curtailment for widespread adoption of high-pressure processing in infant formula manufacturing.

It is noteworthy that, in addition to pressure-based interventions, an array of emerging and re-emerging technologies such as utilization of ohmic, microwave, radio frequency, ultrasonic, or infrared heat; pulsed X-rays; pulsed electric field; and oscillating magnetic field could potentially exhibit promising applications for microbiologically efficacious and economically feasible treatment of infant formula against planktonic cells and biofilms of *C. sakazakii*. These exploratory applications require microbial challenge and safety validation studies as well as feasibility assessments. As an example, mild temperature of up to 50 °C coupled with an ultrasonic treatment at amplitude of up to 61 μm, could yield microbiological reductions comparable to traditional heat treatments for inactivation of *C. sakazakii* [77].

7. Conclusions

There have been a few recent outbreaks and sporadic cases of *C. sakazakii* infections in the country and around the world associated with infant mortality and morbidity. This is almost certainly an underestimation of the public health burden of the pathogen since unlike the vast majority of main foodborne pathogens, *C. sakazakii* infections are not currently a reportable disease in nearly all states.

Strict regulatory standards for nutritional quality of powdered infant formula and heat sensitivity of micronutrient additives of the product lead to the need to addition of the heat-labile ingredients of the formula after heat treatment that creates a potential route of contamination of the product with this bacterium. The ability of this bacterium to survive osmotic stress and low water activity environments for as long as two years, further provides an acceleration in the likelihood of this disease occurrence in the vulnerable population. The *C. sakazakii* has the potential to survive and persist on various biotic and abiotic surfaces such as preparation area in healthcare facilities and form biofilm communities that are more resistant to antimicrobial interventions. These characteristics add another layer of complexity for elimination and prevention of *C. sakazakii* from the manufacturing facilities, hospitals, and domestic environments. Considering the nature of contamination of products in food manufacturing that is mostly heterogeneous in nature as a fraction of larger batches, sampling alone could not assure the safety of infant formula and could lead to false sense of security for manufacturers. Use of supplier's chain verification programs, relying on food safety management system such as those articulated in Food Safety Modernization Act or Hazard Analysis Critical Control Point-based regulations, as well as the use of emerging and validated technologies such as utilization of elevated hydrostatic pressure could assure the safety of the infants and powdered infant formula products. Following the articulated recommendations for the preparation of reconstituted infant formula in healthcare and domestic settings, and reliance on breastfeeding when medically possible are the main preventive approaches that could be implemented by parents and healthcare providers to minimize the risk of infection with this opportunistic bacterium.

Author Contributions: M.H.: Master of Science Candidate and Graduate Research Assistant, Public Health Microbiology Laboratory, Tennessee State University, co-wrote the first version of the manuscript in partial fulfillment of her degree thesis. A.F.: Assistant Professor and Director of Public Health Microbiology Laboratory, secured extramural funding, co-wrote, revised, and edited the manuscript.

Funding: Financial support in part from the National Institute of Food and Agriculture of the United States Department of Agriculture (projects 2017-07534; 2017-07975) and Pressure BioScience Inc. is acknowledged gratefully.

Acknowledgments: Comments and feedback of the three anonymous reviewers and editorial team of *Microorganisms* is gratefully appreciated. This manuscript is supported by a Publication Scholarship from the AGSC 5540 (Food Policies and Regulations; Fall 2018) graduate course of the Public Health Microbiology Laboratory of Tennessee State University for promoting open access publication.

Conflicts of Interest: The authors declare no conflict of interest. The funding sponsors had no role in the design of the study; in the collection, analyses, or interpretation of data; and in the writing of the manuscript. The content of the current publication does not necessarily reflect the views of the funding agencies.

References

1. Hunter, C.J.; Bean, J.F. *Cronobacter*: An emerging opportunistic pathogen associated with neonatal meningitis, sepsis and necrotizing enterocolitis. *J. Perinatol.* **2013**, *33*, 581. [CrossRef] [PubMed]
2. Centers for Disease Control and Prevention. *Cronobacter*: Expanded Information. 2015. Available online: https://www.cdc.gov/cronobacter/technical.html (accessed on 3 February 2019).
3. Healy, B.; Cooney, S.; O'Brien, S.; Iversen, C.; Whyte, P.; Nally, J.; Callanan, J.J.; Fanning, S. *Cronobacter* (*Enterobacter sakazakii*): An opportunistic foodborne pathogen. *Foodborne Pathog. Dis.* **2010**, *7*, 339–350. [CrossRef] [PubMed]
4. Iversen, C.; Lehner, A.; Mullane, N.; Bidlas, E.; Cleenwerck, I.; Marugg, J.; Fanning, S.; Stephan, R.; Joosten, H. The taxonomy of *Enterobacter sakazakii*: Proposal of a new genus *Cronobacter* gen. nov. and descriptions of *Cronobacter sakazakii* comb. nov. *Cronobacter sakazakii* subsp. sakazakii, comb. nov., *Cronobacter sakazakii* subsp. malonaticus subsp. nov., *Cronobacter turicensis* sp. nov., *Cronobacter muytjensii* sp. nov., *Cronobacter dublinensis* sp. nov. and *Cronobacter genomospecies*. *BMC Evol. Biol.* **2007**, *7*, 64.
5. Farmer, J.J.; Asbury, M.A.; Hickman, F.W.; Brenner, D.J.; Enterobacteriaceae Study Group. *Enterobacter sakazakii*: A new species of "Enterobacteriaceae" isolated from clinical specimens. *Int. J. Syst. Evol. Microbiol.* **1980**, *30*, 569–584. [CrossRef]

6. Iversen, C.; Mullane, N.; McCardell, B.; Tall, B.D.; Lehner, A.; Fanning, S.; Stephan, R.; Joosten, H. *Cronobacter* gen. nov., a new genus to accommodate the biogroups of *Enterobacter sakazakii*, and proposal of *Cronobacter sakazakii* gen. nov., comb. nov., *Cronobacter malonaticus* sp. nov., *Cronobacter turicensis* sp. nov., *Cronobacter muytjensii* sp. nov., *Cronobacter dublinensis* sp. nov., *Cronobacter genomospecies* 1, and of three subspecies, *Cronobacter dublinensis* subsp. dublinensis subsp. nov., *Cronobacter dublinensis* subsp. lausannensis subsp. nov. and *Cronobacter dublinensis* subsp. lactaridi subsp. nov. *Int. J. Syst. Evol. Microbiol.* **2008**, *58*, 1442–1447.

7. Joseph, S.; Cetinkaya, E.; Drahovska, H.; Levican, A.; Figueras, M.J.; Forsythe, S.J. *Cronobacter condimenti* sp. nov., isolated from spiced meat and *Cronobacter universalis* sp. nov., a novel species designation for 2 *Cronobacter* sp. genomospecies 1, recovered from a leg infection, 3 water, and food ingredients 4. *Int. J. Syst. Evol. Microbiol* **2012**, *62*, 1277–1283. [CrossRef]

8. Weisbecker, A. The Naming of *Cronobacter Sakazakii*. 2009. Available online: https://www.foodsafetynews.com/2009/09/the-naming-of-cronobacter-sakazakii/ (accessed on 3 February 2019).

9. Ray, B.; Bhunia, A. *Fundamental Food Microbiology*, 5th ed.; CRC Taylor and Francis: Boca Raton, FL, USA, 2014; New and Emerging Foodborne Pathogens; pp. 54, 65, 73, 417, 473, 438, 499, 515. ISBN 13 978-1-4665-6444-2.

10. Kandhai, M.C.; Reij, M.W.; Gorris, L.G.; Guillaume-Gentil, O.; van Schothorst, M. Occurrence of *Enterobacter sakazakii* in food production environments and households. *Lancet* **2004**, *363*, 39–40. [CrossRef]

11. Osaili, T.; Forsythe, S. Desiccation resistance and persistence of *Cronobacter* species in infant formula. *Int. J. Food Microbiol.* **2009**, *136*, 214–220. [CrossRef] [PubMed]

12. Forsythe, S.J. Developments in our understanding of *Cronobacter* genus generated from studying an outbreak at a neonatal intensive care unit for 10 years. *Curr. Trends Microbiol.* **2017**, *11*, 23–31.

13. Forsythe, S.J. Updates on the *Cronobacter* genus. *Annu. Rev. Food Sci. Technol.* **2018**, *9*, 23–44. [CrossRef]

14. Holý, O.; Petrželová, J.; Hanulík, V.; Chroma, M.; Matoušková, I.; Forsythe, S.J. Epidemiology of *Cronobacter* spp. isolates from patients admitted to the Olomouc University Hospital (Czech Republic). *Epidemiol. Mikrobiol. Imunol.* **2014**, *63*, 69–72. [PubMed]

15. Patrick, M.E.; Mahon, B.E.; Greene, S.A.; Rounds, J.; Cronquist, A.; Wymore, K.; Boothe, E.; Lathrop, S.; Palmer, A.; Bowen, A. Incidence of *Cronobacter* spp. infections, United States, 2003–2009. *Emerg. Infect. Dis.* **2014**, *20*, 1520–1523. [CrossRef] [PubMed]

16. Alsonosi, A.; Hariri, S.; Kajsík, M.; Oriešková, M.; Hanulík, V.; Röderová, M.; Petrželová, J.; Kollárová, H.; Drahovská, H.; Forsythe, S.; et al. The speciation and genotyping of *Cronobacter* isolates from hospitalised patients. *Eur. J. Clin. Microbiol. Infect Dis.* **2015**, *34*, 1979–1988. [CrossRef] [PubMed]

17. Pangalos, G. Sur un bacille chromogène isolé par hémoculture. *C. R. Soc. Biol.* **1929**, *100*, 1097–1098.

18. Sakazaki, R. *Genus Enterobacter Hormaeche and Edwards*, 8th ed.; Buchanan, R.E., Gibbons, N.E., Eds.; Willams & Wilkins Co.: Baltimore, MD, USA, 1974; Bergey's Manual of Determinative Bacteriology; pp. 324–325.

19. Urmenyi, A.M.C.; Franklin, A.W. Neonatal death from pigmented coliform infection. *Lancet* **1961**, *1*, 313–315. [CrossRef]

20. Jøker, R.N.; Nørholm, T.; Siboni, K.E. A case of neonatal meningitis caused by a yellow enterobacter. *Dan. Med. Bull.* **1965**, *12*, 128–130.

21. Monroe, P.W.; Tift, W.L. Bacteremia associated with *Enterobacter sakazakii* (yellow, pigmented Enterobacter cloacae). *J. Clin. Microbiol.* **1979**, *10*, 850–851.

22. Muytjens, H.L.; Zanen, H.C.; Sonderkamp, H.J.; Kollée, L.A.; Wachsmuth, I.K.; Farmer, J.J. Analysis of eight cases of neonatal meningitis and sepsis due to *Enterobacter sakazakii*. *J. Clin. Microbiol.* **1983**, *18*, 115–120. [PubMed]

23. Kleiman, M.B.; Allen, S.D.; Neal, P.; Reynolds, J. Meningoencephalitis and compartmentalization of the cerebral ventricles caused by *Enterobacter sakazakii*. *J. Clin. Microbiol.* **1981**, *14*, 352–354. [PubMed]

24. Adamson, D.M.; Rogers, J.R. *Enterobacter sakazakii* meningitis with sepsis. *Clin. Microbiol. Newslett.* **1981**, *3*, 19–20. [CrossRef]

25. Arseni, A.; Malamou-Ladas, E.; Koutsia, C.; Xanthou, M.; Trikka, E. Outbreak of colonization of neonates with *Enterobacter sakazakii*. *J. Hosp. Infect.* **1987**, *9*, 143–150. [CrossRef]

26. Biering, G.; Karlsson, S.I.G.F.U.S.; Clark, N.C.; Jônsdôttir, K.E.; Ludvigsson, P.; Steingrimsson, O. Three cases of neonatal meningitis caused by *Enterobacter sakazakii* in powdered milk. *J. Clin. Microbiol.* **1989**, *27*, 2054–2056. [PubMed]

27. Bowen, A.B.; Braden, C.R. Invasive *Enterobacter sakazakii* disease in infants. *Emerg. Infect. Dis.* **2006**, *12*, 1185–1189. [CrossRef] [PubMed]

28. Simmons, B.P.; Gelfand, M.S.; Haas, M.; Metts, L.; Ferguson, J. *Enterobacter sakazakii* infections in neonates associated with intrinsic contamination of a powdered infant formula. *Infect. Control Hosp. Epidemiol.* **1989**, *10*, 398–401. [CrossRef] [PubMed]
29. Lecour, H.; Seara, A.; Cordeiro, J.; Miranda, M. Treatment of childhood bacterial meningitis. *Infection* **1989**, *17*, 343–346. [CrossRef]
30. Gallagher, P.G.; Ball, S.W. Cerebral infarctions due to CNS infection with *Enterobacter sakazakii*. *Pediatr. Radiol.* **1991**, *21*, 135–136. [CrossRef]
31. Block, C.; Peleg, O.; Minster, N.; Bar-Oz, B.; Simhon, A.; Arad, I.; Shapiro, M. Cluster of neonatal infections in Jerusalem due to unusual biochemical variant of *Enterobacter sakazakii*. *Eur. J. Clin. Microbiol. Infect. Dis.* **2002**, *21*, 613–616. [CrossRef]
32. Caubilla-Barron, J.; Hurrell, E.; Townsend, S.; Cheetham, P.; Loc-Carrillo, C.; Fayet, O.; Prere, M.F.; Forsythe, S.J. Genotypic and phenotypic analysis of *Enterobacter sakazakii* strains from an outbreak resulting in fatalities in a neonatal intensive care unit in France. *J. Clin. Microbiol.* **2007**, *45*, 3979–3985. [CrossRef]
33. van Acker, J.; de Smet, F.; Muyldermans, G.; Bougatef, A.; Naessens, A.; Lauwers, S. Outbreak of necrotizing enterocolitis associated with *Enterobacter sakazakii* in powdered milk formula. *J. Clin. Microbiol.* **2001**, *39*, 293–297. [CrossRef]
34. Himelright, I.; Harris, E.; Lorch, V.; Anderson, M.; Univ of Tennessee Medical Center at Knoxville; Jones, T.; Craig, A.; Tennessee Dept of Health; Kuehnert, M.; Forster, T.; et al. *Enterobacter sakazakii* infections associated with the use of powdered infant formula—Tennessee, 2001. *MMWR* **2002**, *51*, 297.
35. Ray, P.; Das, A.; Gautam, V.; Jain, N.; Narang, A.; Sharma, M. *Enterobacter sakazakii* in infants: Novel phenomenon in India. *Indian J. Med. Microbiol.* **2007**, *25*, 408. [CrossRef]
36. Alerte, C.E. Infections sévères à *Enterobacter sakazakii* chez des nouveau-nés ayant consommé une préparation en poudre pour nourrissons, France, octobre-décembre 2004. NUMÉRO THÉMATIQUE Risques infectieux: Approches. *Méthodologiques de la Veille et de L'aide à la Décision en Santé Publique* **2005**, *353*, 10.
37. Conde, A.A.; Legorburu, A.P.; Urcelay, I.E.; Zárate, Z.H.; Zugazabeitia, J.A. Sepsis neonatal por *Enterobacter sakazakii*. *Anales de Pediatria*. **2007**, *66*, 196–197. [CrossRef]
38. Jackson, E.E.; Flores, J.P.; Fernandez-Escartin, E.; Forsythe, S.J. Reevaluation of a suspected *Cronobacter sakazakii* outbreak in Mexico. *J. Food Prot.* **2015**, *78*, 1191–1196. [CrossRef]
39. Andrews, J. *Cronobacter*: FDA, CDC Find No Connection to Infant Formula. 2011. Available online: http://www.foodsafetynews.com/2011/12/cronobacter-fda-and-cdc-find-no-connection-to-formula/#.W1jQm9JKiUm (accessed on 3 February 2019).
40. McMullan, R.; Menon, V.; Beukers, A.G.; Jensen, S.O.; van Hal, S.J.; Davis, R. *Cronobacter sakazakii* infection from expressed breast milk, Australia. *Emerg. Infect. Dis.* **2018**, *24*, 393–394. [CrossRef]
41. Bowen, A.; Wiesenfeld, H.C.; Kloesz, J.L.; Pasculle, A.W.; Nowalk, A.J.; Brink, L.; Elliot, E.; Martin, H.; Tarr, C.L. Notes from the Field: *Cronobacter sakazakii* infection associated with feeding extrinsically contaminated expressed human milk to a premature infant—Pennsylvania, 2016. *MMWR* **2017**, *66*, 761. [CrossRef]
42. Jason, J. Prevention of invasive *Cronobacter* infections in young infants fed powdered infant formulas. *Pediatrics* **2011**, *130*, 1–9. [CrossRef]
43. Minnesota Department of Public Health. Reporting *Cronobacter* (*Enterobacter*) *sakazakii*. 2018. Available online: http://www.health.state.mn.us/divs/idepc/dtopics/reportable/enterobacters.html (accessed on 3 February 2019).
44. U.S. Food and Drug Administration. FDA Takes Final Step on Infant Formula Protections. 2018. Available online: https://www.fda.gov/ForConsumers/ConsumerUpdates/ucm048694.htm (accessed on 3 February 2019).
45. U.S. Food and Drug Administration. Powdered Infant Formula: An Overview of Manufacturing Processes. 2013. Available online: https://www.fda.gov/ohrms/dockets/ac/03/briefing/3939b1_tab4b.html (accessed on 3 February 2019).
46. Anonymous. How Products Are Made: Baby Formula. 2018. Available online: http://www.madehow.com/Volume-4/Baby-Formula.html (accessed on 3 February 2019).
47. Centers for Disease Control and Prevention. Nutrition, Physical Activity, and Obesity: Data, Trends and Maps. 2018. Available online: https://nccd.cdc.gov/dnpao_dtm/rdPage.aspx?rdReport=DNPAO_DTM.ExploreByLocation&rdRequestForwarding=Form (accessed on 3 February 2019).

48. World Health Organization. Infant and Young Child Feeding. 2018. Available online: http://www.who.int/en/news-room/fact-sheets/detail/infant-and-young-child-feeding (accessed on 3 February 2019).

49. World Health Organization. Breastfeeding Infographics. 2018. Available online: http://www.who.int/topics/breastfeeding/infographics/en/ (accessed on 3 February 2019).

50. Centers for Disease Control and Prevention. Hospital Support for Breastfeeding. 2011. Available online: https://www.cdc.gov/VitalSigns/BreastFeeding/ (accessed on 3 February 2019).

51. Anonymous. Baby Food and Infant Formula. 2018. Available online: https://www.foodsafety.gov/keep/types/babyfood/index.html (accessed on 3 February 2019).

52. Prolacta Bioscience. Breast Milk Donation Process. 2013. Available online: http://www.helpinghandsbank.com/how-to-donate-breast-milk (accessed on 3 February 2019).

53. Centers for Disease Control and Prevention. When & How to Wash Your Hands. 2016. Available online: https://www.cdc.gov/handwashing/when-how-handwashing.html (accessed on 3 February 2019).

54. World Health Organization. How to Prepare Formula for Bottle-Feeding at Home. 2007. Available online: http://www.who.int/foodsafety/publications/micro/PIF_Bottle_en.pdf (accessed on 3 February 2019).

55. World Health Organization. Safe Preparation, Storage and Handling of POWDERED infant Formula: Guidelines. 2007. Available online: http://www.who.int/foodsafety/publications/micro/pif_guidelines.pdf (accessed on 3 February 2019).

56. United States Department of Agriculture. National Nutrient Database for Standard Reference Legacy Release. 2018. Available online: https://ndb.nal.usda.gov/ndb/foods/show/299794?manu=&fgcd=&ds=&q=Infant%20formula,%20MEAD%20JOHNSON,%20Enfamil%20Reguline%20Powder,%20with%20ARA%20and%20DHA,%20not%20reconstituted (accessed on 3 February 2019).

57. Farmer, J.J. My 40-year history with *Cronobacter/Enterobacter sakazakii*–lessons learned, myths debunked, and recommendations. *Front. Pediatr.* **2015**, *3*, 84. [CrossRef]

58. Nazarowec-White, M.; Farber, J.M. Incidence, survival, and growth of *Enterobacter sakazakii* in infant formula. *J. Food Prot.* **1997**, *60*, 226–230. [CrossRef]

59. Kim, K.P.; Klumpp, J.; Loessner, M.J. *Enterobacter sakazakii* bacteriophages can prevent bacterial growth in reconstituted infant formula. *Int. J. Food Microbiol.* **2007**, *115*, 195–203. [CrossRef]

60. Edelson-Mammel, S.G.; Porteous, M.K.; Buchanan, R.L. Survival of *Enterobacter sakazakii* in a dehydrated powdered infant formula. *J. Food Prot.* **2005**, *68*, 1900–1902. [CrossRef]

61. Barron, J.C.; Forsythe, S.J. Dry stress and survival time of *Enterobacter sakazakii* and other Enterobacteriaceae in dehydrated powdered infant formula. *J. Food Prot.* **2007**, *70*, 2111–2117. [CrossRef]

62. Gurtler, J.B.; Kornacki, J.L.; Beuchat, L.R. *Enterobacter sakazakii*: A coliform of increased concern to infant health. *Int. J. Food Microbiol.* **2005**, *104*, 1–34. [CrossRef]

63. Breeuwer, P.; Lardeau, A.; Peterz, M.; Joosten, H.M. Desiccation and heat tolerance of *Enterobacter sakazakii*. *J. Appl. Microbiol.* **2003**, *95*, 967–973. [CrossRef]

64. Kumar, C.G.; Anand, S.K. Significance of microbial biofilms in food industry: A review. *Int. J. Food Microbiol.* **1998**, *42*, 9–27. [CrossRef]

65. Hurrell, E.; Kucerova, E.; Loughlin, M.; Caubilla-Barron, J.; Forsythe, S.J. Biofilm formation on enteral feeding tubes by *Cronobacter sakazakii*, Salmonella serovars and other Enterobacteriaceae. *Int. J. Food Microbiol.* **2009**, *136*, 227–231. [CrossRef]

66. Kim, H.; Ryu, J.H.; Beuchat, L.R. Attachment of and biofilm formation by *Enterobacter sakazakii* on stainless steel and enteral feeding tubes. *Appl. Environ. Microbiol.* **2006**, *72*, 5846–5856. [CrossRef]

67. Oh, S.W.; Chen, P.C.; Kang, D.H. Biofilm formation by *Enterobacter sakazakii* grown in artificial broth and infant milk formula on plastic surface. *J. Rapid Methods Autom. Microbiol.* **2007**, *15*, 311–319. [CrossRef]

68. Fouladkhah, A.; Geornaras, I.; Sofos, J. Biofilm formation of O157 and non-O157 Shiga toxin-producing Escherichia coli and multidrug-resistant and susceptible Salmonella Typhimurium and Newport and their inactivation by sanitizers. *J. Food Sci.* **2013**, *78*, M880–M886. [CrossRef]

69. Kim, H.; Ryu, J.H.; Beuchat, L.R. Effectiveness of disinfectants in killing *Enterobacter sakazakii* in suspension, dried on the surface of stainless steel, and in a biofilm. *Appl. Environ. Microbiol.* **2007**, *73*, 1256–1265. [CrossRef]

70. Henry, M.; Allison, A.; Chowdhury, S.; Fouladkhah, A. High-pressure Pasteurization for Inactivation of Rifampicin-resistant *Cronobacter sakazakii* in Reconstituted Infant Formula. 2018. Available online: https://iafp.confex.com/iafp/2018/meetingapp.cgi/Person/28028 (accessed on 3 February 2019).

71. Allison, A.; Chowdhury, S.; Fouladkhah, A. Biofilm Formation of Wild-type and Pressure-stressed *Cronobacter sakazakii* and Salmonella Serovars and Their Sensitivity to Sodium Hypochlorite. 2018. Available online: https://iafp.confex.com/iafp/2018/meetingapp.cgi/Paper/18392 (accessed on 3 February 2019).

72. Edelson-Mammel, S.G.; Buchanan, R.L. Thermal inactivation of *Enterobacter sakazakii* in rehydrated infant formula. *J. Food Prot.* **2004**, *67*, 60–63. [CrossRef]

73. Wang, C.Y.; Hsiao-Wen, H.; Chiao-Ping Hsu, H.; Binghuei, B.Y. Recent advances in food processing using high hydrostatic pressure technology. *Crit. Rev. Food Sci Nutr.* **2016**, *56*, 527–540. [CrossRef]

74. Tauscher, B. Pasteurization of food by hydrostatic high pressure: Chemical aspects. *Z. Lebensm. Unters. Forsch.* **1995**, *200*, 3–13. [CrossRef]

75. Goyal, A.; Sharma, V.; Upadhyay, N.; Sihag, M.; Kaushik, R. High pressure processing and its impact on milk proteins: A review. *Res. Rev. J. Dairy Sci. Technol.* **2018**, *2*, 12–20.

76. Arroyo, C.; Cebrián, G.; Mackey, B.M.; Condón, S.; Pagán, R. Environmental factors influencing the inactivation of *Cronobacter sakazakii* by high hydrostatic pressure. *Int. J. Food Microbiol.* **2011**, *147*, 134–143. [CrossRef]

77. Adekunte, A.; Valdramidis, V.P.; Tiwari, B.K.; Slone, N.; Cullen, P.J.; Donnell, C.P.O.; Scannell, A. Resistance of *Cronobacter sakazakii* in reconstituted powdered infant formula during ultrasound at controlled temperatures: A quantitative approach on microbial responses. *Int. J. Food Microbiol.* **2010**, *142*, 53–59. [CrossRef]

microorganisms

MDPI

Article

Interactions of Carvacrol, Caprylic Acid, Habituation, and Mild Heat for Pressure-Based Inactivation of O157 and Non-O157 Serogroups of Shiga Toxin-Producing *Escherichia coli* in Acidic Environment

Md Niamul Kabir [1], Sadiye Aras [1], Abimbola Allison [1], Jayashan Adhikari [1], Shahid Chowdhury [1] and Aliyar Fouladkhah [1,2,*]

[1] Public Health Microbiology Laboratory, Tennessee State University, Nashville, TN 37209, USA;
     mkabir492@gmail.com (M.N.K.); sadiyearas47@gmail.com (S.A.); abimbolaallison20@gmail.com (A.A.);
     adkjason99@gmail.com (J.A.); schowdh1@tnstate.edu (S.C.)
[2] Cooperative Extension Program, Tennessee State University, Nashville, TN 37209, USA
*   Correspondence: aliyar.fouladkhah@aya.yale.edu or afouladk@tnstate.edu; Tel.: +1-970-690-7392

Received: 2 May 2019; Accepted: 21 May 2019; Published: 23 May 2019

Abstract: The current study investigated synergism of elevated hydrostatic pressure, habituation, mild heat, and antimicrobials for inactivation of O157 and non-O157 serogroups of Shiga toxin-producing *Escherichia coli*. Various times at a pressure intensity level of 450 MPa were investigated at 4 and 45 °C with and without carvacrol, and caprylic acid before and after three-day aerobic habituation in blueberry juice. Experiments were conducted in three biologically independent repetitions each consist of two replications and were statistically analyzed as a randomized complete block design study using ANOVA followed by Tukey- and Dunnett's-adjusted mean separations. Under the condition of this experiment, habituation of the microbial pathogen played an influential ($p < 0.05$) role on inactivation rate of the pathogen. As an example, O157 and non-O157 serogroups were reduced ($p < 0.05$) by 1.4 and 1.6 Log CFU/mL after a 450 MPa treatment at 4 °C for seven min, respectively, before habituation. The corresponding log reductions ($p < 0.05$) after three-day aerobic habituation were: 2.6, and 3.3, respectively at 4 °C. Carvacrol and caprylic acid addition both augmented the pressure-based decontamination efficacy. As an example, *Escherichia coli* O157 were reduced ($p < 0.05$) by 2.6 and 4.2 log CFU/mL after a seven-min treatment at 450 MPa without, and with presence of 0.5% carvacrol, respectively, at 4 °C.

Keywords: Shiga toxin-producing *Escherichia coli*; habituation; carvacrol; caprylic acid; high-pressure pasteurization

1. Introduction

The 2015–2020 dietary guidelines of the United States Department of Agriculture recommends an increase in consumption of fruits and vegetables [1]. Over the last two decades, consumption of fresh and processed produce has also been increasing [2]. Contamination of plant-based products prior to consumption is practically unavoidable due to the ubiquitous nature of microbial pathogens and complexity of producing and processing operations [3,4], leading to an array of health and economic complications such as foodborne illnesses, hospitalizations, and death episodes, as well as recalls of food products and foodborne disease outbreaks [4–6].

Contamination with *Escherichia coli* O157:H7 and non-O157 serogroups of Shiga toxin-producing *E. coli* are one of the leading concerns of foodborne illnesses linked with muscle- and plant-based foods [7–9]. In addition to the Shiga toxin-producing *E. coli* O157:H7 (STEC) that has historically been linked to an array of food recalls and outbreaks since 1990s [9], non-O157 serogroups of Shiga toxin-producing *E. coli* (nSTEC) have been gaining increasing public health significance recently due to their emergence in food chain [9,10]. The serogroups O26, O45, O103, O111, O121, and O145 (also known as the 'Big Six') are considered as the most epidemiologically significant foodborne serogroups of public health concern among nSTEC [11,12].

Data derived from active surveillance programs of Centers for Disease Control and Prevention [13] indicates that in the United States 3704 and 1579 laboratory confirmed cases occur annually associated with STEC and nSTEC, respectively [13]. It is further estimated that every year in the United States, STEC and nSTEC are responsible for 63,153 and 112,752 domestic foodborne infections, respectively. Among these cases, 68% of STEC and 82% of nSTEC cases are foodborne in nature [5]. From 1998 to 2017, at least 590 foodborne outbreaks in the United Sates, including 14 foodborne outbreaks in the state of Tennessee were associated with STEC and/or nSTEC [13].

Although acidification or use of acidic foods are commonly associated with limited multiplication of microorganisms [14], microbial pathogens could survive and proliferate under acidic conditions [15–17]. Particularly, STEC had been involved in several outbreaks of foodborne diseases in different acidic foods, for example: yoghurt [18], mayonnaise [19] and apple cider [20]. It is also observed that acid adaptation can enhance STEC ability to survive in acidic juices for example in asparagus juice (pH = 3.6) and in mango juice (pH = 3.2) [21]. As an indigenous fruit crop of North America, blueberries have particularly low pH [22], have been associated with a seven-month STEC outbreak in Massachusetts [13], and thus, could be used as a model for investigating validation studies against STEC and nSTEC in acidic environment.

A viable alternative for pasteurization of products in manufacturing is application of elevated hydrostatic pressure [23]. Unlike traditional thermal processing methods that are typically associated with undesirable physiochemical and organoleptic changes in treated products [24], pressure-based pasteurization could be utilized for assuring safety of the products while minimally affecting their sensory and nutritional composition [25,26]. A pressure-based pasteurization could utilizes hydrostatic pressure of 100 to 1000 MPa, pressure-intensity level of around 600 MPa (87 K PSI) for about three min are currently the most common treatment in the private industry [27]. The main challenge for further adaption of pressure-based pasteurization treatments is slightly higher processing costs associated with the technology, thus, application of pressure treatments at intensity levels below 600 MPa, augmented with mild heat and natural antimicrobials could be a desirable approach for the food industry [27].

Caprylic acid is an eight-carbon fatty acid, which could be naturally found in several foods (coconut oil, bovine milk, palm oil, etc.) and is *Generally Recognized as Safe* by the U.S., Food and Drug Administration as a food additive [28,29]. Caprylic acid ($C_8H_{16}O_2$) could be an effective antimicrobial compound against Gram-negative and Gram-positive foodborne pathogens such as *E. coli* O157: H7, *Listeria monocytogenes* and *Salmonella* serovars [28,30–32]. Carvacrol ($C_{10}H_{14}O$), found primarily in oregano, is another natural bioactive compound with reported antimicrobial properties [33] and is broadly known for its effective antioxidant and antimicrobial activity [34,35].

The purpose of this study was to investigate the role of mild heat and addition of caprylic acid and carvacrol on decontamination efficacy of a pressure-based pasteurization treatment against STEC and nSTEC. Habituation of the pathogen, as further delineated in Section 2.1, in an acidic food vehicle were also investigated as an important element for maximizing external validity of a decontamination hurdle validation study.

2. Materials and Methods

2.1. Escherichia coli Strains, Preparation of Culture, Habituation, and Inoculation

A six-strain mixture of Shiga toxin-producing *E. coli* O157:H7 (STEC) (ATCC®, Manassas, VA, USA, numbers BAA 460, 43888, 43894, 35150, 43889 and 43890) and a six-strain mixture of 'Big

Six' non-O157 Shiga toxin-producing *E. coli* (nSTEC) strains, including O26:H11, O45:H2, O103:H2, O111:NM, O121:H19, and O145 (ATCC® numbers BAA 2196, BAA 2193, BAA 2215, BAA 2440, BAA 2219 and BAA 2192 respectively) were used in this study for inoculation of sterilized (autoclaved at 121 °C, for 15 min, under 15 PSI) blueberry juice. The STEC and nSTEC strains with public health significance and those derived from our previously published strain selection trials were selected for this study [9].

The cultures for each of the above-mentioned strains, obtained from American Type Culture Collection (Manassas, VA, USA), were grown on Tryptic Soy Agar (Difco, Becton Dickinson, Franklin Lakes, NJ, USA) supplemented with 0.6% yeast extract (TSA + YE) and for 24 h incubated at 37 °C. Forty eight hours before each experiment, a loopful of single colony of each STEC or nSTEC strains was aseptically transferred for activation into 10 mL Tryptic Soy Broth (Difco, Becton Dickinson, Franklin Lakes, NJ, USA) supplemented with 0.6% yeast extract (TSB + YE). Use of this media and the supplement minimizes acid stress of the bacterial cells during incubation at 37 °C for 20–24 h [23,27,36]. After incubation for 20–24 h at 37 °C, 100-µL aliquot of the culture was individually and aseptically sub-cultured into another 10 mL of TSB + YE, for 22–24 h at 37 °C, for each of the 12 strains, separately.

Each overnight sub-cultured strain (2 mL per strain) was then harvested by centrifugation (Model 5424, Eppendorf North America, Hauppauge, NY, USA; Rotor FA-45-24-11) at 6000 RPM (3548 *g* for 88 mm rotor) for 15 min. Bacterial pellets were then re-suspended in 2 mL Phosphate Buffered Saline (VWR International, Radnor, PA, USA) and washed twice by centrifugation with the above-mentioned intensity and time to remove growth media, excreted secondary metabolites, and sloughed cell components. Two separate six-strain bacterial cocktails (for STEC and nSTEC) were made by combining the washed and re-suspended strains into PBS (VWR International, Radnor, PA, USA), and were used as the inocula for this study. Non-habituated samples were prepared by 10-fold dilution of each of the STEC and nSTEC cocktails in PBS followed by inoculating sterilized blueberry juice samples for target population of 5–6 Log CFU/mL. The habituated samples were prepared by adding 10 mL of STEC and nSTEC cocktails (separately for each strain mixture) to 40 mL of sterilized blueberry juice, followed by a 72 h aerobic storage at 4 °C [23]. Habituation allows pathogen acclimatization to intrinsic factor and temperature of the food product and could impact external validity of a microbial challenge study [37–39]. Levels of inoculation for habituated and non-habituated samples and below-mentioned temperatures and concentrations of antimicrobials were selected after conduct of preliminary trials.

2.2. Preparation of Antimicrobials, and Mild Heat and Pressure-Based Pasteurization

Two naturally occurring antimicrobial compounds (carvacrol and caprylic acid) were used in this study for inactivation of 72-h habituated STEC and nSTEC in sterilized blueberry juice at two temperatures and at an elevated hydrostatic pressure level of 450 MPa. The temperature of the trials were precisely controlled using a water jacket surrounding the treatment chamber, connected to a circulating water bath and monitored by k-type thermocouples as delineated in details in our recent open access publications [23,27]. For 4 °C experiments, 0.5% (7.5 µL of antimicrobial in 1.5 mL of inoculated product (*v/v*)) and for 45 °C experiment, 0.1% concentration (1.5 µL of antimicrobial in 1.5 mL of inoculated product (*v/v*)) of carvacrol and caprylic acid were used based on the above-mentioned preliminary trials. In each experiment, the concentration of antimicrobials was prepared aseptically in sterilized blueberry juice. Inoculated blueberry juice were then exposed to 450 Megapascal (MPa), i.e., c. 65,000 pounds per square inch (PSI) hydrostatic pressure (Barocycler Hub880 Explorer, Pressure Bioscience Inc., South Easton, MA, USA) at 4 and 45 °C for the time intervals of 0 (untreated control) to 7 min. Samples containing antimicrobials were also tested immediately after addition of the antimicrobial and prior to pressure treatment (treated control). The treatments were carried out in no-disk PULSE (Pressure BioScience Inc., South Easton, MA, USA) containing 1.5 mL of inoculated blueberry juice. The PULSE tubes were then used for hydrostatic pressure treatment with 1, 3, 5 and 7 min holding time, in addition to the above-mentioned controls. Pressure and temperature of trials were monitored and recorded automatically every 3 s using HUB Explorer PBI (Version 1.0.8, Pressure BioScience Inc., South Easton, MA, USA) software.

2.3. The pH, Neutralization, and Microbiological Analyses

Each treated sample was neutralized using 5 mL of D/E neutralizing broth (Difco, Becton Dickinson, Franklin Lakes, NJ, USA) to reduce the effect of food vehicle's intrinsic factors before microbiological analyses. The detection limit of microbiological analyses was, thus, 0.48 log CFU/mL. After neutralization, to enhance the recovery of injured cells, samples were 10-fold serially diluted in Maximum Recovery Diluent (Difco, Becton Dickinson, Franklin Lakes, NJ, USA) and then plated on TSA media supplemented with 0.6% yeast extract (TSA + YE). All plates were incubated for 24–48 h at 37 °C. After incubation, colony forming units were counted manually and converted into log values for further statistical analyses. The pH of treated samples was measured two times (after treatment and before neutralization, as well as after neutralization) using a digital pH meter (Mettler Toledo AG, Grelfensee, Switzerland) calibrated at pH levels of 4, 7 and 10 before measurements.

2.4. Statistical Analyses and Experimental Design

The sample size of this study was determined to be at least 5 repetitions per treatment to achieve statistical power of 80%. This sample size was obtained from a previous *a priori* power analysis using Proc Power of SAS software (version 9.2, SAS Institute, Cary, NC, USA) using existing pressure-treated products in the public health microbiology laboratory [40]. The present study was conducted at two temperatures of 4 and 45 °C using two inocula of STEC and nSTEC. At each temperature, the study contained three biologically independent repetitions (three blocks), each consisted of 2 replications. Each replication was also microbiologically analyzed in duplicate (microbiological replications). Thus each reported value is a mean of 12 individual analyses (i.e., 3 blocks, 2 replications, and 2 microbiological repetitions). Initial data arrangement, log transformations and descriptive analysis of the data were completed using Microsoft Excel. The study was considered as a randomized complete block design, and log-transformed microbial counts were statistically analyzed using generalized liner model of SAS for conduct of ANOVA followed by Tukey- and Dunnett's-adjusted mean separations at type I error level of 5% (alpha= 0.05). In order to calculate inactivation indices (D-value and K_{max}) Microsoft Excel and GInaFiT (version 1.7, Katholieke Universiteit, Leuven, Belgium) [41] software were used, respectively.

3. Results and Discussion

As previously delineated in Section 2.2, the experiments were conducted under controlled temperatures to assure microbial inactivation could be attributed to the intrinsic and extrinsic factors of interest rather than temperature fluctuations. Samples treated at 4 and 45 °C, had similar ($p \geq 0.05$) temperature values (mean ± SD) before and after the treatments. Across all treatments at 4 °C, the values before treatments were 4.8 ± 0.2 °C and were 4.9 ± 0.2 °C after the treatments. Values were ranging from 4.3 to 5.2 °C and 4.3 to 5.3 °C, before and after treatments, respectively. For samples treated at 45 °C as well, temperature recordings were similar ($p < 0.05$) before and after treatments. The temperature values were 44.5 ± 0.3 and 44.8 ± 0.4 °C, before and after treatments, respectively. The range for the recordings were from 43.7 to 45.0 °C and 43.7 to 45.2 °C for samples prior and after treatments, respectively. Extent of precision in control of temperature could be further delineated through calculation of coefficient of variation (CV) associated with the temperature recordings. The CVs associated with 4 °C samples were 4.51% and 4.57% and for samples treated at 45 °C were 0.58% and 0.76%, before and after treatments, respectively.

The pH levels of the samples were also similar ($p \geq 0.05$) before and after treatments. For samples treated at 4 °C, and prior to neutralization, the pH value (mean ± SD) and range were 3.16 ± 0.0 and 3.12 to 3.22, respectively. After neutralization, these values were expectedly increased ($p < 0.05$) to 5.56 ± 0.27, ranging from 5.25 to 6.02. Similarly, for samples treated at 45 °C, these values were

3.33 ± 0.1 and 5.54 ± 0.1, before and after neutralization. These values were ranging from 3.24 to 3.44 and 5.37 to 5.69 before and after neutralization, respectively. The CVs associated with pH measurements were 0.42% (4 °C samples, without neutralization), 4.89% (neutralized 4 °C samples), 1.61% (45 °C samples, without neutralization), and 1.70% (neutralized, 45 °C samples).

3.1. Pressure-Based Pasteurization of O157 and Non-O157 Serogroups of Shiga Toxin-Producing Escherichia coli at 4 °C, Before and After Habituation

As further delineated in Section 2.1, this study utilized two separate inoculated products for the pressure-based microbial challenge studies using a six-strain mixture of O157 Shiga toxin-producing *Escherichia coli* (STEC) and a six-strain non-O157 mixture of O26, O45, O103, O111, O121, and O145 Shiga toxin-producing *Escherichia coli* (nSTEC). Data associated with the current study is also provided as a supplementary file. At 4 °C and after the habituation, the STEC and nSTEC counts (mean ± SD) of blueberry juice were 6.32 ± 0.5 and 6.12 ± 0.6 Log CFU/mL, respectively (Figure 1A). Hydrostatic pressure treatment of 450 MPa (c. 65 K PSI), for 1, 3, 5, and 7 min, reduced the STEC by 1.7 to 2.6 log CFU/mL and specifically reduced ($p < 0.05$) the STEC counts to 4.60 ± 0.8, 4.45 ± 0.9, 4.51 ± 0.8, 3.68 ± 1.1, respectively (Figure 1A). Sensitivity of nSTEC were similar to STEC- the treatments for 1, 3, 5, and 7 min at the above-referenced pressure and intensity level lead to 1.1, 2.7, 2.6, and 3.3 log reductions of nSTEC samples (Figure 1A). Under the condition of our experiment, habituation played an influential role on sensitivity of both STEC and nSTEC serogroups to pressure-based treatments at 4 °C (Figure 1A,B). For non-habituated samples at 4 °C, STEC and nSTEC counts were 5.55 ± 0.6 and 5.00 ± 0.1 prior to treatments, respectively. The STEC were reduced ($p < 0.05$) to 4.35 ± 0.3, 4.26 ± 0.5, 4.57 ± 0.9, and 4.13 ± 0.2 log CFU/mL, after treatments for 1, 3, 5, and 7 min at 450 MPa, respectively (Figure 1B). These reductions were considerably less that reductions of the habituated STEC. In other words, the habituated STEC were more sensitive to pressure-based treatments at this temperature relative to the non-habituated phenotype. As an example, 7 min of treatment at 450 MPa at 4 °C reduced the habituation STEC ($p < 0.05$) by 3.7 log CFU/mL (Figure 1A), while the same treatment were only capable of reducing ($p < 0.05$) the non-habituated STEC for 1.4 log CFU/mL (Figure 1B). This trend was also observed for habituated and non-habituated nSTEC (Figure 1A,B).

This considerable difference in sensitivity of the pathogen before and after habituation had been discussed in the microbiology literature in the past. While studies, similar to our current study, had observed that post-stress, pathogens exhibit more sensitivity to a decontamination treatment. Some studies also indicate certain stressors could lead to cross-protective effects, i.e., increasing the tolerance of a pathogen post-stress [23,38,42,43]. If a manufacturer is relying on validation studies with non-habituated inoculated pathogen, the validation data could be an overestimation or underestimation of the treatment decontamination efficacy, and thus, leading to false sense of treatment efficacy or a treatment that is overly conservative. This could also lead to over- or under-estimation of microbial reductions in risk assessment analyses throughout the supply chain. It is thus recommended that habituation for each specific product-pathogen-treatment combination be considered as an important factor of a validation study to assure data obtained from a microbial challenge study has external validity and is conducted in an environment that is as close as possible to actual processing condition of a product. This could assure economic feasibility of a treatment as well as providing assurance that a treatment is safeguarding the public health. Currently, there is a knowledge gap about sensitivity of acid-adapted and acid-stressed foodborne pathogens of public health concern to various pressure-based treatments relative to their wild-type phenotypes.

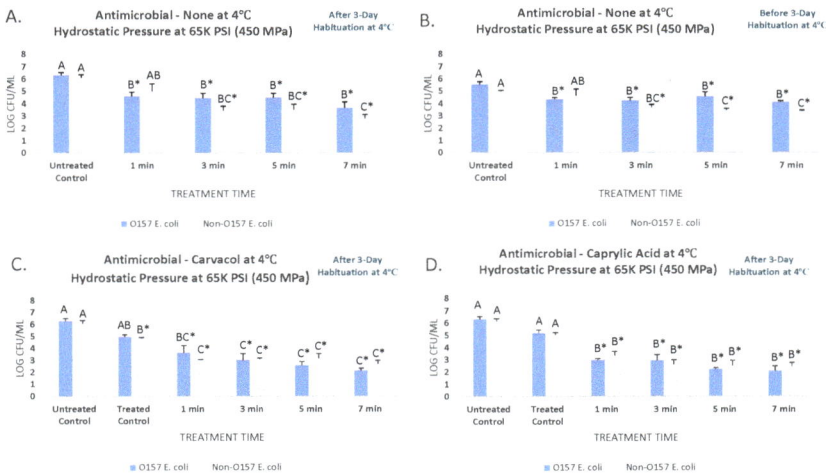

Figure 1. Inactivation of six-strain cocktail of habituated and non-habituated *E. coli* O157:H7 (ATCC® numbers BAA 460, 43888, 43894, 35150, 43889, 43890) and the 'Big Six' non-O157 *E. coli* mixtures (ATCC® numbers BAA 2196, BAA 2193, BAA 2215, BAA 2440, BAA 2219, BAA 2192) in sterilized blueberry juice, treated by carvacrol (0.5%), caprylic acid (0.5%) and elevated hydrostatic pressure at 450 MPa (Barocycler Hub880 Explorer, Pressure Bioscience Inc., South Easton, MA, USA) for 0, 1, 3, 5, and 7 min at 4 °C. In each graph, and for each pathogen mixture separately, columns of each time interval followed by different uppercase letters are representing log CFU/mL values (Mean ± SE) that are statistically ($p < 0.05$) different (Tukey-adjusted ANOVA). Uppercase letters followed by * sign are statistically ($p < 0.05$) different than the untreated control (not treated with antimicrobial) (Dunnett's-adjusted ANOVA). (**A**) After 3 days of habituation, treated by no antimicrobial at 4 °C; (**B**) Before 3 days of habituation, treated by no antimicrobial at 4 °C; (**C**) After 3 days of habituation, treated by 0.5% carvacrol at 4 °C; (**D**) After 3 days of habituation, treated by 0.5% caprylic acid at 4 °C.

3.2. Augmenting the Efficacy of High Pressure Pasteurization using Carvacrol and Caprylic Acid at 4 °C

Under the condition of our experiments, we observed the selected two natural antimicrobials could appreciably augment the efficacy of the pressure-based pasteurization of STEC and nSTEC at 4 °C. It is noteworthy that the synergism of elevated hydrostatic pressure and carvacrol and caprylic acid were investigated on inoculated samples with three-day aerobic habituation that, as discussed in Section 3.1, yields more realistic outcome with higher external validity. Data and graphical representations obtained and reported for these experiments were similar in structure to those elaborated in Section 3.1 with the exception that the microbial reductions immediately after exposure to 0.5% antimicrobial were also determined, thus graphs contain untreated controls as well as treated controls (e.g., samples that are immediately neutralized and enumerated after exposure to the antimicrobial).

The STEC and nSTEC counts (mean ± SD) for untreated controls were 6.32 ± 0.5 and 6.12 ± 0.6 log CFU/mL, respectively at 4 °C. Immediately after exposure to 0.5% carvacrol, these counts were reduced ($p < 0.05$) to 4.99 ± 0.4 and 4.86 ± 0.1 log CFU/mL, for STEC and nSTEC samples, respectively (Figure 1C). Carvacrol were able to enhance ($p < 0.05$) the efficacy of the treatment. As an example, treatments of STEC samples for 5 and 7 min at 450 MPa at 4 °C lead to 3.8 and 4.2 log CFU/mL reductions ($p < 0.05$) while same treatment at the same temperature and intensity level without presence of carvacrol resulted in 1.0 and 1.4 log CFU/mL reductions ($p < 0.05$) in habituated samples, respectively (Figure 1A,C). In vast majority of tested time intervals, STEC and nSTEC serogroups exhibited comparable sensitivity to high hydrostatic pressure (Figure 1A–D). Caprylic acid, at 0.5% concentration, were similarly effective to augment the decontamination efficacy of the pressure-based treatments at 4 °C. The nSTEC counts, as an example, were 6.12 ± 0.6 log CFU/mL prior to treatment and prior to exposure to caprylic acid (untreated control). These counts were reduced ($p < 0.05$) to 5.02 ± 0.5 log CFU/mL immediately after exposure to 0.5% caprylic acid (treated controls) and were further reduced ($p < 0.05$) to 3.35 ± 0.8, 2.60 ± 0.9, 2.49 ± 1.1, 2.44 ± 0.8 log CFU/mL after 1-, 3-, 5-, and 7-min treatments at 450 MPa at 4 °C (Figure 1D). These reductions were appreciably higher than those obtained from elevated hydrostatic pressure alone for both STEC and nSTEC. As an example, the above-reference 7-min treatment reduced ($p < 0.05$) the STEC and nSTEC for 4.2 and 3.7 log CFU/mL in presence of 0.5% caprylic acid, respectively, while the same treatment resulted in 1.4 and 1.6 log CFU/mL reductions ($p < 0.05$) for the habituated samples without caprylic acid (Figure 1A,D).

These results could be of practical importance for the private industry with a high-pressure processing plant. At current times, slightly higher operation costs of many pressure-treated products relative to existing heat-treated commodities in the market are the main curtailment for further expanding the utilization of this technology in the food processing industry [23,27]. Main costs of the operation are associated with maintenance and energy expenditure associated with use of high levels of hydrostatic pressure. Our study indicates that lower levels of pressure could lead to similar decontamination efficacy in presence of natural antimicrobials such as carvacrol and caprylic acid.

3.3. Pressure-Based Pasteurization of the Pathogen at 45 °C as Affected by Habituation, Carvacrol and Caprylic Acid

The pressure treatments discussed in Sections 3.1 and 3.2, coupled with mild heat were appreciably more efficacious for decontamination of the product from STEC and nSTEC (Figure 2A–C). This thermal-assisted pressure-based treatment at 450 MPa and 45 °C were able to reduce ($p < 0.05$) the STEC counts by 3.8, 4.0, 4.8, and 5.4 log CFU/mL for habituated samples (Figure 2A). This decontamination efficacy were also observed with similar trends for the nSTEC samples, leading to 3.3 to 4.8 log CFU/mL reductions for treatments of up to 7 min (Figure 2A). Effects of habituation at this temperature were less pronounced relative to the experiment conducted at 4 °C (Figure 2A). As an example, counts of non-habituated STEC and nSTEC samples were 5.96 ± 0.3 and 5.88 ± 0.5 before treatments and were reduced ($p < 0.05$) to 0.66 ± 0.2 and 0.91 ± 0.7 log CFU/mL after 7-min treatments at 450 MPa and 45 °C, respectively. Counts for habituated STEC and nSTEC were reduced ($p < 0.05$) by 5.4 and 4.8 log values, similar to the reductions obtained by treatment of non-habituated samples (Figure 2A,B). Our data indicates, habituation could have a more pronounced effect on external validity of a pressure-based validation study at 4 °C while may have only modest effects on validity of a thermal-assisted high-pressure processing.

At elevated temperature, effects of carvacrol and caprylic acid at 0.1% were also less pronounced in augmenting the decontamination efficacy of the treatments (Figure 2C,D). This indicates that while these antimicrobials might be efficacious alone, or coupled with pressure-based treatments at lower temperature, at 0.5% concentrations, but these do not augment the efficacy of a treatment at higher temperature when tested at 0.1%. Similar effects were observed in the past when acidic acid was not able to augment efficacy of a heat treatment at elevated temperature while efficacious at ambient environment [44]. As an example, STEC counts of habituated samples treated without antimicrobial, with 0.1% carvacrol, and with 0.1% caprylic acid for 3 min at 450 MPa were similar ($p \geq 0.05$) and were 3.09 ± 1.3, 3.91 ± 0.4, and 3.79 ± 0.6 log CFU/mL, respectively (Figure 2A,C,D). Similar to treatments at lower temperature, STEC and nSTEC counts were comparable for the vast majority of time and pressure treatments, prior and after habituation, and in presence or absence of the antimicrobials (Figure 2A–D). Our results, thus indicate that mild elevated heat and natural antimicrobial could augment efficacy of a pressure-based pasteurization with similar effectiveness against STEC and nSTEC, but utilization of both mild heat and antimicrobials simultaneously does not necessarily provide added decontamination benefit.

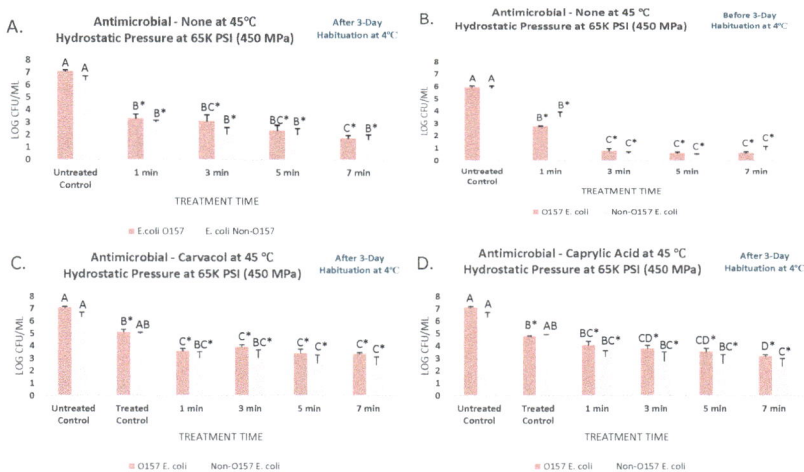

Figure 2. Inactivation of six-strain cocktail of habituated and non-habituated *E. coli* O157:H7 (ATCC® numbers BAA 460, 43888, 43894, 35150, 43889, 43890) and the 'Big Six' non-O157 *E. coli* strain mixtures (ATCC® numbers BAA 2196, BAA 2193, BAA 2215, BAA 2440, BAA 2219, BAA 2192) in sterilized blueberry juice, treated by carvacrol (0.1%), caprylic acid (0.1%) and elevated hydrostatic pressure at 450 MPa (Barocycler Hub880 Explorer, Pressure Bioscience Inc., South Easton, MA, USA) for 0, 1, 3, 5, and 7 min at 45 °C. In each graph, and for each pathogen mixture separately, columns of each time interval followed by different uppercase letters are representing log CFU/mL values (mean ± SE) that are statistically ($p < 0.05$) different (Tukey-adjusted ANOVA). Uppercase letters followed by * sign are statistically ($p < 0.05$) different than the untreated control (not treated with antimicrobial) (Dunnett's-adjusted ANOVA). (**A**) After three days of habituation, treated by no antimicrobial at 45 °C; (**B**) Before three days of habituation, treated by no antimicrobial at 45 °C; (**C**) After three days of habituation, treated by 0.1% carvacrol at 45 °C; (**D**) After three days of habituation, treated by 0.1% caprylic acid at 45 °C.

3.4. Linear and Non-Leaner Inactivation Indices for High Pressure Pasteurization of O157 and Non-O157 Serogroups of Shiga Toxin-Producing Escherichia coli at 4 and 45 °C

Effects of habituation and synergism of heat, carvacrol and/or caprylic acid with the pressure-based pasteurization could be further discussed by interpretation of linear and non-linear inactivation indices (Figures 3 and 4). D-value was the linear model utilized in this study that could be interpreted as the time required at the specific condition of the experiment to achieve 90% reduction of the inoculated pathogen (i.e., one-log reduction). A non-linear model had also been utilized in this study using GInaFiT version 1.7 software [41]. The reported k_{max} values are in unit of 1/min thus smaller K_{max} values indicate longer time required for reduction of the pathogen, in contrast to D-value that is in unit of min.

The D-value for STEC for habituated and non-habituated samples (Figure 3A,C) emphasizes on importance of this practice on outcome of a challenge study. The D-value associated with habituated STEC were 13.70 min while for non-habituated samples this inactivation index was 7.76 min (Figure 3A,C). This effect was not observed at higher temperature. At 45 °C, the D-values were similar for habituated and non-habituated STEC samples and were 1.65 and 1.51 min, respectively (Figure 3A,C).

Carvacrol was able to augment the efficacy of pressure-based pasteurization of the pathogen as evidenced by inactivation indices. As an example, nSTEC required 8.03 min of treatment at 450 MPa and 4 °C for one-log reduction e.g., D-value = 8.03 min (Figure 3B). In presence of 0.5% carvacrol, same treatment required only 2.92 min for one-log reduction (Figure 3F). The k_{max} values also delivered similar trend, having values of 2.77 and 13.19 1/min for nSTEC samples without carvacrol, and those treated with presence of 0.5% carvacrol (Figure 3B,F). At 4 °C, 0.5% caprylic acid was also capable of reducing the time for one-log reduction of both STEC and nSTEC as exhibited in Figure 3A,B,G,H.

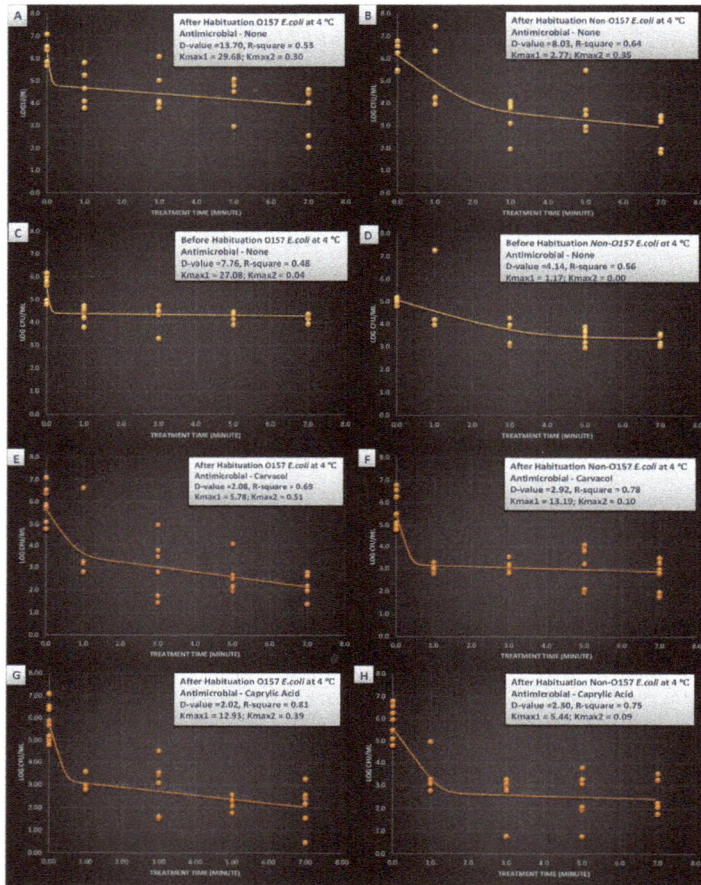

Figure 3. Inactivation rates for six-strain habituated and non-habituated mixture of *E. coli* O157:H7 (ATCC® numbers BAA 460, 43888, 43894, 35150, 43889, 43890) and the 'Big Six' non-O157 *E. coli* strain mixtures (ATCC® numbers BAA 2196, BAA 2193, BAA 2215, BAA 2440, BAA 2219, BAA 2192) exposed to 0.5% carvacrol, 0.5% caprylic acid, and elevated hydrostatic pressure at 450 MPa (Barocycler Hub 440, Pressure BioScience Inc., South Easton, MA) in sterilized blueberry juice at 4 °C. Using the GInaFiT software, the provided K_{max} values are selected from the best-fitted model (goodness-of-fit indicator of R^2 values, $\alpha = 0.05$). K_{max} values indicate the expressions of number of log cycles of reduction in 1/min unit for each pressure/temperature combinations. Presented D-values are calculated based on best-fitted linear model, showing time required for one log (90%) of microbial cell reductions of the microbial cell mixture. (**A**) Habituated *E. coli* O157 treated by no antimicrobial at 4 °C with $R^2 = 0.53$; (**B**) Habituated *E. coli* non-O157 treated by no antimicrobial at 4 °C with $R^2 = 0.64$; (**C**) Non-habituated *E. coli* O157 treated by no antimicrobial at 4 °C with $R^2 = 0.48$; (**D**) Non-habituated *E. coli* non-O157 treated by no antimicrobial at 4 °C with $R^2 = 0.56$; (**E**) Habituated *E. coli* O157 treated by carvacrol (0.5%) at 4 °C with $R^2 = 0.69$; (**F**) Habituated *E. coli* non-O157 treated by carvacrol (0.5%) at 4 °C with $R^2 = 0.78$; (**G**). Habituated *E. coli* O157 treated by caprylic acid (0.5%) at 4 °C with $R^2 = 0.81$; (**H**). Habituated *E. coli* non-O157 treated by caprylic acid (0.5%) at 4 °C with $R^2 = 0.75$.

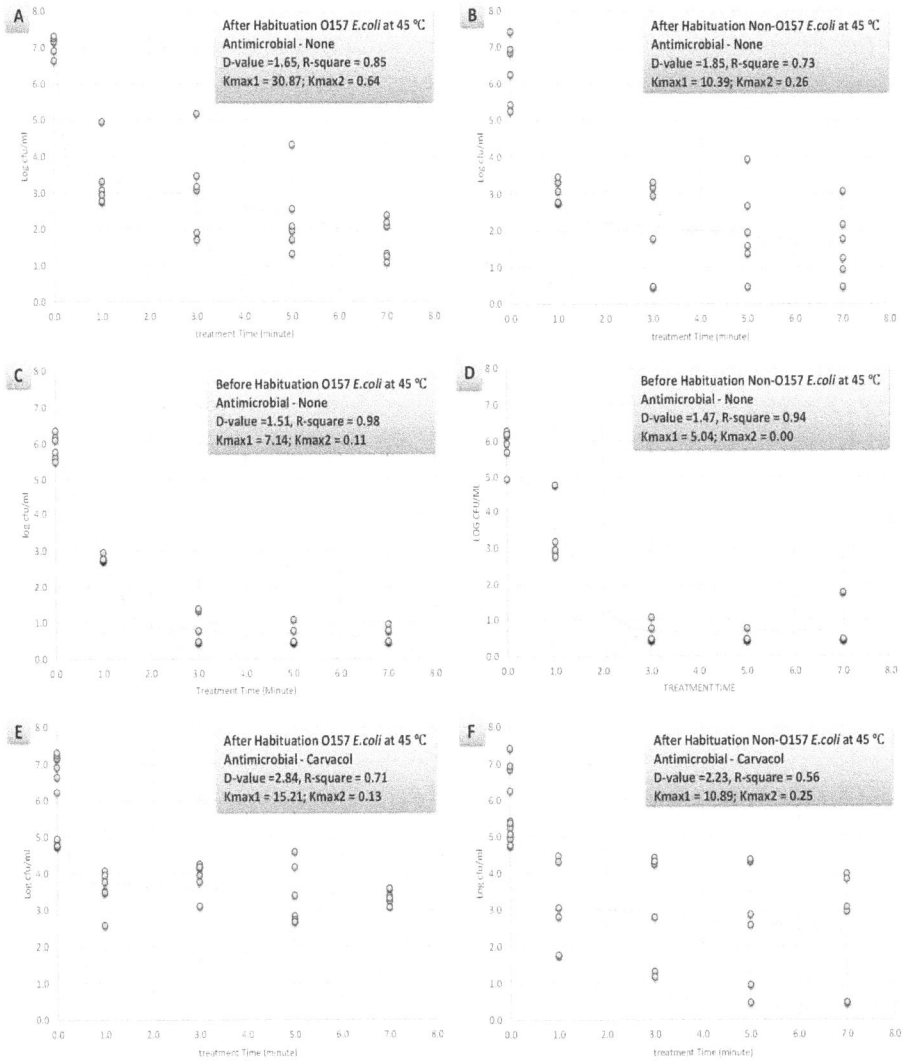

Figure 4. *Cont.*

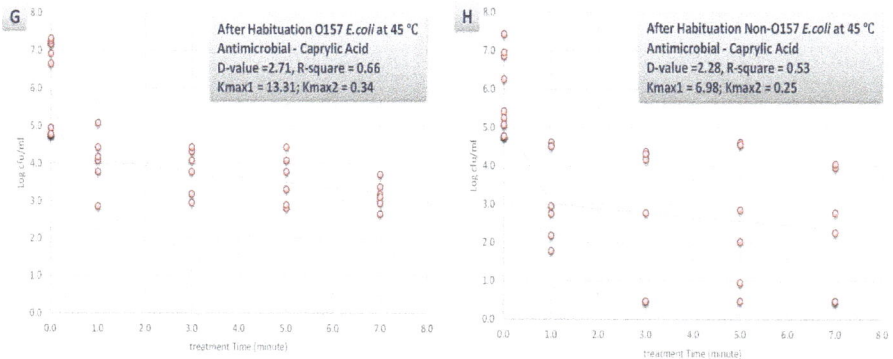

Figure 4. Inactivation rates for six-strain habituated and non-habituated mixture of *E. coli* O157:H7 (ATCC® numbers BAA 460, 43888, 43894, 35150, 43889, 43890) and the 'Big Six' non-O157 *E. coli* strain mixtures (ATCC® numbers BAA 2196, BAA 2193, BAA 2215, BAA 2440, BAA 2219, BAA 2192) exposed to 0.1% carvacrol, 0.1% caprylic acid, and elevated hydrostatic pressure at 450 MPa (Barocycler Hub 440, Pressure BioScience Inc., South Easton, MA) in sterilized blueberry juice. Using the GInaFiT software, the provided K_{max} values are selected from the best-fitted model (goodness-of-fit indicator of R^2 values, $\alpha = 0.05$). K_{max} values indicate the expressions of number of log cycles of reduction in 1/min unit for each pressure/temperature combinations. Presented D-values are calculated based on best-fitted linear model, showing time required for one log (90%) of microbial cell reductions of the habituated microbial cell mixture. (**A**) Habituated *E. coli* O157 treated by no antimicrobial at 45 °C with $R^2 = 0.53$; (**B**) Habituated *E. coli* non-O157 treated by no antimicrobial at 45 °C with $R^2 = 0.64$; (**C**) Non-habituated *E. coli* O157 treated by no antimicrobial at 45 °C with $R^2 = 0.48$; (**D**) Non-habituated *E. coli* non-O157 treated by no antimicrobial at 45 °C with $R^2 = 0.56$; (**E**) Habituated *E. coli* O157 treated by carvacrol (0.1%) at 45 °C with $R^2 = 0.69$; (**F**) Habituated *E. coli* non-O157 treated by carvacrol (0.1%) at 45 °C with $R^2 = 0.78$; (**G**) Habituated *E. coli* O157 treated by caprylic acid (0.1%) at 45 °C with $R^2 = 0.81$; (**H**) Habituated *E. coli* non-O157 treated by caprylic acid (0.1%) at 45 °C with $R^2 = 0.75$.

The synergistic effects of the tested antimicrobial (0.1% concentration) and habituation were less pronounced at elevated temperature of 45 °C (Figure 4). For example, the D-values for habituated with no antimicrobial, non-habituated with no antimicrobial, habituated and treated with 0.1% carvacrol, and habituated and treated with 0.1% caprylic acid for STEC samples were similar ($p < 0.05$) and were 1.65, 1.51, 2.84, and 2.71 min, respectively (Figure 4A,C,E,G). This indicates that additional of antimicrobials could appreciable enhance the decontamination efficacy of a pressure-based intervention at 4 °C while could have minor to no effects for augmenting the efficacy of a thermal-assisted high pressure pasteurization.

As further discussed in the introduction, antimicrobials used in the current study have *Generally Recognized as Safe* status in the United States regulatory landscape [28,45] and the concentrations utilized are similar to those used previously in literature [46]. As for any product development project, incorporation of these antimicrobials in a product formula for enhancing safety of the product, requires product specific and close attention to organoleptic properties of the product with and without the antimicrobials.

It is also noteworthy that this study utilized a six-strain mixture of *E. coli* O157:H7 and a six-strain mixture of non-O157 Shiga toxin-producing *E. coli*. As delineated in Section 2.1, these were selected based on our previously published screening trials as well as the strains' public health significance. Acid tolerance, sensitivity to intrinsic and extrinsic factors of a product, and reduction as a result of a thermal or non-thermal treatment could vary immensely among the plethora of Shiga toxin-producing isolates of the pathogen. Conducting experiments with similar design to the current study in future, using an array of individual strains followed by further analyses of the survivors after the treatments could be experiments of utmost importance and a complement to the current study for better assimilation of sensitivity of this pathogen of public health concern to pressure-based interventions under various intrinsic and extrinsic conditions of a product and processing conditions.

4. Conclusions

Under the condition of our experiments, for the vast majority of tested time and pressure intervals in presence or absence of two antimicrobials, O157 and non-O157 serogroups of Shiga toxin-producing *Escherichia coli* exhibited similar sensitivity to elevated hydrostatic pressure. Thus, if a pressure-based treatment is validated and is efficacious for decontamination of O157 serogroups of *Escherichia coli*, it would almost certainly exhibit comparable efficacy for reduction of non-O157 serogroups of the pathogen as well. We also observed that, particularly for treatments at 4 °C, habituation of samples could meaningfully alter the results of a microbial challenge study and thus would need to be carefully considered for maximizing the external validity of a validation study. Reducing the cost of pressure-based treatments are currently the major curtailment for further adaption of this emerging technology. Our study indicates that application of natural antimicrobials could augment the decontamination efficacy of this technology, allowing the practitioners to benefit from synergism of natural antimicrobials and elevated hydrostatic pressure, to utilize lower intensity of the treatment with the same level of microbiological safety. This could be a practical solution for ultimately reducing high-pressure processing operation costs and increasing the competitiveness of products manufactured with this technology. This could also lead to enhanced preservation of nutritional and sensory properties of the products since mild hydrostatic pressure treatments are typically associated with no or minimal deleterious effects on physiochemical and organoleptic properties of food products.

Supplementary Materials: Supplementary materials can be found at http://www.mdpi.com/2076-2607/7/5/145/s1.

Author Contributions: M.N.K.: Conducted the main and preliminary trials, assisted in data management and analyses, co-wrote the first version of the manuscript. S.A.: Assisted in conduct of the trials, data collection, and data management. A.A.: Assisted in conduct of the trials, data collection, and data management. J.A.: Assisted in conduct of the trials, data collection, and data management. S.C.: Assisted in conduct of the trials, data collection, data management, and supervision of the students. A.F.: Secured extramural funding, designed the experiments and preliminary trials, prepared the inferential analyses code, supervised the students, co-wrote, revised and edited the manuscript.

Funding: Financial support in part from the National Institute of Food and Agriculture of the United States Department of Agriculture (Projects 2017-07534; 2017-07975; 2017-06088) and Pressure BioScience Inc. is acknowledged gratefully by the corresponding author.

Acknowledgments: Technical assistance and contributions of the members of the Public Health Microbiology Laboratory is sincerely appreciated by the authors. Authors also appreciate the feedback of anonymous reviewers and editorial team of *Microorganisms*.

Conflicts of Interest: The authors declare no conflict of interest. The funding sponsors have no role in design, data collection, microbiological and statistical analyses, or data interpretation and writing of the manuscript. The content of the current publication does not necessarily reflect views of the funding agencies.

References

1. United States Department of Agriculture, 2015–2020 Dietary Guidelines for Americans. Available online: https://health.gov/dietaryguidelines/2015/ (accessed on 1 May 2019).
2. Olaimat, A.N.; Holley, R.A. Factors influencing the microbial safety of fresh produce: A review. *Food Microbiol.* **2012**, *32*, 1–19. [CrossRef]
3. Warriner, K.; Huber, A.; Namvar, A.; Fan, W.; Dunfield, K. Recent advances in the microbial safety of fresh fruits and vegetables. *Food Nutr. Res.* **2009**, *57*, 155–208.
4. Fouladkhah, A. The Need for Evidence-Based Outreach in the Current Food Safety Regulatory Landscape. Commentary section. *J. Ext.* **2017**, *55*, 2COM1.
5. Scallan, E.; Hoekstra, R.M.; Angulo, F.J.; Tauxe, R.V.; Widdowson, M.A.; Roy, S.L.; Jones, J.L.; Griffin, P.M. Foodborne illness acquired in the United States—Major pathogens. *Emerg. Infect. Dis.* **2011**, *17*, 7–15. [CrossRef] [PubMed]

6. Crim, S.M.; Griffin, P.M.; Tauxe, R.; Marder, E.P.; Gilliss, D.; Cronquist, A.B.; Cartter, M.; Tobin-D'Angelo, M.; Blythe, D.; Smith, K.; et al. Preliminary incidence and trends of infection with pathogens transmitted commonly through food—Foodborne Diseases Active Surveillance Network, 10 US sites, 2006–2014. *MMWR* **2015**, *64*, 495. [PubMed]

7. Johnson, K.E.; Thorpe, C.M.; Sears, C.L. The emerging clinical importance of non-O157 Shiga toxin-producing *Escherichia coli*. *Clin. Infect. Dis.* **2006**, *43*, 1587–1595.

8. Bosilevac, J.M.; Arthur, T.M.; Bono, J.L.; Brichta-Harhay, D.M.; Kalchayanad, N.; King, D.A.; Shackelford, S.D.; Wheeler, M.L.; Koohmaraie, M. Prevalence and enumeration of *Escherichia coli* O157: H7 and *Salmonella* in U.S. abattoirs that process fewer than 1,000 head of cattle per day. *J. Food Prot.* **2009**, *72*, 1272–1278. [CrossRef]

9. Fouladkhah, A.; Geornaras, I.; Yang, H.; Sofos, J. Lactic Acid Resistance of Shiga Toxin-Producing *Escherichia coli* and Multidrug-resistant and Susceptible *Salmonella* Typhimurium and *Salmonella* Newport in Meat Homogenate. *Food Microbiol.* **2013**, *36*, 260–266. [CrossRef]

10. Brooks, J.T.; Sowers, E.G.; Wells, J.G.; Greene, K.D.; Griffin, P.M.; Hoekstra, R.M.; Strockbine, N.A. Non-O157 Shiga toxin–producing Escherichia coli infections in the United States, 1983–2002. *J. Infect. Dis.* **2005**, *192*, 1422–1429. [CrossRef]

11. Pihkala, N.; Bauer, N.; Eblen, D.; Evans, P.; Johnson, R.; Webb, J.; Williams, C.; FSIS. Risk Profile for Pathogenic Non-O157 Shiga Toxin-producing *Escherichia coli*. Available online: https://www.fsis.usda.gov/shared/PDF/Non_O157_STEC_Risk_Profile_May2012.pdf (accessed on 22 May 2019).

12. Cutter, C.; Depasquale, D.; Hayes, J.; Raines, C.; Seniviranthne, R. Meat Science Review: HPP, Ground Beef and the 'Big 6' STEC. The National Provisioner. Available online: http://www.provisioneronline.com/articles/98113-meat-science-reviewehppeground-beef-and-theebig-6estec (accessed on 21 April 2013).

13. Centers for Disease Control and Prevention (CDC), National Outbreak Reporting System (NORS). 2018. Available online: https://wwwn.cdc.gov/norsdashboard/ (accessed on 1 May 2019).

14. Brown, M.H.; Booth, I.R. Acidulants and low pH. In *Food Preservatives*; Russell, N.J., Gould, G.W.Ž, Eds.; Springer: Berlin/Heidelberg, Germany, 1991; pp. 22–43.

15. O'Driscoll, B.; Gahan, C.G.M.; Hill, C. Adaptive acid tolerance response in *Listeria monocytogenes*: Isolation of an acid-tolerant mutant, which demonstrates, increased virulence. *Appl. Environ. Microbiol.* **1996**, *62*, 1693–1698.

16. Cheng, H.Y.; Chou, C.C. Survival of acid-adapted *Escherichia coli* O157: H7 in some acid foods and subsequent environmental stress. MS thesis, National Taiwan University, Taipei, Taiwan, 1999.

17. Foster, J.W.; Hall, H.K. Adaptive acidification tolerance response of *Salmonella typhimurium*. *J. Bacteriol.* **1990**, *172*, 771–778. [CrossRef]

18. Morgan, D.; Newman, C.P.; Hutchinson, D.N.; Walker, A.M.; Rowe, B.; Majid, F. Verotoxin producing *Escherichia coli* O157: H7 infections associated with the consumption of yoghurt. *Epidemiol. Infect.* **1993**, *111*, 181–187. [CrossRef]

19. Weagant, S.D.; Bryant, J.L.; Bark, D.H. Survival of *Escherichia coli* O157: H7 in mayonnaise-based sauces at room and refrigerated temperatures. *J. Food Prot.* **1994**, *57*, 629–631. [CrossRef]

20. Besser, R.E.; Lett, S.M.; Weber, J.T.; Doyle, M.P.; Barrett, T.J.; Wells, J.G.; Griffin, P.M. An outbreak of diarrhea and hemolytic uremic syndrome from *Escherichia coli* O157: H7 in fresh-pressed apple cider. *J. Am. Med. Assoc.* **1993**, *269*, 2217–2220. [CrossRef]

21. Cheng, H.Y.; Chou, C.C. Acid adaptation and temperature effect on the survival of Escherichia coli O157: H7 in acidic fruit juice and lactic fermented milk product. *Int. J. Food Microbiol.* **2001**, *70*, 189–195.

22. Luna, R.E.; Mody, R. Non-O157 Shiga Toxin-Producing *E. coli* (STEC) Outbreaks, United States. Available online: http://blogs.cdc.Gov/publichealthmatters/files/2010/05/nono157stec_obs_052110.pdf (accessed on 22 April 2013).

23. Allison, A.; Daniels, E.; Chowdhury, S.; Fouladkhah, A. Effects of elevated hydrostatic pressure against mesophilic background microflora and habituated *Salmonella* serovars in orange juice. *Microorganisms* **2018**, *6*, 23. [CrossRef]

24. Cao, X.M.; Zhang, Y.; Zhang, F.S.; Wang, Y.T.; Yi, J.Y.; Liao, X.J. Effects of high hydrostatic pressure on enzymes, phenolic compounds, anthocyanins, polymeric color and color of strawberry pulps. *J. Sci. Food Agric.* **2011**, *91*, 877–885. [CrossRef] [PubMed]

25. Han, Y.; Jiang, Y.; Xinglian, S.; Xinsheng, B.; Zhou, G. Effect of high pressure treatment on microbial populations of sliced vacuum—packed cooked ham. *Meat Sci.* **2011**, *88*, 682–688. [CrossRef] [PubMed]

26. Hiremath, H.D.; Ramaswamy, H.S. High-pressure destruction kinetics of spoilage and pathogenic microorganisms in mango juice. *J. Food Process. Preserv.* **2012**, *36*, 113–125. [CrossRef]

27. Allison, A.; Chowdhury, S.; Fouladkhah, A. Synergism of Mild Heat and High-Pressure Pasteurization Against Listeria monocytogenes and Natural Microflora in Phosphate-Buffered Saline and Raw Milk. *Microorganisms* **2018**, *6*, 102. [CrossRef] [PubMed]

28. Vasudevan, P.; Marek, P.; Nair, M.K.M.; Annamalai, T.; Darre, M.; Khan, M.; Venkitanarayanan, K. In vitro inactivation of *Salmonella* Enteritidis in autoclaved chicken cecal contents by caprylic acid. *J. Appl. Poult. Res.* **2005**, *14*, 122–125. [CrossRef]

29. The U.S. Food and Drug Administration. 21CFR184.1025: Caprylic acid. Available online: https://www. accessdata.fda.gov/scripts/cdrh/cfdocs/cfcfr/CFRSearch.cfm?fr=184.1025 (accessed on 20 May 2019).

30. Annamalai, T.; Nair, M.K.M.; Marek, P.; Vasudevan, P.; Schreiber, D.; Knight, R.; Hoagland, T.; Venkitanarayanan, K. In vitro inactivation of *Escherichia coli* O157: H7 in bovine rumen fluid by caprylic acid. *J. Food Prot.* **2004**, *67*, 884–888. [CrossRef] [PubMed]

31. Nair, M.K.M.; Vasudevan, P.; Hoagland, T.; Venkitanarayanan, K. Inactivation of *Escherichia coli* O157: H7 and *Listeria monocytogenes* in milk by caprylic acid and monocaprylin. *Food Microbiol.* **2004**, *21*, 611–616. [CrossRef]

32. Chang., S.-S.; Redondo-Solano, M.; Thippareddi, H. Inactivation of *Escherichia coli* O157: H7 and *Salmonella* spp. on alfalfa seeds by caprylic acid and monocaprylin. *Int. J. Food Microbiol.* **2010**, *144*, 141–146. [CrossRef] [PubMed]

33. Burt, S. Essential oils: Their antibacterial properties and potential applications in foods—a review. *Int. J. Food Microbiol.* **2004**, *94*, 223–253. [CrossRef] [PubMed]

34. Ultee, A.; Slump, R.A.; Steging, G.; Smid, E.I. Antimicrobial activity carvacrol toward Bacillus cereus on rice. *J. Food Prot.* **2000**, *63*, 620–624. [CrossRef] [PubMed]

35. Lu, Y.; Wu, C. Reduction of *Salmonella enterica* contamination on grape tomatoes by washing with thyme oil, thymol, and carvacrol as compared with chlorine treatment. *J. Food Prot.* **2010**, *73*, 2270–2275. [CrossRef]

36. Allison, A.; Chowdhury, S.; Fouladkhah, A. Effects of Lactic Acid and Elevated Hydrostatic Pressure against Wild-Type and Rifampicin-Resistant O157 and Non-O157 Shiga Toxin-Producing *Escherichia coli* in Meat Homogenate. In Proceedings of the 2018 Annual Meeting of Institute of Food Technologists, Chicago, IL, USA. Available online: https://ift.planion.com/Web.User/AbstractDet?ACCOUNT=IFT&ABSID=21564& CONF=IFT18&ssoOverride=OFF&CKEY= (accessed on 22 May 2019).

37. Fouladkhah, A.; Geornaras, I.; Sofos, J.N. Effects of Reheating against *Listeria monocytogenes* Inoculated on Cooked Chicken Breast Meat Stored Aerobically at 7 °C. *Food Prot. Trends* **2012**, *32*, 697–704.

38. Koutsoumanis, K.P.; Sofos, J.N. Comparative acid stress response of *Listeria monocytogenes*, *Escherichia coli* O157: H7 and *Salmonella Typhimurium* after habituation at different pH conditions. *Lett. Appl. Microbiol.* **2004**, *38*, 321–326. [CrossRef]

39. Fouladkhah, A.; Geornaras, I.; Nychas, G.J.; Sofos, J.N. Antilisterial properties of marinades during refrigerated storage and microwave oven reheating against post—Cooking inoculated chicken breast meat. *J. Food Sci.* **2013**, *78*, M285–M289. [CrossRef]

40. Allison, A.; Fouladkhah, A. Sensitivity of *Salmonella* serovars and natural microflora to high-pressure pasteurization: Open access data for risk assessment and practitioners. *Data Brief* **2018**, *21*, 480–484. [CrossRef] [PubMed]

41. Geeraerd, A.H.; Valdramidis, V.P.; Van Impe, J.F. GInaFiT, a freeware tool to assess non-log-linear microbial survivor curves. *Int. J. Food Microbiol.* **2005**, *102*, 95–105. [CrossRef]

42. Ryu, J.H.; Beuchat, L.R. Influence of acid tolerance responses on survival, growth, and thermal cross-protection of *Escherichia coli* O157: H7 in acidified media and fruit juices. *Int. J. Food Microbiol.* **1988**, *45*, 185–193. [CrossRef]

43. Haberbeck, L.U.; Wang, X.; Michiels, C.; Devlieghere, F.; Uyttendaele, M.; Geeraerd, A.H. Cross-protection between controlled acid-adaptation and thermal inactivation for 48 *Escherichia coli* strains. *Int. J. Food Microbiol.* **2017**, *241*, 206–214. [CrossRef]

44. Fouladkhah, A.; Avens, J.S. Effects of combined heat and acetic acid on natural microflora reduction on cantaloupe melons. *J. Food Prot.* **2010**, *73*, 981–984. [CrossRef] [PubMed]

45. The U.S. Food and Drug Administration, Code of Federal Regulation Title 21. Available online: https://www.accessdata.fda.gov/scripts/cdrh/cfdocs/cfcfr/CFRSearch.cfm?CFRPart=172&showFR=1 (accessed on 20 May 2019).

46. Moschonas, G.; Geornaras, I.; Stopforth, J.D.; Wach, D.; Woerner, D.R.; Belk, K.E.; Smith, G.C.; Sofos, J.N. Activity of caprylic acid, carvacrol, ε-polylysine and their combinations against *Salmonella* in not-ready-to-eat surface-browned, frozen, breaded chicken products. *J. Food Sci.* **2012**, *77*, M405–M411. [CrossRef]

microorganisms

MDPI

Article

Presence of Shiga Toxin-Producing *Escherichia coli* (STEC) in Fresh Beef Marketed in 13 Regions of ITALY (2017)

Bianca Maria Varcasia, Francesco Tomassetti, Laura De Santis, Fabiola Di Giamberardino, Sarah Lovari, Stefano Bilei and Paola De Santis *

Istituto Zooprofilattico Sperimentale Lazio e Toscana, "M. Aleandri", 00178 Rome, Italy;
biancamaria.varcasia@izslt.it (B.M.V.); francesco.tomassetti@izslt.it (F.T.); laura.desantis@izslt.it (L.D.S.);
fabiola.digiamberardino@izslt.it (F.D.G.); sarah.lovari@izslt.it (S.L.); stefano.bilei@izslt.it (S.B.)
*   Correspondence: paola.desantis@izslt.it; Tel.: +39-06-72596047

Received: 30 September 2018; Accepted: 5 December 2018; Published: 6 December 2018

Abstract: The aim of this study was to determine the prevalence of Shiga toxin-producing *Escherichia coli* in fresh beef marketed in 2017 in 13 regions of Italy, to evaluate the potential risk to human health. According to the ISO/TS 13136:2012 standard, 239 samples were analysed and nine were STEC positive, from which 20 strains were isolated. The STEC-positive samples were obtained from Calabria ($n = 1$), Campania ($n = 1$), Lazio ($n = 2$), Liguria ($n = 1$), Lombardia ($n = 1$) and Veneto ($n = 3$). All STEC strains were analysed for serogroups O26, O45, O55, O91, O103, O104, O111, O113, O121, O128, O145, O146 and O157, using Real-Time PCR. Three serogroups were identified amongst the 20 strains: O91 ($n = 5$), O113 ($n = 2$), and O157 ($n = 1$); the O-group for each of the 12 remaining STEC strains was not identified. Six *stx* subtypes were detected: *stx1a*, *stx1c*, *stx2a*, *stx2b*, *stx2c* and *stx2d*. Subtype *stx2c* was the most common, followed by *stx2d* and *stx2b*. Subtype *stx2a* was identified in only one *eae*-negative strain and occurred in combination with *stx1a*, *stx1c* and *stx2b*. The presence in meat of STEC strains being potentially harmful to human health shows the importance, during harvest, of implementing additional measures to reduce contamination risk.

Keywords: *Escherichia coli* (STEC); beef; serogroups; *stx*-genes; *stx*-subtypes

1. Introduction

Shiga toxins (Stx) are potent cytotoxins encoded by lambdoid phages and integrated into the bacterial chromosome of a large and complex group of pathogenic *Escherichia coli* (STEC) strains that cause disease in humans [1,2]. Shiga toxins are immunologically distinct [3] and based on this antigenic diversity are divided into two groups, Stx1 and Stx2 [4]. Epidemiological studies [3–5] have shown that some Stx1 and Stx2 subtypes often are associated with severe human STEC illnesses [4]. Three Stx1 variants have been identified: Stx1a, Stx1c and Stx1d [2,4]. Usually, subtypes Stx1c and Stx1d are found in the meat of sheep, deer, and wildlife [6–9] and are rarely associated with disease in humans [2]. Subtypes Stx2a, Stx2b, Stx2c, Stx2d, Stx2e, Stx2f and Stx2g, along with the recently discovered Stx2h, are the eight Stx2 subtypes known to exist [4,10–14]. Subtypes Stx2a, Stx2c and Stx2d are associated with STEC infections in humans [11,15–17], while Stx2e, Stx2f, Stx2g are mainly found in animals [2,10,11]. Some studies have suggested that STEC strains producing Stx2f can cause diarrhoea in humans [18]; however, recently, STEC strains carrying the *stx2f* gene have been isolated from patients with hemolytic uremic syndrome (HUS) [18,19].

The majority of STEC strains associated with disease in humans possess adherence factors that facilitate their attachment to the intestinal epithelial cells [20]. The principal adherence factor is the intimin protein encoded by the *eae* gene, and responsible for what is known as the "attaching and effacing" (A/E) lesion of the intestinal mucosa [4,21–23]. The simultaneous presence of *eae* and *stx2* genes is considered a reliable indicator of a particular STEC strain's ability to cause severe disease in humans [24]. However, STEC strains that lack the *eae* gene can also cause severe disease by utilising alternative adherence mechanisms, as evidenced recently during a large outbreak of HUS in Germany in 2011 and caused by an enteroaggregative haemorrhagic *Escherichia coli* (EAHEC) O104:H4 carrying the *aggR* and *aaiC* genes in combination with *stx2a* [25].

Serological identification, based on the somatic (O) and flagellar (H) antigens, has to date resulted in the identification of ~470 STEC serotypes [8,26], all able to produce any one of the twelve known Stx subtypes or combinations of these subtypes [27]. The European Food and Safety Authority (EFSA) has identified STEC encoding the *stx* and *eae* genes that belong to serogroups O26, O103, O111, O145, O157, the so-called "big five", as those of major concern to human health in Europe [28]. Following the O104:H4 outbreak in Germany, this serotype was incorporated into the screening protocol for all *eae*-negative STEC isolated from food (Regulation EU No. 209/2013) [29].

STEC O157 is the most frequently reported serogroup worldwide [17,30,31]. The incidence of STEC O157 has however decreased in recent years, whereas the so-called non-O157 STEC serogroups are increasingly associated with haemorrhagic colitis (HC) and HUS in humans [27,31]; the most frequently encountered non-O157 serogroups are O26, O103, O111, O121, O145, O45, O118, O71 and O186 [31,32]. In 2015, as reported by EFSA and the European Centre for Disease Prevention (ECDC), in Europe, the STEC serogroups most commonly isolated from beef were O157 and O26, followed by O148, O145, O8, O113, O91, O130, O174 and O113. Many of these STEC serogroups were linked to human illnesses, confirming the epidemiological involvement of beef in STEC infections [33]. In 2016, in Europe, the STEC serogroup most frequently isolated from bovine meat was O157, followed by O113, O26, O145 and O174 [34]. In Italy, STEC O26 was the predominant serogroup in 2012 and responsible for about half of STEC cases in humans, followed by STEC O157 and STEC O111 [35]. In Europe, as a consequence of only a handful of countries doing any monitoring, few data exist on the isolation of STEC from beef [1,21,34].

The main reservoirs of STEC are ruminants, including wildlife. STEC can colonize the gut asymptomatically, their excretion into the environment [21] serving as a significant route of infection in humans [3]. Other studies have demonstrated that the hides of cattle represent an important source of STEC, resulting in carcass contamination during harvest [21,36]. Transmission to cattle may take place on-farm or during transportation to the abattoir [5,36]. STEC prevalence in cattle appears to be influenced by the age of the animal, the season, and probably, also feed composition [5,17]. Pathways along which humans may become infected include faecal-oral contamination during harvest, direct contact with faeces, STEC cross-contamination and multiplication during the preparation and handling of animal-derived foodstuffs, and human-to-human transmission [21,37]. European legislation (Regulation EU No. 2073/2005 and its amendments Regulation EU No. 1441/2007) [38,39] did not include the screening of STEC from meat products because, originally, very few data were available on the health risks associated with STEC-contaminated food [1,21].

In 2012, the International Organisation for Standardisation (ISO) issued the ISO/TS 13136:2012 method for the detection of STEC with a focus on the *stx1* and *stx2* virulence genes and on the *eae* adhesion factor gene, as these are associated with the "big five" serogroups [40]. The method is based on the Real-Time PCR screening of enrichment cultures, followed by serogroup identification and characterisation of isolated strains. The initial enrichment step, by increasing concentrations of the target bacteria, not only enhances the sensitivity of the method but also ensures the viability of bacterial cells from which positive results are obtained [41]. To date, as stated in the ESFA and ECDC 2017 report on trends and sources of zoonoses, zoonotic agents and food-borne outbreaks, 91.5% of the samples tested during 2016 by the European Member States, were analysed using ISO/TS 13136:2012 [34]. Some studies have suggested that the culture conditions involving media formulations and incubation temperature, as currently recommended in the ISO, be modified to further enhance STEC growth [41–44]. While improvements to the current ISO standard are possible, food authorities will always promote the use of a standardized method so results from different countries remain comparable [34].

The aim of this study was to determine the prevalence of Shiga toxin-producing *Escherichia coli* in fresh beef marketed in 13 regions of Italy in 2017, to evaluate the potential risk to human health.

2. Materials and Methods

2.1. Sampling

Between January and December 2017, 239 samples of refrigerated fresh beef were obtained from the retail market in 13 regions of Italy. The samples were collected originally to monitor antimicrobial resistance in zoonotic bacteria from food-producing animals and meat, under Decision 2013/652/EC [45]. The antimicrobial resistance aspects do not form part of this study, but they provided us with the opportunity to assay samples that were representative of most of Italy. The 13 regions account for >90% of the total animals harvested in Italy. The samples were arbitrarily chosen from supermarkets and traditional butcheries, and were obtained at least once monthly throughout the year to cover all four seasons. A single sample was collected from each lot of origin, either domestic or imported; frozen meat was excluded. The samples included meat either sliced or diced, vacuum-wrapped, or packaged under a controlled atmosphere. The 239 samples were obtained from the regions of Abruzzo ($n = 8$), Calabria ($n = 7$), Campania ($n = 3$), Emilia Romagna ($n = 23$), Friuli Venezia Giulia ($n = 6$), Lazio ($n = 34$), Liguria ($n = 6$), Lombardia ($n = 53$), Marche ($n = 7$), Piemonte ($n = 25$), Puglia ($n = 22$), Toscana ($n = 20$) and Veneto ($n = 25$).

2.2. Screening of Enrichment Cultures

The samples were analysed following the ISO/TS 13136:2012 standard [40]. Twenty-five grams of meat homogenised with 225 mL of modified Tryptone Soya Broth (mTSB) (Biolife Italiana srl, Milan, Italy) supplemented with 16 mg/mL of novobiocin (Biolife Italiana srl, Milan, Italy) and incubated at 37 °C for 18–24 h. DNA was extracted from 1 mL of each enrichment culture, using an automated nucleic acid purification system (MagPurix® 12S, Resnova, Rome, Italy), following the manufacturer's instructions. The extraction method provided approximately 100 ng/µL of DNA eluted in nuclease-free water at a final volume of 200 µL. DNA extracts were tested for the *stx1*, *stx2* and *eae* genes by Real-Time PCR, following the ISO standard procedure given above. PCR amplifications were done maintaining a final volume of 20 µL that contained 3 µL of DNA template (standardized at a concentration of 20 ng/µL), 1× qPCR Master Mix (Kapa Biosystems, Resnova, Rome, Italy), 300 nM of each primer, and 125 nM of each probe (Eurofins Genomics, Milan, Italy). All the reactions included an internal amplification control (Exo IPC kit) (Eurogentec, Italy). PCR conditions comprised an enzyme activation step of 95 °C for 5 min, followed by 40 cycles: 95 °C for 3 s (denaturation) and 60 °C for 30 s (annealing/extension/data acquisition). All the reference material used as reaction positive controls were provided by the European Union Reference Laboratory for *E. coli* (EU-RL VTEC). All *stx*-positive and *eae*-positive enrichment broths were screened for serogroups O26, O111, O103, O145 and O157 [40], while the *stx*-positive but *eae*-negative broths were screened also for the O104 serogroup following an additional protocol provided by the EURL VTEC [46]. All the serogroup Real-Time PCRs were done using the same reagent formulas and PCR conditions described above for the *stx* and *eae* genes; only for serogroup O103 was the annealing/extension temperature lowered to 55 °C.

2.3. Isolation of STEC Strains

For STEC strain isolation, the *stx*-positive enrichment broths were cultured on Tryptone Bile X-Glucuronide (TBX) agar (Biolife Italiana srl, Milan, Italy) or, if screening of the enrichment broths indicated the presence of serogroup O26, were cultured also on Rhamnose MacConkey (RMAC) agar and incubated at 37 °C for 18–24 h. Then, of many colonies that phenotypically resembled *E. coli*, fifty were selected arbitrarily and re-analysed singly for the presence (or absence) of the *stx* and *eae* genes using Real-Time PCR. Based on the original enrichment broth results, STEC colonies were tested for one or more of the "big five" serogroups. Those *stx*-positive colonies that tested negative for the "big five", were then analysed for the O45, O55, O91, O113, O121, O128 and O146 serogroups, using a method provided by the EURL VTEC [46].

2.4. Stx Subtyping

The *stx* subtype of each STEC strain was identified using the PCR-based subtyping protocol of the Statens Serum Institut, WHO Collaborating Centre for Reference and Research on *Escherichia* and *Klebsiella* [9,11]. The PCR was done using the 2GFast Master mix (Resnova, Rome, Italy), 280 nM of each primer (Eurofins, Milan, Italy) and 5 µL of template DNA (20 ng/µL). Each reaction was adjusted to a final volume of 25 µL in nuclease-free water. The annealing temperature was 66 °C for subtyping *stx1a-c*, 62 °C for *stx2a-c*, and 64 °C for *stx2d-g*. Agarose gel electrophoresis was used to visualize the PCR products. A molecular weight marker (Euroclone S.p.a., Milan, Italy) was used to assign the molecular weight to amplicons produced. The samples were run in Tris-Borate-EDTA running buffer (VWR International Srl, Milan, Italy) at a constant voltage (100 V for the first ten minutes and 60 V until the end of electrophoresis). All Statens Serum Institut reference material used as reaction positive controls was provided through the EURL VTEC.

3. Results

3.1. Real-Time PCR Screening of Enrichment Cultures, and Isolation of STEC Strains

During initial Real-Time PCR screening of enrichment cultures, *stx* genes were detected in 20 (8.4%) of the 239 samples. STEC was not isolated from 11 of these 20 *stx*-positive enrichment cultures, hence, based on the ISO/TS 13136:2012, in these samples only the "presumptive" presence of STEC could be determined. The *eae* gene was detected in eight of the eleven "presumptive" enrichment cultures, *stx1* in seven, and *stx2* in ten. One or more of the "big five" serogroups occurred singly, or in combination, in five enrichment cultures, as follows: O104 ($n = 1$), O103 ($n = 1$), O104 + O111 ($n = 1$), O26 + O103 + O157 ($n = 2$). For the six remaining "presumptive" STEC positive cultures, the serogroup was not identified. The regions of Italy from which "presumptive" positive STEC samples were obtained, are provided along with the sampling month in Table 1.

Table 1. STEC "presumptive" presence. Intimin (*eae*), Shiga-toxin *stx1* and *stx2* genes and serogroups detected in enrichment broth cultures obtained from fresh beef samples collected within 13 regions of Italy (2017).

Region of Italy	STEC Presumptive Presence (No)	Sampling Month	Sample ID	STEC Virulence Gene Profile			E. coli Serogroup
				eae	*stx1*	*stx2*	
Abruzzo	1	May	42696	+	+	+	nd [1]
Lazio	2	September	78963		+	+	nd [1]
		October	82856			+	O104
Liguria	1	June	51045	+	+	+	O103
Lombardia	4	August	64370			+	nd [1]
		September	72350	+		+	O104-O111
		September	75247	+	+	+	O26-O103-O157
		November	96150	+	+	+	nd [1]
Marche	1	November	97189	+		+	nd [1]
Puglia	1	July	57025	+	+		nd [1]
Veneto	1	October	87734	+	+	+	O26-O103-O157

[1] nd (not determined): enrichment broth cultures that tested negative to the "big five" serogroups analysed (O26, O103, O111, O145, O157) and O104.

Twenty STEC strains were isolated from nine of the 20 *stx*-positive enrichment cultures, hence the samples are classified as STEC "presence". The STEC-positive samples were obtained from the regions of Calabria (*eae* + *stx1* + *stx2*; *n* = 1), Campania (*stx2*; *n* = 1), Lazio (*eae* + *stx1* + *stx2* and *stx1* + *stx2*; *n* = 2), Liguria (*stx1* + *stx2*; *n* = 1), Lombardia (*stx2*; *n* = 1) and Veneto (*eae* + *stx1* + *stx2*; *n* = 3). Three different serogroups were identified amongst the 20 strains isolated: O91 (*n* = 5), O113 (*n* = 2), and O157 (*n* = 1). The remaining 12 STEC strains tested negative for all the serogroups analysed (O26, O45, O55, O91, O103, O104, O111, O113, O121, O128, O145, O146 and O157).

3.2. Stx Subtyping

Subtyping of the *stx* genes detected in the 20 STEC isolates, displayed various *stx1* and *stx2* subtype profiles. Strains belonging to serogroup O91 (*n* = 5) carried five different virulence gene profiles (*eae* + *stx1a*; *eae* + *stx1c*; *stx2c*; *eae* + *stx1a* + *stx1c* + *stx2c*; *eae* + *stx1a* + *stx1c* + *stx2c* + *stx2d*). Two O113 isolates were *eae*-negative and contained subtypes *stx2c* and *stx2c* + *stx2d*, respectively. One O157 strain was *eae*-positive and comprised *stx* subtypes *stx1a* + *stx2c* + *stx2d*. Twelve STEC isolates tested negative to all 13 serogroups analysed (nd, Table 2); amongst these isolates various combinations of *eae* and *stx* subtypes were detected, with *stx2c* found most frequently, followed by *stx2d*, *stx2b* and *stx2a*, respectively. Finally, subtypes *stx1d*, *stx2e*, *stx2f* and *stx2g* were not detected in any of the 20 STEC isolates obtained (Table 2).

3.3. Discussion

Cattle are considered a major reservoir for virulent strains of Shiga toxin-producing *Esherichia coli* (STEC) and the most important source of human infections through the consumption of contaminated beef products. The aim of this study was to identify and characterise the STEC strains found to occur in fresh beef obtained from 13 regions in Italy. A culture method involving selective and non-selective media, and following an initial enrichment step, was used to isolate STEC strains [1,47]. Specific PCR assays were used to identify pathogenicity factors (*eae* and *stx* genes), serogroups, and *stx* subtypes [1,17,40,48,49]. Initial enrichment yielded 20 (8.4%) *stx*-positive cultures, while STEC strains were only isolated from nine cultures. The failure to isolate STEC from a *stx*-positive enrichment culture has been reported upon previously [21,50–52]. To isolate STEC from food can be challenging because the number of STEC cells are likely to be low; other hurdles include sublethal cell injury, or cell growth suppressed in the presence of a large population of competing microflora [4,21,53]. For these reasons, enrichment cultures are essential to augment sensitivity, thereby promoting the isolation of STEC strains needed to confirm the presence of the *stx* genes in the live cell, while excluding the presence of free DNA or free prophages in the cultures [28]. Recently, various authors have reported the reduced sensitivity of mTSB enrichment broths supplemented with novobiocin (16 mg/L), suggesting that a decrease in novobiocin concentration might improve detection of O111 and other non-O157 serogroups [41–44]. While it is possible that reduced concentrations of novobiocin facilitate the isolation of non-O157 serogroups, 19 of the 20 isolates obtained represented non-O157 STEC strains.

Table 2. Subtyping of Shiga-toxin stx1 and stx2 genes and serogroups associated with STEC strains isolated from fresh beef samples collected within 13 regions of Italy (2017).

Region of Italy	Samples Positive (No)	Sampling Month	Strain ID	stx Subtypes[1]						*eae*	Serogroup
				stx1a	*stx1c*	*stx2a*	*stx2b*	*stx2c*	*stx2d*		
Calabria	1	November	94200-01	+						+	O91
			94200-02		+					+	O91
			94200-03	+	+			+	+	+	nd[2]
			94200-04	+	+			+	+	+	O91
Campania	1	December	107402-01					+			nd[2]
Lazio	2	June	50231-01	+				+			O91
			50231-02					+	+	+	O157
			50231-03					+			O113
		November	107400-01					+	+		O113
			107400-02		+		+				nd[2]
			107400-03	+				+	+		nd[2]
			107400-04					+	+		nd[2]
Liguria	1	December	104076-01	+	+	+	+				nd[2]
Lombardia	1	July	58927-01				+	+			nd[2]
Veneto	3	March	27847-01				+				nd[2]
		March	22774-01					+	+		nd[2]
			22774-02					+			nd[2]
		November	94197-01	+	+			+		+	O91
			94197-02	+	+			+		+	nd[2]
			94197-03	+	+			+	+	+	nd[2]

[1] *stx* subtypes included *stx1a-d* and *stx2a-g*; subtypes *stx1d*, *stx2e*, *stx2f* and *stx2g*, were not detected in any of the isolated STEC strains. [2] nd (not determined): strains that tested negative to all serogroups analysed (O26, O45, O55, O91, O103, O104, O111, O113, O121, O128, O145, O146 and O157).

In this study, overall STEC contamination in beef was 3.8%, a prevalence rate that agrees only partly with rates obtained previously in Italy. A frequency rate of 0.42% for STEC O157 matches that obtained during a nationwide survey conducted by Conedera et al. [54] and who reported STEC in four (0.43%) of 931 minced beef samples. These were screened only for serogroup O157. In the region of Piemonte, Rantsiou et al. [55] found six (5.9%) STEC strains in 101 mixed meat products using a method developed in-house. In the Emilia Romagna region, Bardasi et al. [56], following the ISO/TS 13136:2012 protocol, demonstrated an STEC presence in four (0.6%) of 689 meat samples (representing pork, bovine and poultry). In a more recent study, the same ISO protocol was used to test 675 pork samples (comprising both fresh and dried products) collected in the Umbria and Marche regions of Italy [57]; these authors reported the presumptive presence of *stx*-genes in 2.8% of the products, but were unable to isolate any STEC strains. The discrepant STEC prevalence rates obtained may find causes in various factors, including geographic compartmentalization of the *E. coli* population amongst food animals, laboratory techniques and protocols employed [54,55], and the wide range in meat products analysed. According to EFSA, in Europe, the overall presence of STEC in 18,975 food samples assayed was 2.5%, the highest proportion found in meat, particularly that from small ruminants [34]. In Switzerland, Fantelli et al. reported the presence of STEC in 2.3% of 211 minced beef samples tested [58]; in France, 4% of 411 beef samples were STEC positive [59]. Our STEC prevalence rates are comparable to some of those obtained for beef previously in Europe [34,58,59].

Real-Time PCR, based on the O-antigen synthesis genes (*wzx* and *wzy*), is widely used to serogroup STEC strains [60]. However, the Real-Time PCR methods currently used do not cover all known serogroups, hence many serogroups to which a STEC strain may belong remain unidentified [33]; consequently, 12 of our 20 STEC strains could not be identified to serogroup. While serogroups and serotypes are not virulence factors and not predictive of a virulence profile, they nevertheless remain useful for conducting surveillance and for investigating outbreaks [61]. Serogroups O26, O103, O104, O111 and O157, along with the *eae* and *stx2* virulence genes, were detected in 11 enrichment broths; the failure to isolate STEC strains from these 11 broths is the reason why the corresponding samples were classified as "presumptive" positive. The potential risk to human health that "presumptives" represent, means the responsible authorities must continue to monitor for STEC.

An association between Stx subtype and severity of disease in humans has been observed [51,62]. In this study, 20 STEC strains were isolated from nine beef samples and carried *stx* subtypes in various combinations. The *stx1* subtypes detected were *stx1a* and *stx1c*. The Stx1a toxin subtype is often produced by strains that are *eae*-positive and known to cause severe disease in man [32]. In this study, nine STEC strains, either *eae*-positive or *eae*-negative, had *stx1a* alone or in combination with *stx1c* or *stx2a*, *stx2b*, *stx2c* and *stx2d*. The Stx1c toxin subtype is reported mainly in *eae*-negative strains causing mild infections [63]; we found *stx1c* both in identified (serogroup O91) and unidentified serogroups that were *eae*-positive (Table 2). With regard to toxin type Stx2, the subtypes *stx2a*, *stx2c* and *stx2d* have been linked to HC and HUS in humans [11,16]. In this study STEC isolates carrying *stx2c* were the *stx* subtypes most commonly found, followed by *stx2d* and *stx2b*. Subtype *stx2a* was identified in only one *eae*-negative strain and occurred in combination with *stx1a*, *stx1c* and *stx2b*. Of the eight *eae*-positive STEC strains obtained, four belonged to serogroup O91, one to O157, while three represented unidentified serogroups. Two strains of O113 and one of O91 were *eae*-negative; the nine remaining strains were not identified to serogroup (Table 2). Contrary to other reports [64], we found four O91 STEC strains to be *eae*-positive; this finding is unusual and we cannot explain it satisfactorily. Subtype *stx2d* was found in eight STEC strains, of which four were *eae*-negative, while the other four were *eae*-positive; the latter group included a strain O157 that was also *stx1a*- and *stx2c*-positive. The *stx2d* subtype is usually associated with *eae*-negative strains and severe disease in humans [65]; recently, in Spain, Sanchez et al. [66] reported upon a O157:H7 strain that was *eae*- and *stx2d*-positive and isolated from a 2-year-old child with BD. This unusual virulence combination, though rare, has been reported also from several HUS-affected patients in France and separately involving STEC O26:H11 [67] and STEC O80:H2 [68].

4. Conclusions

In Italy, the isolation of STEC strains from fresh meat samples signals the recurring threat that beef, consumed either undercooked or raw, poses to human health. The variety of *stx* types and subtypes and multiple STEC serogroups detected, are amongst those found elsewhere in the world and where, in humans, they have been demonstrated to be involved in severe diseases, such as BD, HC and HUS. The presence in meat of potentially harmful STEC strains emphasizes the importance, during harvest, of implementing additional measures to reduce contamination risk. Linked to this, an efficient surveillance strategy for STECs in retail foodstuffs, remains a national priority. The laboratory diagnostic protocols needed to isolate and accurately identify STEC strains are laborious, expensive, and time-consuming. However, they continue to remain pivotal to assessing the strain of pathogenic *E. coli* involved, and for identifying the possible source of infection. This knowledge is needed to enable the competent authorities to respond precisely and rapidly. Improvements to current isolation techniques, and the validation and standardization of molecular protocols, remain a matter of urgency. It is foreseen that in the future new high-power methodologies, such as Next-Generation Sequencing (NGS), will become more widely utilised and that these will lead to further improvements in the currently used standards for diagnosing STEC in foods.

Author Contributions: Conceptualization, P.D.S. and B.M.V.; Methodology, B.M.V., F.T., L.D.S., F.D.G. and S.L.; Data curation, B.M.V.; Writing—original draft preparation, B.M.V.; Writing—review and editing, P.D.S.; Supervision, P.D.S.; project administration, S.B.; funding acquisition, S.B.

Funding: This research received no external funding.

Acknowledgments: The authors thank the staff of the Laboratory of Food Microbiology (Istituto Zooprofilattico Sperimentale Lazio e Toscana "M. Aleandri", Italy) for the collaboration provided to this study. The authors also wish to thank the National Reference Laboratory for Antimicrobial Resistance (Istituto Zooprofilattico Sperimentale Lazio e Toscana "M. Aleandri", Italy) for kindly providing the samples used in this study. Special thanks to Rudy Meiswinkel.

Conflicts of Interest: The authors declare no conflict of interest.

References and Note

1. Bonardi, S.; Alpigiani, I.; Tozzoli, R.; Vismarra, A.; Zecca, V.; Greppi, C. Shiga toxin-producing *Escherichia coli* O157, O26 and O111 in cattle faeces and hides in Italy. *Vet. Rec. Open* **2015**, *2*, 1–9. [CrossRef] [PubMed]
2. Melton-Celsa, A.R. Shiga Toxin (STx) Classification, Structure, and Function. *Microbiol. Spectr.* **2014**, *2*, 1–21. [CrossRef] [PubMed]
3. Bertin, Y.; Boukhors, K.; Pradel, N.; Livrelli, V.; Martin, C. Stx2 Subtyping of Shiga Toxin-Producing *Escherichia coli* Isolated from Cattle in France: Detection of a New Stx2 Subtype and Correlation with Additional Virulence Factors. *J. Clin. Microbiol.* **2001**, *39*, 3060–3065. [CrossRef]
4. Farrokh, C.; Jordan, K.; Auvray, F.; Glass, K.; Oppegaard, H.; Raynaud, S.; Thevenot, D.; Condron, R.; De Reu, K.; Govaris, A.; et al. Review of Shiga-toxin producing *Escherichia coli* and their significance in dairy production. *Int. J. Food Microbiol.* **2013**, *162*, 190–212. [CrossRef] [PubMed]
5. Hussein, H.S. Prevalence and pathogenicity of Shiga toxin-producing Escherichia coli in beef cattle and their products. *J. Anim. Sci.* **2014**, *85*, E63–E72. [CrossRef] [PubMed]
6. Brett, K.N.; Ramachandran, V.; Hornitzky, M.A.; Bettelheim, K.A.; Walker, M.J.; Djordjevic, S.P. stx$_{1c}$ is the most common Shiga toxin 1 subtype among Shiga toxin-producing *Escherichia coli* isolates from sheep but not among isolates from cattle. *J. Clin. Microbiol.* **2003**, *41*, 926–936. [CrossRef] [PubMed]
7. Hofer, E.; Cernela, N.; Stephan, R. Shiga Toxin Subtypes Associated with Shiga Toxin–Producing *Escherichia coli* Strains Isolated from Red Deer, Roe Deer, Chamois, and Ibex. *Foodborne Pathog Dis.* **2012**, *9*, 792–795. [CrossRef]
8. Mora, A.; López, C.; Dhabi, G.; López-Beceiro, A.M.; Fidalgo, L.E.; Díaz, E.A.; Martínez-Carrasco, C.; Mamani, R.; Herrera, A.; Blanco, J.E.; et al. Seropathotypes, Phylogroups, Stx Subtypes, and Intimin Types of Wildlife-Carried, Shiga Toxin-Producing *Escherichia coli* Strains with the Same Characteristics as Human-Pathogenic Isolates. *Appl. Environ. Microbiol.* **2012**, *78*, 2578–2585. [CrossRef]

9. Scheutz, F.; Teel, L.D.; Beutin, L.; Piérard, D.; Buvens, G.; Karch, H.; Mellmann, A.; Caprioli, A.; Tozzoli, A.; Morabito, S.; et al. Multicenter evaluation of a sequence-based protocol for subtyping Shiga toxins and standardizing Stx nomenclature. *J. Clin. Microbiol.* **2012**, *50*, 2951–2963. [CrossRef]

10. Bielaszewska, M.; Prager, R.; Zhang, W.; Friedrich, A.W.; Mellmann, A.; Tschäpe, H.; Karch, H. Chromosomal dynamism in progeny of outbreak-related sorbitol-fermenting enterohemorrhagic *Escherichia coli* O157: NM. *Appl. Environ. Microbiol.* **2006**, *72*, 1900–1909. [CrossRef]

11. Persson, S.; Olsen, K.E.; Ethelberg, S.; Scheutz, F. Subtyping method for *Escherichia coli* Shiga toxin (Verocytotoxin) 2 variants and correlation to clinical manifestations. *J. Clin. Microbiol.* **2007**, *45*, 2020–2024. [CrossRef] [PubMed]

12. Hussein, H.S.; Bollinger, L.M. Prevalence of Shiga toxin-producing *Escherichia coli* in beef cattle. *J. Food Prot.* **2005**, *68*, 2224–2241. [CrossRef] [PubMed]

13. Schmidt, H.; Scheef, J.; Morabito, S.; Caprioli, A.; Wieler, L.H.; Karch, H. A new Shiga toxin 2 variant (Stx2f) from *Escherichia coli* isolated from pigeons. *Appl. Environ. Microbiol.* **2000**, *66*, 1205–1208. [CrossRef] [PubMed]

14. Leung, P.H.; Yam, W.C.; Ng, W.W.; Peiris, J.S. The prevalence and characterization of verotoxin-producing *Escherichia coli* isolated from cattle and pigs in an abattoir in Hong Kong. *Epidemiol. Infect.* **2001**, *126*, 173–179. [CrossRef]

15. Eklund, M.; Leino, K.; Siitonen, A. Clinical *Escherichia coli* strains carrying stx genes: Stx variants and stx-positive profiles. *J. Clin. Microbiol.* **2002**, *40*, 4585–4593. [CrossRef] [PubMed]

16. Friedrich, A.W.; Bielaszewska, M.; Zhang, W.; Pulz, M.; Kuczius, T.; Ammon, A.; Karch, H. *Escherichia coli* harboring Shiga toxin 2 gene variants: Frequency and association with clinical symptoms. *J. Infect. Dis.* **2002**, *185*, 74–84. [CrossRef] [PubMed]

17. Caprioli, A.; Morabito, S.; Brugere, H.; Oswald, E. Enterohaemorrhagic *Escherichia coli*: Emerging issues on virulence and modes of transmission. *Vet. Res.* **2005**, *36*, 289–311. [CrossRef] [PubMed]

18. Friesema, I.H.M.; Keijzer-Veen, M.G.; Koppejan, M.; Schipper, H.S.; van Griethuysen, A.J.; Heck, M.E.O.; van Pelt, W. Hemolytic Uraemic Syndrome associated with *Escherichia coli* O8:H19 and shiga toxin 2f gene. *Emerg. Infect. Dis.* **2015**, *79*, 1329–1337.

19. Grande, L.; Michelacci, V.; Giugliacci, F.; Badouei, M.A.; Schlager, S.; Minelli, F.; Caprioli, A.; Morabito, S. Whole-genome Characterization and strain comparison of VT2f-producing *Escherichia coli* causing haemolytic uraemic syndrome. *Emerg. Infect. Dis.* **2016**, *22*, 2078–2086. [CrossRef]

20. Kaper, J.B.; Nataro, J.P.; Mobley, H.L.T. Pathogenic *Escherichia coli*. *Nat. Rev. Microbiol.* **2004**, *2*, 123–140. [CrossRef]

21. Nobili, G.; Franconieri, I.; La Bella, G.; Basanisi, M.G.; La Salandra, G. Prevalence of Verocytotigenic *Escherichia coli* strains isolated from raw beef in southern Italy. *Int. J. Food Microbiol.* **2017**, *257*, 201–205. [CrossRef]

22. Duffy, G.; Burgess, C.M.; Bolton, D.J. A review of factors that affect the transmission and survival of verocytotoxigenic *Escherichia coli* in the European farm to fork chain. *Meat Sci.* **2014**, *97*, 375–383. [CrossRef]

23. Caprioli, A.; Maugliani, A.; Michelacci, V.; Morabito, S. Molecular typing of verocytotoxin-producing *E. coli* (VTEC) strains isolated from food, feed and animals: State of play and standard operating procedures for pulsed field gel electrophoresis (PFGE) typing, profiles interpretation and curation. *EFSA Support. Publ.* **2014**, *704*, 1–55.

24. Boerlin, P.; McEwen, S.A.; Boerlin-Petzold, F.; Wilson, J.B.; Johnson, R.P.; Gyles, C.L. Associations between Virulence Factors of Shiga Toxin-Producing *Escherichia coli* and Disease in Humans. *J. Clin. Microbiol.* **1999**, *37*, 497–503.

25. Beutin, L.; Martin, A. Outbreak of Shiga Toxin–Producing *Escherichia coli* (STEC) O104:H4 Infection in Germany Causes a Paradigm Shift with Regard to Human Pathogenicity of STEC Strains. *J. Food Prot.* **2012**, *75*, 408–418. [CrossRef]

26. Bettelheim, K.A. The non-O157 Shiga-toxigenic (verocytotoxigenic) *Escherichia coli*: Under-rated pathogens. *Crit. Rev. Microbiol.* **2007**, *33*, 67–87. [CrossRef]

27. Food and Agriculture Organization of the United Nations (FAO); World Health Organization (WHO). *Shiga Toxin-Producing Escherichia coli (STEC) and Food: Attribution, Characterization, and Monitoring: Report*; Microbiological Risk Assessment Series; FAO: Rome, Italy; WHO: Geneva, Switzerland, 2018.

28. European Food Safety Authority. Scientific opinion on VTEC-seropathotype and scientific criteria regarding pathogenicity assessment. *EFSA J.* **2013**, *11*, 1–106.
29. Commission Regulation (EU) No 209/2013. 2013 of 11 March 2013 amending Regulation (EC) No 2073/2005 as regards microbiological criteria for sprouts and the sampling rules for poultry carcases and fresh poultry meat. Available online: http://data.europa.eu/eli/reg/2013/209/oj (accessed on 12 March 2013).
30. Kiranmayi, Ch.B.; Krishnaiah, N.; Mallika, E.N. Escherichia coli O157:H7—An Emerging Pathogen in foods of Animal Origin. *Vet. World.* **2010**, *3*, 382–389. [CrossRef]
31. Centre for Disease Prevention and Control (CDC). *National Enteric Disease Surveillance: Shiga Toxin-Producing Escherichia coli (STEC) Annual Report, 2015*; CS 282919-A; CDC: Atlanta, GA, USA, 2017.
32. Brooks, J.T.; Sowers, E.G.; Wells, J.G.; Greene, K.D.; Griffin, P.M.; Hoekstra, R.M.; Strockbine, N.A. Non-O157 Shiga toxin-producing *Escherichia coli* infections in the United States, 1983–2002. *J. Infect. Dis.* **2005**, *192*, 1422–1429. [CrossRef]
33. European Food Safety Authority (EFSA); European Centre for Disease Prevention and Control (ECDC). The European Union summary report on trends and sources of zoonoses, zoonotic agents and food-borne outbreaks in 2015. *EFSA J.* **2016**, *14*, 4634.
34. European Food Safety Authority (EFSA); European Centre for Disease Prevention And Control (ECDC). The European Union summary report on trends and sources of zoonoses, zoonotic agents and food-borne outbreaks in 2016. *EFSA J.* **2017**, *15*, 5077.
35. European Food Safety Authority. Technical specifications on randomised sampling for harmonized monitoring of antimicrobial resistance in zoonotic and commensal bacteria. *EFSA J.* **2014**, *12*, 3686.
36. Arthur, T.M.; Bosilevac, J.M.; Nou, X.; Shackelford, S.D.; Wheeler, T.L.; Kent, M.P.; Jaroni, D.; Pauling, B.; Allen, D.M.; Koohmaraie, M. *Escherichia coli* O157 prevalence and enumeration of aerobic bacteria, *Enterobacteriaceae*, and *Escherichia coli* O157 at various steps in commercial beef processing plants. *J. Food Prot.* **2004**, *67*, 658–665. [CrossRef]
37. Buncic, S.; Nychas, G.J.; Lee, M.R.; Koutsoumanis, K.; Hébraud, M.; Desvaux, M.; Antic, D. Microbial pathogen control in the beef chain: Recent research advances. *Meat Sci.* **2014**, *97*, 288–297. [CrossRef]
38. No. 1441/2007 of 5 December 2007 Amending Regulation (EC) No 2073/2005 on Microbiological Criteria for Foodstuffs. *OJ L 322*, 7.12.2007, p 12–29. Available online: http://data.europa.eu/eli/reg/2007/1441/oj (accessed on 23 December 2014).
39. Commission Regulation (EC), No. 2073/2005 of 15 November 2005 on Microbiological Criteria for Foodstuffs. *OJ L 338*, 22.12.2005, p 1–26. Available online: http://data.europa.eu/eli/reg/2005/2073/oj (accessed on 23 December 2014).
40. International Organization for Standardization, ISO/TS 13136:2012, 2012. Microbiology of food and animal feed—Real-time polymerase chain reaction (PCR)-based method for the detection of food-borne pathogens —Horizontal method for the detection of Shiga toxin-producing *Escherichia coli* (STEC) and the determination of O157, O111, O26, O103 and O145 serogroups.
41. Amagliani, G.; Rotundo, L.; Carlonia, E.; Omiccioli, E.; Magnania, M.; Brandia, G.; Fratamico, P. Detection of Shiga toxin-producing *Escherichia coli* (STEC) in ground beef and bean sprouts: Evaluation of culture enrichment conditions. *Food Res. Int.* **2018**, *103*, 398–405. [CrossRef]
42. Fratamico, P.M.; Wasilenko, J.L.; Garman, B.; Demarco, D.R.; Varkey, S.; Jensen, M.; Rhoden, K.; Tice, G. Evaluation of a multiplex real-time PCR method for detecting shiga toxin-producing *Escherichia coli* in beef and comparison to the U.S. Department of Agriculture Food Safety and Inspection Service Microbiology laboratory guidebook method. *J. Food Prot.* **2014**, *77*, 180–188. [CrossRef]
43. Fratamico, P.M.; Bagi, L.K.; Cray, W.C., Jr.; Narang, N.; Yan, X.; Medina, M.; Liu, Y. Detection by multiplex real-time polymerase chain reaction assays and isolation of Shiga toxin-producing *Escherichia coli* serogroups O26, O45, O103, O111, O121, and O145 in ground beef. *Foodborne Pathog. Dis.* **2011**, *8*, 601–607. [CrossRef]
44. Vimont, A.; Delignette-Muller, M.-L.; Vernozy-Rozand, C. Supplementation of enrichment broths by novobiocin for detecting Shiga toxin-producing *Escherichia coli* from food: A controversial use. *Lett. Appl. Microbiol.* **2007**, *44*, 326–331. [CrossRef]
45. 2013/652/EU: Commission Implementing Decision of 12 November 2013 on the monitoring and reporting of antimicrobial resistance in zoonotic and commensal bacteria (notified under document C(2013) 7145) Text with EEA relevance. *OJ L 303*, 14.11.2013, p 26–39. Available online: http://data.europa.eu/eli/dec_impl/2013/652/oj (accessed on 12 November 2013).

46. European Union Reference Laboratory for *E. coli.*. Identification of the STEC serogroups mainly associated with human infections by Real-Time PCR amplification of O-associated genes. Available online: http://old.iss.it/vtec/index.php?lang=2&anno=2014&tipo=3 (accessed on 25 March 2018).

47. Wang, L.; Wakushima, M.; Aota, T.; Yoshida, Y.; Kita, T.; Maehara, T. Specific properties of enteropathogenic *Escherichia coli* isolates from diarrheal patients and comparison to strains from foods and faecal specimens from cattle, swine, and healthy carriers in Osaka City, Japan. *Appl. Environ. Microbiol.* **2013**, *79*, 1232–1240. [CrossRef]

48. Paddock, Z.; Shi, X.; Bai, J.; Nagaraja, T.G. Applicability of a multiplex PCR to detect O26, O45, O103, O111, O121, O145, and O157 serogroups of *Escherichia coli* in cattle feces. *Vet. Microbiol.* **2012**, *156*, 381–388. [CrossRef]

49. Scheutz, F.; Strockbine, N.A. Escherichia. In *Bergey's Manual of Systematic Bacteriology*, 2nd ed.; Garrity, G.M., Brenner, D.J., Krieg, N.R., Staley, J.T., Eds.; Springer: New York, NY, USA, 2005; Volume 2, pp. 607–624. ISBN 978-0-387-29298-4.

50. Meng, Q.; Bai, X.; Zhao, A.; Lan, R.; Du, H.; Wang, T.; Shi, C.; Yuan, X.; Bai, X.; Ji, S.; et al. Characterization of Shiga toxin-producing *Escherichia coli* from healthy pigs in China. *BMC Microbiol.* **2014**, *14*, 5. [CrossRef]

51. Bai, X.; Wang, H.; Xin, Y.; Wei, R.; Tang, X.; Zhao, A.; Sun, H.; Wang, Z.; Wang, J.; Xu, Y.; et al. Prevalence and characteristics of Shiga toxin-producing *Escherichia coli* isolated from retail raw meats in China. *Int. J. Food Microbiol.* **2015**, *200*, 31–38. [CrossRef]

52. Quiros, P.; Martinez-Castillo, A.; Muniesa, M. Improving detection of Shiga toxin-producing *Escherichia coli* by molecular methods by reducing the interference of free Shiga toxin-encoding bacteriophages. *Appl. Environ. Microbiol.* **2015**, *81*, 415–421. [CrossRef]

53. Brusa, V.; Aliverti, V.; Aliverti, F.; Ortega, E.E.; De la Torre, J.H.; Linares, L.H.; García, P.P. Shiga toxin-producing *Escherichia coli* in beef retail markets from Argentina. *Front. Cell. Infect. Microbiol.* **2012**, *2*, 171. [CrossRef]

54. Conedera, G.; Dalvit, P.; Martini, M.; Galiero, G.; Gramaglia, M.; Goffredo, E.; Loffredo, G.; Morabito, S.; Ottaviani, D.; Paterlini, F.; et al. Verocytotoxin-producing *Escherichia coli* O157 in minced beef and dairy products in Italy. *Int J. of Food Microb.* **2004**, *96*, 67–73. [CrossRef]

55. Rantsiou, K.; Alessandria, V.; Cocolin, L. Prevalence of Shiga toxin-producing *Escherichia coli* in food products of animal origin as determined by molecular methods. *Int. J. Food Microb.* **2012**, *154*, 37–43. [CrossRef]

56. Bardasi, L.; Taddei, R.; Nocera, L.; Ricchi, M.; Merialdi, G. Shiga toxin-producing *Escherichia coli* in meat and vegetable products in Emilia Romagna Region, years 2012–2013. *Ital. J. Food Saf.* **2015**, *4*, 4511. [CrossRef]

57. Ercoli, L.; Farneti, S.; Zivaco, A.; Mencaroni, G.; Blasi, G.; Striano, G.; Scuota, S. Prevalence and characteristics of verotoxigenic Escherichia coli strains isolated from pigs and pork products in Umbria and Marche regions of Italy. *Int J. Food Microbiol.* **2016**, *232*, 7–14. [CrossRef]

58. Fantelli, K.; Stephan, R. Prevalence and characteristics of shiga toxin-producing *Escherichia coli* and *Listeria monocytogenes* strains isolated from minced meat in Switzerland. *Int J. Food Microbiol.* **2001**, *70*, 63–69. [CrossRef]

59. Pradel, N.; Livrelli, V.; De Champs, C.; Palcoux, J.B.; Reynaud, A.; Scheutz, F.; Sirot, J.; Joly, B.; Forestier, C. Prevalence and characterization of Shiga toxin-producing *Escherichia coli* isolated from cattle, food, and children during a one-year prospective study in France. *J. Clin. Microbiol.* **2000**, *38*, 1023–1031.

60. Lin, A.; Sultan, O.; Lau, H.K.; Wong, E.; Hartman, G.; Lauzon, C.R. O serogroup specific real time PCR assays for the detection and identification of nine clinically relevant non-O157 STECs. *Food Microbiol.* **2011**, *28*, 478–483. [CrossRef]

61. Conrad, C.C.; Stanford, K.; McAllister, T.A.; Thomas, J.; Reuter, T. Further development of sample preparation and detection methods for O157 and the top 6 non-O157 STEC serogroups in cattle feces. *J. Microbiol. Methods* **2014**, *105*, 22–30. [CrossRef]

62. Bosilevac, J.M.; Koohmaraie, M. Prevalence and characterization of non-O157 shiga toxin-producing *Escherichia coli* isolates from commercial ground beef in the United States. *Appl. Environ. Microbiol.* **2011**, *77*, 2103–2112. [CrossRef]

63. Friedrich, A.W.; Borell, J.; Bielaszewska, M.; Fruth, A.; Tschäpe, H.; Karch, H. Shiga toxin 1c-producing *Escherichia coli* strains: Phenotypic and genetic characterization and association with human disease. *J. Clin. Microb.* **2003**, *41*, 2448–2453. [CrossRef]

64. Feng, P.C.H.; Delannoy, S.; Lacher, D.W.; Bosilevac, J.M.; Fach, P.; Beutin, L. Shiga toxin-producing serogroup O91 *Escherichia coli* strains isolated from food and environmental samples. *Appl. Environ. Microbiol.* **2017**, *83*, e01231-17. [CrossRef]

65. Bielaszewska, M.; Friedrich, A.W.; Aldick, T.; Schürk-Bulgrin, R.; Karch, H. Shiga toxin activatable by intestinal mucus in *Escherichia coli* isolated from humans: Predictor for a severe clinical outcome. *Clin Infect. Dis.* **2006**, *43*, 1160–1167. [CrossRef]

66. Sánchez, S.; Llorente, M.; Herrera-León, L.; Ramiro, R.; Nebreda, S.; Remacha, M.; Herrera-León, S. Mucus-Activatable Shiga Toxin Genotype *stx2d* in *Escherichia coli* O157:H7. *Emerg. Infect. Dis.* **2017**, *23*, 1431–1433. [CrossRef]

67. Delannoy, S.; Mariani-Kurkdjian, P.; Bonacorsi, S.; Liguori, S.; Fach, P. Characteristics of emerging human-pathogenic *Escherichia coli* O26:H11 strains isolated in France between 2010 and 2013 and carrying the *stx*2d gene only. *J. Clin. Microbiol.* **2015**, *53*, 486–492. [CrossRef]

68. Mariani-Kurkdjian, P.; Lemaître, C.; Bidet, P.; Perez, D.; Boggini, L.; Kwon, T.; Bonacorsi, S. Haemolytic-uraemic syndrome with bacteraemia caused by a new hybrid *Escherichia coli* pathotype. *New Microbes New Infect.* **2014**, *2*, 127–131. [CrossRef]

microorganisms

MDPI

Article

Biofilm Formation by Shiga Toxin-Producing *Escherichia coli* on Stainless Steel Coupons as Affected by Temperature and Incubation Time

Zhi Ma [1,2,3], Emmanuel W. Bumunang [2,3], Kim Stanford [3], Xiaomei Bie [1,*], Yan D. Niu [4] and Tim A. McAllister [2,*]

[1]   College of Food Science and Technology, Nanjing Agriculture University, Nanjing 210095, China;
      mazhi19900504@sina.com
[2]   Agriculture and Agri-Food Canada, Lethbridge, AB T1J 4B1, Canada; emmanuel.bumunang@canada.ca
[3]   Alberta Agriculture and Forestry, Lethbridge, AB T1J 4V6, Canada; kim.stanford@gov.ab.ca
[4]   Faculty of Veterinary Medicine, University of Calgary, Calgary, AB T2N 4Z6, Canada;
      dongyan.niu@ucalgary.ca
*   Correspondence: bxm43@njau.edu.cn (X.B.); tim.mcallister@canada.ca (T.A.M.)

Received: 2 March 2019; Accepted: 27 March 2019; Published: 31 March 2019

Abstract: Forming biofilm is a strategy utilized by Shiga toxin-producing *Escherichia coli* (STEC) to survive and persist in food processing environments. We investigated the biofilm-forming potential of STEC strains from 10 clinically important serogroups on stainless steel at 22 °C or 13 °C after 24, 48, and 72 h of incubation. Results from crystal violet staining, plate counts, and scanning electron microscopy (SEM) identified a single isolate from each of the O113, O145, O91, O157, and O121 serogroups that was capable of forming strong or moderate biofilms on stainless steel at 22 °C. However, the biofilm-forming strength of these five strains was reduced when incubation time progressed. Moreover, we found that these strains formed a dense pellicle at the air-liquid interface on stainless steel, which suggests that oxygen was conducive to biofilm formation. At 13 °C, biofilm formation by these strains decreased ($P < 0.05$), but gradually increased over time. Overall, STEC biofilm formation was most prominent at 22 °C up to 24 h. The findings in this study identify the environmental conditions that may promote STEC biofilm formation in food processing facilities and suggest that the ability of specific strains to form biofilms contributes to their persistence within these environments.

Keywords: Shiga toxin-producing *Escherichia coli* (STEC); biofilm formation; temperature; stainless steel

1. Introduction

Currently, biofilm formation has gained considerable attention in food processing environments. The attachment of microorganisms and subsequent development of biofilms in these environments may be a leading cause of the adulteration of food, which results from the biofouling of pipelines and processing equipment [1]. In addition, biofilms of spoilage and pathogenic microflora that form on contact surfaces are often responsible for the contamination of food during post-processing production [2,3]. It has been shown that, even with diligent cleaning and sanitation, microorganisms within biofilms can remain viable on equipment surfaces [4].

Bacteria can readily bind to stainless steel and polymer surfaces in food production systems and form biofilms where cells are embedded within a matrix made up of proteins, carbohydrates, and extracellular DNA [5,6]. Portions of mature biofilm often detach and are able to colonize downstream environments [7]. Biofilm formation in food processing environments increases the resistance of cells to a number of stressors including starvation, heat, cold, and sanitizers [8,9].

Shiga toxin–producing *Escherichia coli* (STEC) are foodborne pathogens responsible for human enteric infections [10]. They are associated with important public health concerns worldwide. Symptoms associated with STEC infections range from abdominal cramps and bloody diarrhea to post-infection complications arising from hemolytic-uremic syndrome [11]. According to the Public Health Agency of Canada [12], more than 652 cases of STEC infections occur in Canada each year. The rate of STEC O157:H7 has remained relatively constant at 1.2 cases per 100,000 people per year since 2010. For STEC non-O157, the incidence rate increased slightly between 2010 and 2016 from 0.25 to 0.6 cases per 100,000 people per year. STEC O157:H7 is the most predominant serotype causing outbreaks, but other STEC serogroups, such as O26, O45, O91, O103, O111, O113, O121, O128, and O145, have also been linked to severe illness [13,14]. Although the first reported infections by STEC were associated with contaminated meat, foods such as cheese, vegetables, and drinking water have also been implicated in STEC outbreaks [15–17]. STEC isolates of different origins (i.e., animal, food, and human) can form strong biofilms on various food-contact surfaces [17]. The extracellular matrix of STEC biofilms mainly consists of proteins, poly-*N*-acetylglucosamine, cellulose, and colanic acid [5]. Although biofilm formation by STEC isolates is influenced by temperature and time [9], researchers have found that the number of STEC O157:H7 reached up to 5 log CFU/mL in beef juice at 4 °C over 72 h [6]. Thus, STEC biofilms are a potential threat to food hygiene and may become a source of infectious disease in both farm and food-processing environments.

Numerous studies have evaluated the impact of STEC O157 biofilms on food safety, as well as understanding the mechanisms and genetic basis for biofilm formation by this pathogen [18,19]. In contrast, there are relatively few reports on the ability of non-O157 STEC to form biofilms on food-contact surfaces. Adator et al. demonstrated that 12 non-O157 strains remained viable within dry-surface biofilms on stainless steel for at least 30 days and were able to contaminate fresh lettuce within 2 min of exposure [17]. Furthermore, Rong et al. showed that O26:H11 and O111:H8 exhibited a superior ability to form biofilms at the air-liquid interface on glass surfaces and be insensitive to sanitizers [20]. These studies demonstrated that non-O157 can adhere to food contact surfaces and subsequently result in contamination of vegetables and meat.

Thirty-six non-O157 STEC strains from nine serogroups (O113, O145, O91, O26, O121, O128, O103, O45, and O111) were investigated for biofilm formation on polystyrene in our previous study [21]. Of these strains, EC20020170 O113:H21, EC19990166 O145:H25, EC20010076 O91:H21, EC19970119 O26:H11, EC19990161 O121:H19, EC19960949 O128:NM, EC19970327 O103:H2, EC19940040 O45:H2, and EC20030053 O111:NM from each serogroup formed strong biofilms on polystyrene at 22 °C and 37 °C [21]. This finding coupled with the previous studies motivated us to further explore their biofilm-forming abilities on stainless steel, since it is the most common contact surface used in food-processing plants. In addition, we included a representative O157 strain of phage type 14a, which is the predominant phage type isolated from humans in Canada. Therefore, the objective of this study was to investigate the biofilm forming potential of STEC over time on stainless steel surfaces at different temperatures.

2. Materials and Methods

2.1. Bacterial Strains and Cultivation

EC20020170 O113:H21, EC19990166 O145:H25, EC20010076 O91:H21, EC19970119 O26:H11, EC19990161 O121:H19, EC19960949 O128:NM, EC19970327 O103:H2, EC19940040 O45:H2, EC20030053 O111:NM, and EC2011007 O157:H7 were kindly provided by Dr. Roger Johnson of the Public Health Agency of Canada (Guelph, ON, USA). All strains were streaked onto Lysogeny broth (LB) agar (Sigma-Aldrich, Oakville, ON, USA) and incubated at 37 °C for 18 h. An isolated colony was then inoculated into 10 mL of Minimal Salt (M9) broth (Sigma-Aldrich) supplemented with 0.4% glucose, 0.02% MgSO$_4$, and 0.001% CaCl$_2$ (w/v) and grown at 37 °C, on a rocker platform at 180 rpm for 18 h.

2.2. Biofilm Formation

Type-304 stainless steel coupons (No. 2b finish, 2.54 cm × 7.62 cm × 0.081 cm, Biosurface, Bozeman, MT, USA) were used to assess biofilm formation. Prior to use, coupons were soaked in 10% bleach (0.5% hypochlorite) for 24 h. This was followed by rinsing three times with sterile distilled water to remove residual hypochlorite and then dried in the biosafety cabinet. Coupons were then treated with 70% ethanol and air-dried for 5 min at room temperature. Lastly, the coupons were autoclaved at 121 °C for 15 min.

To assess biofilm formation, a conventional static assay was used with minor modifications [22]. Briefly, overnight STEC cultures were diluted in M9 medium to achieve a final concentration of 7 log CFU/mL. Subsequently, 20 mL of the diluted cultures were introduced into 50 mL falcon tubes containing a sterile stainless steel coupon. The tubes were loosely capped and incubated at 13 °C or 22 °C for 24, 48, or 72 h at which point biofilm formation was assessed. Two replicate coupons for each strain were evaluated, and coupons in un-inoculated medium were used as negative controls. Data are presented as the average of three independent trials.

2.3. Crystal Violet Staining

Following incubation, the coupons were carefully removed from the growth medium using sterile forceps, gently tapped against the side of the falcon tube to remove excess liquid, and rinsed three times with 25 mL of sterile filtered water to remove loosely-adherent bacteria. Subsequently, the coupons with attached bacterial cells were fixed with 25-mL absolute methanol (analytical grade, >99%, Sigma-Aldrich) for 15 min. Coupons were air-dried for 2 min, and stained with 0.5 % (w/v) crystal violet solution (Sigma-Aldrich) for 15 min, which was followed by three washes with distilled water and air-drying. The dye bound to the biofilm was then dissolved with 25 mL of 33% glacial acetic acid (Sigma-Aldrich) and measured at a wavelength of 590 nm using a spectrophotometer as described previously [21].

2.4. Enumeration of the Planktonic and Attached Cells

To enumerate the planktonic cells after each period, 1 mL of the cell suspension was pipetted from the falcon tubes, serially diluted with 10 mM phosphate buffered saline (PBS, pH 7.4), plated on LB agar, and incubated at 37 °C for 18 h. To recover the attached bacterial cells, the coupons were rinsed three times with sterile water, immersed into 25 mL of sterile PBS, and sonicated at 20 kHz for 10 min. After sonication, the tubes containing coupons were vigorously vortexed for 1 min, and 1 mL of the bacterial suspension was serially diluted, plated on LB agar, and incubated at 37 °C for 18 h. Bacteria were enumerated and the results were expressed as the average of the data from three independent assays.

2.5. SEM Analysis

Based on the above assays, three strong biofilm formers (strains O113, O145 and O91) were further observed by scanning electron microscopy (SEM), as described previously [23]. Coupons with biofilms were rinsed three times as described above, air-dried, and then fixed in 2.5 % glutaraldehyde (v/v) for 24 h. Subsequently, the samples were dehydrated in a series of ethanol dilutions (v/v) (i.e., 10%, 30%, 50%, 70%, 90%, and 100%) and isobutyl alcohol dilutions (v/v) (i.e., 10%, 30%, 50%, 70%, 90%, and 100%). The samples were then treated with 100% (v/v) hexamethyldisilazane for 10 min and coated with gold using a sputter coater and visualized using a SEM (HITACHI S-4800, Japan).

2.6. Statistical Analysis

Results from biofilm formation were compiled from the three independent experiments. Data were reported as the averages of replicates ± the standard deviation. The student's t-test and one-way ANOVA with the SPSS software (19.0, IBM, Armonk, NY, USA) were used to calculate P values among treatment groups. Significant differences were presented at a 95% confidence level ($P \leq 0.05$).

3. Results

3.1. Growth of the Planktonic Cells in M9 Medium

As shown in Figure S1, all strains grew well and exhibited similar growth patterns at 22 °C, which reached the stationary phase within 24 h at about 9 log CFU/mL, but, at 13 °C, it required 72 h for cultures to reach an average of 8.5 log CFU/mL.

3.2. Biofilm Formation by STEC on a Stainless Steel Surface

The STEC isolates differed in their ability to form biofilms on stainless steel (Figure 1A). Based on the OD_{590nm} produced by biofilms, strains were classified as no biofilm, weak, moderate, or strong biofilm producers, as previously described [24]. The cutoff optical density value (ODc) of 0.043 was three standard deviations above the mean OD of negative controls. Isolates were classified as no (A_{590} < 0.043), weak (0.086 > A_{590} > 0. 043), moderate (0.172 > A_{590} > 0.086), or strong biofilm formers (A_{590} > 0.172; Figure 1B). At 22 °C, O113 exhibited the highest biofilm-forming capacity (P < 0.05), followed by O145, O91, O157, and O121, respectively. The isolates from serogroups O26, O103, O128, O111, and O45 formed only weak biofilms. We also found that the attached biomass of strains O113, O145, O91, O157, and O121 decreased (P < 0.05) with incubation time at 22 °C (Figure 1B). Compared to 22 °C, the biofilm formation of isolates from serogroups O113, O145, O91, O157, and O121 decreased (P < 0.05) at 13 °C. Only the isolate from serogroup O113 formed a moderate biofilm at 13 °C after 72 h of incubation.

3.3. Enumeration of the Biofilm Cells

Populations of biofilm cells of strains O113, O145, O91, O157, and O121 all reached approximately 7.0 log CFU/cm^2 at 22 °C after 24 h, but consistently decreased (P < 0.05) to less than 5.9 log CFU/cm^2 by 72 h (Figure 2). The number of biofilm cells for these five isolates ranged from 2.0 to 3.3 log CFU/cm^2 at 13 °C at 24 h, achieving 5.5 to 6.4 log CFU/cm^2 over 72 h. In contrast, there was no difference in the number of biofilm cells of O111, O128, O103, O26, and O45 at 22 °C, with the concentration remaining between 4.1 to 4.9 log CFU/cm^2. However, the populations of these five isolates did increase from 1.1 to 3.1 log CFU/cm^2 at 24 h and from 4.1 to 4.8 log CFU/cm^2 after 72 h at 13 °C.

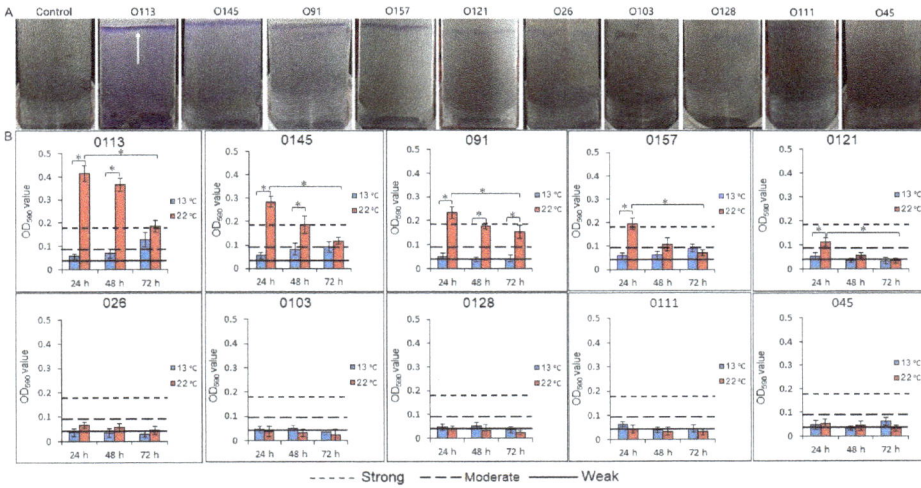

Figure 1. Biofilm formation of 10 STEC strains on a stainless steel surface. (**A**) Biofilms of 10 STEC strains were stained by crystal violet after incubation at 22 °C for 24 h. The arrow shows that some strains formed a dense pellicle at the air-liquid interface. (**B**) Biofilm formation of 10 STEC strains in M9 medium at 22 °C or 13 °C after 24, 48, and 72 h. The vertical axis represents the average of OD values, determined at 590 nm. Horizontal lines represent the cutoff values between weak, moderate, and strong biofilm producers. The cutoff optical density value (ODc) of 0.043 is defined as three standard deviations above the mean OD of the negative controls. Strains were classified as OD $\leq$ ODc (0.043), no biofilm producer, ODc < OD $\leq$ 2 $\times$ ODc, weak biofilm producers, 2 $\times$ ODc < OD $\leq$ 4 $\times$ ODc, moderate biofilm producers, and 4 $\times$ ODc < OD, strong biofilm producers. OD, optical density, STEC, Shiga toxin-producing *E. coli*. Asterisk denotes significant difference ($P < 0.05$).

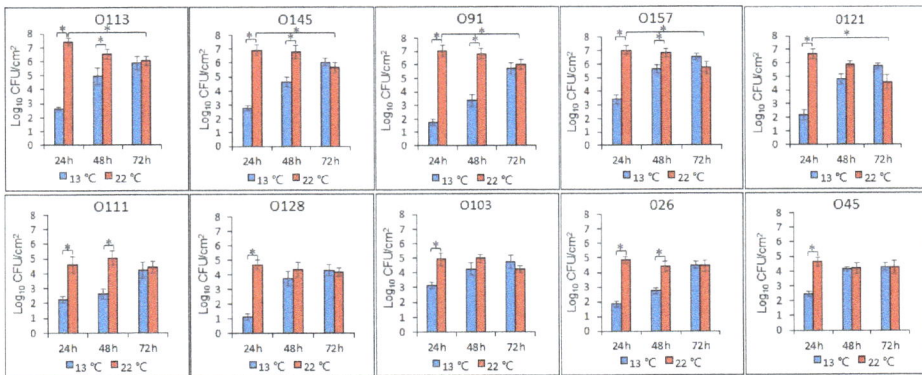

Figure 2. Number of biofilm cells of the 10 STEC isolates after incubation at 22 °C and 13 °C for 24, 48, and 72 h. Asterisk denotes a significant difference ($P < 0.05$).

3.4. SEM Observation

Overall, biofilms formed by O113 (Figure 3), O145 (Figure 4), and O91 (Figure 5) at 13 °C were dramatically different from those formed at 22 °C. At 22 °C, the biofilms of strains O113, O145, and O91 consisted of multiple layers of bacterial cells and completely covered the surface of the stainless steel coupon. However, the biofilms decreased with increasing incubation time (Figure 3, Figure 4, and Figure 5). At 72 h, although there were still some cells attached to the surface, they remained in monolayers and were sparsely distributed on the surface. In contrast, no cell aggregates of O113, O145, and O91 were observed on the stainless steel coupon at 13 °C after 24 h, with only sporadic cell aggregates observed on the surface of stainless steel after 72 h.

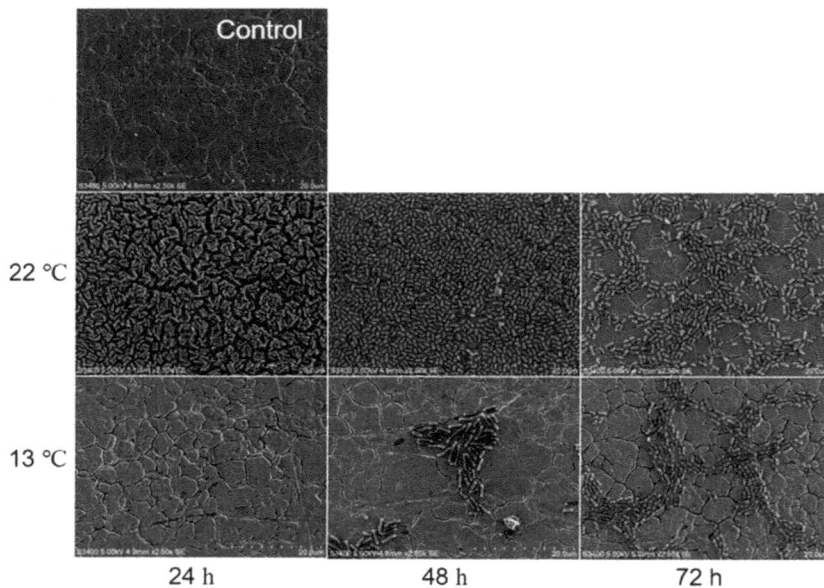

Figure 3. Representative SEM images of O113 biofilm grown in M9 medium at 22 °C and 13 °C for 24, 48, and 72 h on stainless steel coupons. Bar = 20 μm.

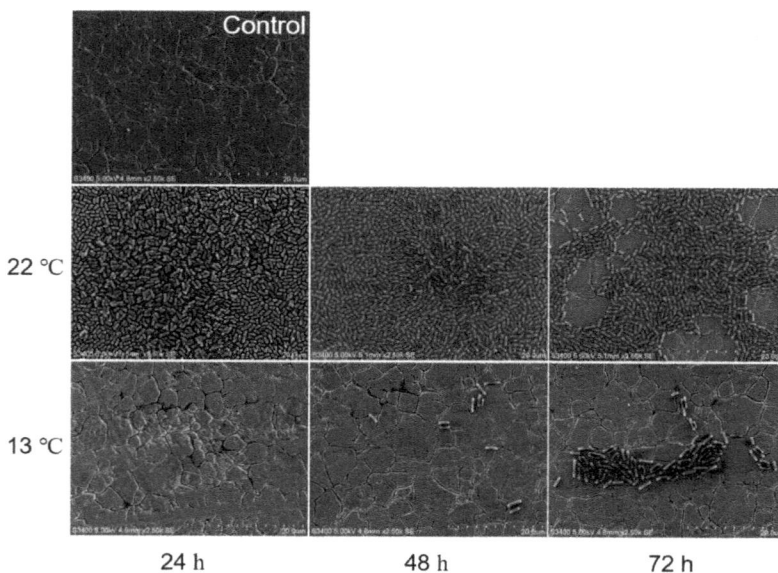

Figure 4. Representative SEM images of O145 biofilm grown in M9 medium at 22 °C and 13 °C for 24, 48, and 72 h on stainless steel coupons. Bar = 20 μm.

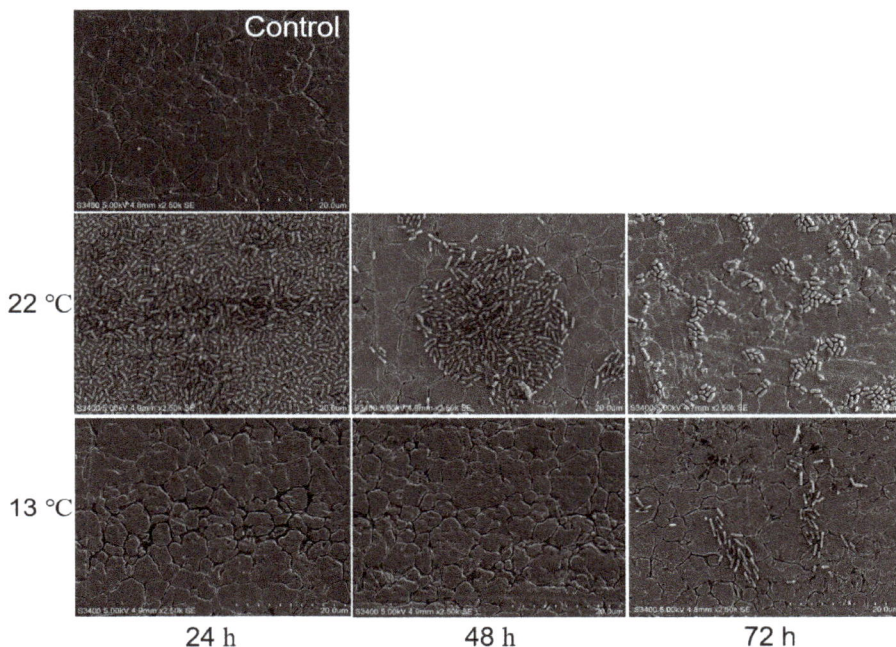

Figure 5. Representative SEM images of O91 biofilm grown in M9 medium at 22 °C and 13 °C for 24, 48, and 72 h on stainless steel coupons. Bar = 20 μm.

4. Discussion

Biofilms are recognized as one of the major strategies that bacteria utilize to support survival under adverse environmental conditions [25]. Bacteria can form biofilms on a wide variety of surfaces and at air-liquid or liquid-solid interfaces [26,27]. Stainless steel surfaces are widely used in food processing plants [28,29]. However, previous studies have shown that the adhesion of Salmonella to stainless steel was significantly higher than to other materials such as rubber and polyurethane surfaces in processing plants [30]. Stainless steel appears smooth to the unaided eye, but it is actually irregular and can harbor bacterial cells when viewed under a microscope [29].

In this study, the capacity of the 10 STEC isolates varied in their ability to form biofilms on stainless steel (Figure 1A). Similar strain-dependent results were obtained from a study that compared the biofilm-forming ability on polystyrene of 18 O157:H7 strains and 33 non-O157 strains belonging to serogroups O26, O111, O103, and O145 [31]. Another feature of biofilm formation in the current study was that the isolates capable of forming biofilms formed a dense pellicle at the air-liquid interface on stainless steel (white arrows in Figure 1A). It is known that STEC biofilm formation is influenced by a variety of factors, including the characteristics of the strains, nutrient availability, temperature, and other environmental conditions [9]. Under our experimental conditions, the combination of oxygen and moisture available at the air-liquid interfaces may have played an important role in biofilm formation, which is an observation supported by others [20].

Of the 10 isolates that were previously identified as capable of forming strong or moderate biofilms on polystyrene at 22 °C [21], five strains (O113, O145, O91, O157, and O121) were able to form strong or moderate biofilms on stainless steel at this temperature. This indicates that these bacterial isolates can survive on various food-contact surfaces, which increases the likelihood that they could contaminate food. However, the lack of ability for the other five strains (O45, O111, O26, O103, and O128) to form strong biofilms on stainless steel may be due to a number of factors. First, the properties of attachment surface may affect the binding strength of bacteria to the substrate. Second, the differences in the nutrient composition of media might be a contributing factor considering the fact that our previous study was conducted using LB broth [21,32], whereas M9 medium was used to grow the bacteria that formed biofilms in the current study.

At 22 °C, the attached biomass of strains O113, O145, O91, O157, and O121 decreased with increasing incubation time. This reduction in biofilm cells was further confirmed by representative SEM images of O113, O145, and O91 (Figure 3, Figure 4, and Figure 5). Based on our previous studies [21], the biofilms of strains O113, O145, O121, O45, and O103 increased on polystyrene with incubation time, so it was expected that the density of biofilm cells of these strains would also increase on stainless steel over time. Previous studies demonstrated a linear increase in *E. coli* biofilm formation on stainless steel at 23 °C [33] and 15 °C [6] up to 24 h. However, some studies observed a reduction in cell density in biofilms after 48 h [33]. These findings suggest that the biofilm medium plays a major role in the biofilm phenotype. Under static conditions, a lack of nutrients and metabolic waste accumulation in M9 during incubation may contribute to a decrease in biofilm surface populations [5]. Moreover, daughter cells of attached bacteria are often released from the surface upon completion of cell division [34]. These released cells may remain in a planktonic state, which leads to a decrease in STEC biofilm formation. Otherwise, the release of cells from STEC biofilm on stainless steel might have occurred at higher cell densities than during biofilm growth on polystyrene since biofilm cell numbers of STEC continued to increase at 22 °C [21]. Biofilms formed on stainless steel dislodge at a faster rate than those on highly hydrophobic acrylic surfaces [35]. Compared to polystyrene, stainless steel surfaces are hydrophilic and negatively charged at a neutral pH. Since bacteria are also typically negatively charged, this surface may be less conducive to microbial colonization. In weakly charged liquids, the repulsive electrostatic forces are significant [32,36].

Low temperatures (5–15°C) are generally maintained in meat processing environments [37]. Therefore, the potential for STEC strains to develop biofilms on stainless steel at 13 °C was examined. Compared to 22 °C, the biofilm formation of strains O113, O145, O91, O157, and O121 was dramatically inhibited at 13 °C. As previously reported, temperature influences the biofilm-forming capacity of isolates [9]. Weak biofilm formation at lower temperatures could be due to the presence of fewer planktonic cells at 13 °C than at 22 °C (Figure S1). Another explanation could be that biofilm cells grew slower at lower temperatures, which results in weaker biofilms at 13 °C than at 22 °C. These findings are in agreement with Dewanti and Wong [38], who observed stronger biofilm formation by STEC O157:H7 on stainless steel at 22 °C than at 10 °C. Furthermore, the surface properties of STEC at 13 °C may be different from that grown at 22 °C. It has been shown that the cell surface hydrophobicity and fimbriae production of *E. coli* and *Salmonella* are temperature-dependent [39–41]. Adrian et al. reported that higher temperature increased the cell surface hydrophobicity level in STEC isolates [40], which is positively correlated with biofilm formation [42,43]. Furthermore, Walker et al. demonstrated that fimbriae in Salmonella were not produced at temperatures below 20 °C [41], which reduced bacterial attachment to commonly used food processing surfaces [44,45]. We also found that, over 72 h of incubation, only strain O113 formed a moderate biofilm at 13 °C. This biofilm-forming capacity of O113 may contribute to its persistence in the processing environment and influence its high relative incidence [46].

5. Conclusions

The findings in this study indicated that STEC isolates form biofilms in a strain-dependent manner and the process was affected by various environmental factors such as temperature, atmosphere, and incubation time. Of the ten isolates previously shown to readily form biofilms on polystyrene, only strains O113, O145, O91, O157, and O121 formed moderate to strong biofilms on stainless steel at 22 °C. At 13 °C, biofilm formation by strains O113, O145, O91, O157, and O121 decreased, which indicates that low temperature environments will reduce STEC biofilm formation on food contact surfaces. Further studies are underway to assess the ability of these strains to form biofilms in meat processing environments and to identify methods to remove and prevent biofilm development as a means of reducing the risk of food contamination.

Supplementary Materials: The following are available online at http://www.mdpi.com/2076-2607/7/4/95/s1.

Author Contributions: Z.M., K.S., Y.D.N., and T.A.M. designed the study. Z.M. performed the experiments. Z.M. and K.S. analyzed the data. Z.M. wrote the manuscript. K.S., Y.D.N., E.W.B., T.A.M., and X.B revised the manuscript. T.A.M., C.N., and K.S. provided technical assistance.

Funding: Funding support was provided from Agriculture and Agri-Food Canada, Alberta Agriculture and Forestry, the China Scholarship Council and Xianqin Yang and Claudia Narvaez through a Beef Cattle Research Council Beef Cluster grant.

Acknowledgments: We acknowledge the technical assistance of R. Ha, C. Conrad, R. Barbieri, S. Trapp, and Y. Graham during the study as well as Roger Johnson (PHAC) for provision of STEC strains.

Conflicts of Interest: The authors declare no conflict of interest.

References

1. Dzieciol, M.; Schornsteiner, E.; Muhterem, M.; Stessl, B.; Wagner, M.; Schmitz-Esser, S. Bacterial diversity of floor drain biofilms and drain waters in a *Listeria monocytogenes* contaminated food processing environment. *Int. J. Food Microbiol.* **2016**, *223*, 33–40. [CrossRef] [PubMed]

2. Al-Adawi, A.S.; Gaylarde, C.C.; Sunner, J.; Beech, I.B. Transfer of bacteria between stainless steel and chicken meat: a CLSM and DGGE study of biofilms. *AIMS Microbiol.* **2016**, *2*, 340–358. [CrossRef]

3. Wang, H.; Zhang, X.; Zhang, Q.; Ye, K.; Xu, X.; Zhou, G. Comparison of microbial transfer rates from *Salmonella* spp. biofilm growth on stainless steel to selected processed and raw meat. *Food Control* **2015**, *50*, 574–580. [CrossRef]

4. Gibson, H.; Taylor, J.; Hall, K.; Holah, J. Effectiveness of cleaning techniques used in the food industry in terms of the removal of bacterial biofilms. *J. Appl. Microbiol.* **1999**, *87*, 41–48. [CrossRef] [PubMed]

5. Vogeleer, P.; Tremblay, Y.D.; Jubelin, G.; Jacques, M.; Harel, J. Biofilm-forming abilities of Shiga toxin-producing Escherichia coli isolates associated with human infections. *Appl. Environ. Microbiol.* **2016**, *82*, 1448–1458. [CrossRef] [PubMed]

6. Dourou, D.; Beauchamp, C.S.; Yoon, Y.; Geornaras, I.; Belk, K.E.; Smith, G.C.; Nychas, G.-J.E.; Sofos, J.N. Attachment and biofilm formation by *Escherichia coli* O157:H7 at different temperatures, on various food-contact surfaces encountered in beef processing. *Int. J. Food Microbiol.* **2011**, *149*, 262–268. [CrossRef] [PubMed]

7. Petrova, O.E.; Sauer, K. Escaping the biofilm in more than one way: desorption, detachment or dispersion. *Curr. Opin. Microbiol.* **2016**, *30*, 67–78. [CrossRef] [PubMed]

8. Ryu, J.H.; Beuchat, L.R. Biofilm formation by Escherichia coli O157:H7 on stainless steel: Effect of exopolysaccharide and curli production on its resistance to chlorine. *Appl. Environ. Microbiol.* **2005**, *71*, 247–254. [CrossRef] [PubMed]

9. Ryu, J.H.; Kim, H.; Beuchat, L.R. Attachment and biofilm formation by Escherichia coli O157:H7 on stainless steel as influenced by exopolysaccharide production, nutrient availability, and temperature. *J. Food Prot.* **2004**, *67*, 2123–2131. [CrossRef] [PubMed]

10. Caprioli, A.; Scavia, G.; Morabito, S. Public health microbiology of Shiga toxin-producing *Escherichia coli*. In *Enterohemorrhagic Escherichia coli and Other Shiga Toxin-Producing E. coli*, 1st ed.; Vanessa, S., Carolyn, J.H., Eds.; American Society of Microbiology: Washington, DC, USA, 2015; pp. 263–295.

11. Fakhouri, F.; Zuber, J.; Frémeaux-Bacchi, V.; Loirat, C. Haemolytic uraemic syndrome. *Lancet* **2017**, *390*, 681–696. [CrossRef]

12. Public Health Agency of Canada. *National Enteric Surveillance Program (NESP): Annual summary 2016*; Public Health Agency of Canada: Ottawa, ON, Canada, 2016.

13. Luna-Gierke, R.; Griffin, P.; Gould, L.; Herman, K.; Bopp, C.; Strockbine, N.; Mody, R. Outbreaks of non-O157 Shiga toxin-producing *Escherichia coli* infection: USA. *Epidemiol. Infect.* **2014**, *142*, 2270–2280. [CrossRef] [PubMed]

14. Park, S.; Kim, S.H.; Seo, J.J.; Kee, H.Y.; Kim, M.J.; Seo, K.W.; Lee, D.H.; Choi, Y.H.; Lim, D.J.; Hur, Y.J. An outbreak of inapparent non-O157 enterohemorrhagic *Escherichia coli* infection. *Korean J. Med.* **2006**, *70*, 495–504.

15. Yatsuyanagi, J.; Saito, S.; Ito, I.J. A case of hemolytic-uremic syndrome associated with Shiga toxin 2-producing *Escherichia coli* O121 infection caused by drinking water contaminated with bovine feces. *Jpn. J. Infect. Dis.* **2002**, *55*, 174–176. [PubMed]

16. Reid, T.M.S. A case study of cheese associated *E. coli* O157 outbreaks in Scotland. In *Verocytotoxigenic Escherichia coli*, 2nd ed.; Duffy, G., Garvey, P., McDowell, D., Eds.; Food and Nutrition: Hartford, CT, USA, 2001; pp. 201–212.

17. Adator, E.H.; Cheng, M.; Holley, R.; McAllister, T.; Narvaez-Bravo, C. Ability of Shiga toxigenic *Escherichia coli* to survive within dry-surface biofilms and transfer to fresh lettuce. *Int. J. Food Microbiol.* **2018**, *269*, 52–59. [CrossRef] [PubMed]

18. Wang, R.; Luedtke, B.E.; Bosilevac, J.M.; Schmidt, J.W.; Kalchayanand, N.; Arthur, T.M. *Escherichia coli* O157:H7 strains isolated from high-event period beef contamination have strong biofilm-forming ability and low sanitizer susceptibility, which are associated with high pO157 plasmid copy number. *J. Food Prot.* **2016**, *79*, 1875–1883. [CrossRef]

19. Sharma, V.K.; Kudva, I.T.; Bearson, B.L.; Stasko, J.A. Contributions of EspA filaments and curli fimbriae in cellular adherence and biofilm formation of enterohemorrhagic *Escherichia coli* O157:H7. *PLoS ONE* **2016**, *11*, e0149745. [CrossRef] [PubMed]

20. Wang, R.; Bono, J.L.; Kalchayanand, N.; Shackelford, S.; Harhay, D.M. Biofilm formation by Shiga toxin–producing *Escherichia coli* O157:H7 and Non-O157 strains and their tolerance to sanitizers commonly used in the food processing environment. *J. Food Prot.* **2012**, *75*, 1418–1428. [CrossRef] [PubMed]

21. Wang, J.; Stanford, K.; McAllister, T.A.; Johnson, R.P.; Chen, J.; Hou, H.; Zhang, G.; Niu, Y.D. Biofilm formation, virulence gene profiles, and antimicrobial resistance of nine serogroups of non-O157 Shiga toxin–producing *Escherichia coli*. *Foodborne Pathog. Dis.* **2016**, *13*, 316–324. [CrossRef] [PubMed]

22. Ramírez, M.D.; Groot, M.N.; Smid, E.J.; Hols, P.; Kleerebezem, M.; Abee, T. Role of cell surface composition and lysis in static biofilm formation by *Lactobacillus plantarum* WCFS1. *Int. J. Food Microbiol.* **2018**, *271*, 15–23. [CrossRef]

23. Dhowlaghar, N.; Bansal, M.; Schilling, M.W.; Nannapaneni, R.J. Scanning electron microscopy of *Salmonella* biofilms on various food-contact surfaces in catfish mucus. *Food Microbiol.* **2018**, *74*, 143–150. [CrossRef]

24. Stepanović, S.; Vuković, D.; Dakić, I.; Savić, B.; Švabić-Vlahović, M.J. A modified microtiter-plate test for quantification of staphylococcal biofilm formation. *J. Microbiol. Methods* **2000**, *40*, 175–179. [CrossRef]

25. Fuente-Núñez, C.; Reffuveille, F.; Fernández, L.; Hancock, R.E. Bacterial biofilm development as a multicellular adaptation: Antibiotic resistance and new therapeutic strategies. *Curr. Opin. Microbiol.* **2013**, *16*, 580–589. [CrossRef] [PubMed]

26. Prado, D.B.; Fernandes, M.S.; Anjos, M.M.; Tognim, M.C.; Nakamura, C.V.; Machinski Junior, M.; Mikcha, J.M.; Abreu Filho, B.A. Biofilm-forming ability of *Alicyclobacillus* spp. isolates from orange juice concentrate processing plant. *J. Food Saf.* **2018**, e12466. [CrossRef]

27. Medrano-Félix, J.A.; Chaidez, C.; Mena, K.D.; Socorro, M.; Castro, N. Characterization of biofilm formation by *Salmonella enterica* at the air-liquid interface in aquatic environments. *Environ. Monit. Assess.* **2018**, *190*, 221. [CrossRef] [PubMed]

28. Hilbert, L.R.; Bagge-Ravn, D.; Kold, J.; Gram, L.J. Influence of surface roughness of stainless steel on microbial adhesion and corrosion resistance. *Int. Biodeterior. Biodegr.* **2003**, *52*, 175–185. [CrossRef]

29. Giaouris, E.; Chorianopoulos, N.; Nychas, G.J. Effect of temperature, pH, and water activity on biofilm formation by *Salmonella enterica* enteritidis PT4 on stainless steel surfaces as indicated by the bead vortexing method and conductance measurements. *J. Food Prot.* **2005**, *68*, 2149–2154. [CrossRef]

30. Chia, T.; Goulter, R.; McMeekin, T.; Dykes, G.; Fegan, N.J. Attachment of different *Salmonella* serovars to materials commonly used in a poultry processing plant. *Food Microbiol.* **2009**, *26*, 853–859. [CrossRef]

31. Biscola, F.T.; Abe, C.M.; Guth, B.E. Determination of adhesin gene sequences in, and biofilm formation by, O157 and Non-O157 Shiga toxin-producing *Escherichia coli* strains isolated from different sources. *Appl. Environ. Microbiol.* **2011**, *77*, 2201–2208. [CrossRef] [PubMed]

32. Dewanti, R.; Wong, A.C. Influence of culture conditions on biofilm formation by *Escherichia coli* O157:H7. *Int. J. Food Microbiol.* **1995**, *26*, 147–164. [CrossRef]

33. Nguyen, H.; Yang, Y.; Yuk, H. Biofilm formation of *Salmonella* Typhimurium on stainless steel and acrylic surfaces as affected by temperature and pH level. *LWT-Food Sci. Technol.* **2014**, *55*, 383–388. [CrossRef]
34. Sinde, E.; Carballo, J.J. Attachment of *Salmonella* spp. and Listeria monocytogenes to stainless steel, rubber and polytetrafluorethylene: The influence of free energy and the effect of commercial sanitizers. *Food Microbiol.* **2000**, *17*, 439–447. [CrossRef]
35. Bakterij, A.J. An overview of the influence of stainless-steel surface properties on bacterial adhesion. *Mater. Tehnol.* **2014**, *48*, 609–617.
36. Hood, S.K.; Zottola, E.A. Adherence to stainless steel by foodborne microorganisms during growth in model food systems. *Int. J. Food Microbiol.* **1997**, *37*, 145–153. [CrossRef]
37. Somers, E.B.; LEE, A.C. Efficacy of two cleaning and sanitizing combinations on *Listeria monocytogenes* biofilms formed at low temperature on a variety of materials in the presence of ready-to-eat meat residue. *J. Food Prot.* **2004**, *67*, 2218–2229. [CrossRef]
38. Frank, J.F. Microbial attachment to food and food contact surfaces. In *Advances in Food and Nutrition Research*; Academic Press: Cambridge, MA, USA, 2001; Volume 43, pp. 319–370.
39. Andersen, T.E.; Kingshott, P.; Palarasah, Y.; Benter, M.; Alei, M.; Kolmos, H.J. A flow chamber assay for quantitative evaluation of bacterial surface colonization used to investigate the influence of temperature and surface hydrophilicity on the biofilm forming capacity of uropathogenic *Escherichia coli*. *J. Microbiol. Methods* **2010**, *81*, 135–140. [CrossRef]
40. Cookson, A.L.; Cooley, W.A.; Woodward, M.J. The role of type 1 and curli fimbriae of Shiga toxin-producing *Escherichia coli* in adherence to abiotic surfaces. *Int. J. Food Microbiol.* **2002**, *292*, 195–205. [CrossRef]
41. Walker, S.L.; SOJKA, M.; Dibb-fuller, M.; Woodward, M.J. Effect of pH, temperature and surface contact on the elaboration of fimbriae and flagella by *Salmonella* serotype Enteritidis. *J. Med. Microbiol.* **1999**, *48*, 253–261. [CrossRef]
42. Bonaventura, G.; Piccolomini, R.; Paludi, D.; Dorio, V.; Vergara, A.; Conter, M.; Ianieri, A.J. Influence of temperature on biofilm formation by Listeria monocytogenes on various food-contact surfaces: Relationship with motility and cell surface hydrophobicity. *J. Appl. Microbiol.* **2008**, *104*, 1552–1561. [CrossRef] [PubMed]
43. Liu, Y.; Yang, S.F.; Li, Y.; Xu, H.; Qin, L.; Tay, J.H. The influence of cell and substratum surface hydrophobicities on microbial attachment. *J. Biotechnol.* **2004**, *110*, 251–256. [CrossRef]
44. Jain, S.; Chen, J.J. Attachment and biofilm formation by various serotypes of *Salmonella* as influenced by cellulose production and thin aggregative fimbriae biosynthesis. *J. Food Prot.* **2007**, *70*, 2473–2479. [CrossRef]
45. Cogan, T.A.; Rgensen, F.; Lappin-Scott, H.M.; Benson, C.E.; Woodward, M.J.; Humphrey, T.J. Flagella and curli fimbriae are important for the growth of *Salmonella enterica* serovars in hen eggs. *Microbiology* **2004**, *150*, 1063. [CrossRef] [PubMed]
46. Paton, A.W.; Woodrow, M.C.; Doyle, R.M.; Lanser, J.A.; Paton, J.C. Molecular characterization of a Shiga toxigenic *Escherichia coli* O113: H21 strain lacking eae responsible for a cluster of cases of hemolytic-uremic syndrome. *J. Clin. Microbiol.* **1999**, *37*, 3357–3361. [PubMed]

microorganisms

MDPI

Article

Inactivation of *Escherichia coli* O157:H7 by High Hydrostatic Pressure Combined with Gas Packaging

Bing Zhou [1,2,3,4], Luyao Zhang [1,2,3,4], Xiao Wang [1,2,3,4], Peng Dong [1,2,3,4], Xiaosong Hu [1,2,3,4] and Yan Zhang [1,2,3,4,*]

[1] College of Food Science and Nutritional Engineering, China Agricultural University, Beijing 100083, China; bingcau110@163.com (B.Z.); zhangly7667@163.com (L.Z.); echowang2014@163.com (X.W.); dongpeng12@hotmail.com (P.D.); huxiaos@263.net (X.H.)
[2] National Engineering Research Center for Fruits and Vegetables Processing, Ministry of Science and Technology, Beijing 100083, China
[3] Key Laboratory of Fruits and Vegetables Processing, Ministry of Agriculture and Rural Affairs, Beijing 100083, China
[4] Beijing Key Laboratory of Food Non-Thermal Processing, Beijing 100083, China
* Correspondence: zhangyan348@163.com; Tel.: +86-10-62737434-23

Received: 19 April 2019; Accepted: 16 May 2019; Published: 28 May 2019

Abstract: The inactivation of *Escherichia coli* O157:H7 (*E. coli*) in physiological saline and lotus roots by high hydrostatic pressure (HHP) in combination with CO_2 or N_2 was studied. Changes in the morphology, cellular structure, and membrane permeability of the cells in physiological saline after treatments were investigated using scanning electron microscopy, transmission electron microscopy, and flow cytometry, respectively. It was shown that after HHP treatments at 150–550 MPa, CO_2-packed *E. coli* cells had higher inactivation than the N_2-packed and vacuum-packed cells, and no significant difference was observed in the latter two groups. Further, both the morphology and intracellular structure of CO_2-packed *E.coli* cells were strongly destroyed by high hydrostatic pressure. However, serious damage to the intracellular structures occurred in only the N_2-packed *E. coli* cells. During HHP treatments, the presence of CO_2 caused more disruptions in the membrane of *E. coli* cells than in the N_2-packed and vacuum-packed cells. These results indicate that the combined treatment of HHP and CO_2 had a strong synergistic bactericidal effect, whereas N_2 did not have synergistic effects with HHP. Although these two combined treatments had different effects on the inactivation of *E. coli* cells, the inactivation mechanisms might be similar. During both treatments, *E. coli* cells were inactivated by cell damage induced to the cellular structure through the membrane components and the extracellular morphology, unlike the independent HHP treatment.

Keywords: high hydrostatic pressure; carbon dioxide; nitrogen; modified atmosphere packaging; *Escherichia coli*

1. Introduction

During the last decade, some mild and efficient food preservation technologies have been developed to satisfy the growing consumer demands for minimally processed and preservative-free food products [1]. Among these technologies, high hydrostatic pressure (HHP) processing technology is a new, commercially successful nonthermal technology that meets these consumer demands to some extent while retaining the sensorial and nutritional properties of freshly prepared foods [2]. Compared with conventional technologies, HHP can inactivate food spoilage and pathogenic microorganisms at room temperature, extend the shelf life of foods, and reduce damage to heat-sensitive food components and the formation of harmful food components such as heterocyclic amines (HCAs) caused by high temperature [3–5]. In addition, unlike thermal processing and other preservations, pressure can penetrate the entire food product to inactivate both surface and internalized microorganisms and acts instantaneously and uniformly throughout foods regardless of size, shape, and geometry [6,7].

However, two main deficiencies of HHP treatment exist that limit its commercial use in low-acid foods. One is the economic costs of the high-pressure equipment that can reach pressures up to 600 MPa or more, which are required to efficiently inactivate microorganisms in food [8]. The other is its weak inactivation of bacterial spores, the most resistant cell type known. At room temperature, high-pressure food processes have been reported to be effective in reducing or inactivating vegetative pathogens, human rotavirus, hepatitis A virus, and calicivirus in foods [5], but even ultra-high pressure levels (1000 MPa) are not effective at inactivating bacterial endospores [9]. In order to overcome these critical drawbacks of HHP, various preservative factors (hurdles) to increase or accompany the efficacy of pressure-induced inactivation of microorganisms have been thoroughly studied, including moderate temperature, pH, modified atmospheres, other nonthermal food processing methods, and antimicrobial agents such as nisin [10–13].

The combination of HHP and modified atmosphere packaging (MAP) as an effective synergistic treatment that has attracted much attention [14]. Modified atmosphere packaging (MAP) is a well-established technology that is generally used for extending the shelf-life of minimally processed foods by replacing the surrounding atmosphere of the food with a gas mixture. The gas mixture usually includes the bactericidal gas CO_2 and the comparatively inert gases of N_2 and O_2. It is reported that microbial growth could be inhibited by compressed gas (CO_2, Kr, Xe, and N_2O) over a range of pressures (1.5 to 5.5 MPa) [15,16]. Also, combining modified atmosphere packaging (50% O_2 + 50% CO_2) with low high pressure (150 MPa) was investigated for shelf-life extension of carrots, and it was found that spoilage microorganisms and pathogens are more susceptible to being inactivated by HHP in the presence of gas [17]. Furthermore, due to the bacteriostatic effect of CO_2, pressurized CO_2, known as high-pressure carbon dioxide (HPCD), whose pressure level is less than 100 MPa, has been widely used to inactivate microorganisms in foods and has become an alternative nonthermal pasteurization technique for foods [18].

In recent years, the synergistic effect of antimicrobial gas (CO_2) and HHP, in which the pressure level is more than 100 MPa, has been extensively reported in the literature [19–21]. The inactivation effect of HHP on bacteria was greatly enhanced by using a new setup to dissolve and retain the concentration of CO_2 in fruit juices [22]. Further, low- or medium-acid fruit and vegetable juices treated with a combination of HHP and dissolved CO_2 were also effectively preserved. More importantly, when using CO_2 in combination with HHP, the treatment pressure could be reduced without compromising a reduction in microbial count.

However, how this treatment synergistically inactivates the valid objective microorganisms is unexplored. Furthermore, the mechanism of this synergic effect remains elusive. Thus, we determined the effect of high nitrogen (N_2) and high carbon dioxide (CO_2) on the inactivation effect of high pressure on *Escherichia coli* O157:H7. We investigated the morphology and cellular structures by scanning electron microscopy (SEM), transmission electron microscopy (TEM), and flow cytometry (FCM) to provide more evidence for the microbial inactivation mechanism of HHP treatment combined with gas.

2. Materials and Methods

2.1. Bacterial Strain and Culture Conditions

A stock culture of *E. coli* (CGMCC1.90), obtained from the China General Microbiological Culture Collection Center (CGMCC, Beijing, China), was maintained on nutrient agar (NA) plates (Beijing Aoboxing Biological Technology, Beijing, China). The *E. coli* inoculum was made by transferring a single colony to 20 mL of nutrient broth (NB, Beijing Aoboxing Biological Technology, Beijing, China), which was shaken at 170 rpm at 37 °C for 12 h to obtain cells in stationary phase. Cultures were inoculated by transferring a 1% (v/v) inoculum to 1.5 L nutrient broth and continuing growth under the same conditions as described above for 2.5 h to obtain cells in the middle exponential phase. Cells were harvested, washed twice in sterile physiological saline (PS, 0.85% NaCl solution, pH 6.80) and resuspended in PS. For some samples, *E. coli* cells were resuspended in sterilized lotus root sauce. Sterilized lotus root sauce was prepared by the following procedure. At first, the lotus root was homogenized by the beating machine. Then, the acquired sauce was autoclaved at 121 °C. The final number of *E. coli* cells generally ranged from 10^7 to 10^8 CFU per milliliter (mL).

2.2. Packing and HHP Processing

A 50 mL *E. coli* cell suspension was transferred to ultraviolet-sterilized polyethylene terephthalate (PET) trays (200 mL) and then conditioned under 100% CO_2 or 100% N_2 using a DT-6A modified atmosphere packaging machine (Dajiang machinery equipment CO., Ltd., Zhejiang, China). The other cell suspensions were transferred to sterile polyethylene plastic bags, vacuum packed, and stored at 4 °C for less than 1 h before treatment.

These samples were treated in a hydrostatic pressurization unit (HHP-750, Baotou Kefa Co., Ltd., Inner Mongolia, China) with a chamber of 30 L capacity. The pressure-transmitting fluid was water. The treatment time in this study did not include the pressurization and depressurization times. The prepared samples were placed in the pressure vessel and treated at 150, 250, 350, 450, and 550 MPa for 1 min at room temperature.

2.3. Determination of Viable Cells

Treated and untreated samples were serially diluted and surface plated on NA agar plates (Oxoid, Basingstoke, UK). Plates were incubated at 37 °C for 24 h, and then the colonies were enumerated. Survival was expressed as the logarithmic viability reduction $\log_{10}$ (N_i/N_0) with N_0 and N_i representing the colony counts before and after HHP treatment, respectively. Survival counts are presented as averages ± standard deviation of three independent experiments.

2.4. SEM and TEM Analysis

According to previous methods [23,24], a suspension of *E. coli* cells was centrifuged at 8000× rpm for 10 min at 4 °C, the supernatant was removed, and then the pellet was resuspended in 2.5% (v/v) glutaraldehyde solution to fix for 12 h. After fixation, the cells in suspension were washed several times with 0.1 M phosphate buffer (PBS, pH 7.2) and fixed again by 1% osmium tetroxide solution (pH 7.2). After 1.5 h, the cells were washed in PBS three times and dehydrated 10 min each with a series of cold ethanol solutions (10%, 30%, 50%, 75%, and 95%). For the SEM assay, the dehydrated cells were rinsed with 50%, 70%, 90%, and 100% isoamyl acetate for 3 min each, critical point dried, and coated with gold–palladium for 60 s. Observations and photomicrographs were carried out with a Hitachi S-3400 N SEM (Hitachi Instruments Inc., Tokyo, Japan) and a JEM-1230 TEM (JEOL Japan Electronics Co., Ltd., Japan).

2.5. FCM Analysis

The FCM analysis of untreated cells (negative control), 75% isopropanol-treated cells (positive control), and the above-treated cells were measured as described by previous studies [23,25]. Cell suspensions were washed twice with physiological saline, resuspended in physiological saline, and adjusted to 10^8–10^9 CFU/mL. Then, 0.15 µL of dye mixture containing equivalent SYTO9 (Sigma-Aldrich, St. Louis, MO, USA) and propidium iodine (PI, Sigma-Aldrich, USA) were added to 50 µL of the cell suspensions and incubated for 20 min at room temperature in the dark. After that, the cells were immediately analyzed with a BD FACSCalibur flow cytometer (BD Biosciences, San Jose, CA, USA) and about 30,000 cells were collected in each run. Forward scatter and side scatter were collected, and the fluorescence signals were collected in the FL1 (green fluorescence of SYTO9 at 502 nm) and FL2 (red fluorescence of PI at 613 nm) channels [26] using Cell Quest software (Becton Dickinson, San Jose, CA, USA).

2.6. Statistical Analysis

All experiments were repeated at least three times. All data were statistically analyzed using Microcal Origin 8.1 (Microcal Software, Inc., Northampton, MA, USA).

3. Results and Discussion

3.1. Inactivation of E. coli in Buffer and Lotus Root Suspension

The inactivation of vacuum-packed, N_2-packed, and CO_2-packed *E. coli* cells subjected to high pressure at 150–550 MPa at room temperature for 1 min is shown in Figure 1a. When the pressure was at 150 MPa, there was no significant difference in inactivation levels among these cells, as indicated by less than 2 logs of inactivation for them (Figure 1a). This indicated that the synergistic effect of low high pressure and gas on the inactivation of *E. coli* cells did not occur. However, the reduction in cell counts of CO_2-packed *E. coli* cells was about 1, 4, and 2 logs more than that of vacuum-packed and N_2-packed cells at 250 MPa, 350 MPa, and 450 MPa, respectively (Figure 1a). Furthermore, a reduction of more than 8 log units, the detection limit, was achieved at 350 MPa for CO_2-packed cells; the pressure for a similar inactivation effect for vacuum-packed and N_2-packed cells was 550 MPa (Figure 1a). Thus, we conclude that the combined treatment of moderately high pressure (250–450 MPa) and CO_2 showed a strong synergistic bactericidal effect.

Next, we sought to determine whether the similar inactivation kinetics of HHP combined with gases also exist in *E. coli* cells suspended in lotus root. Comparing the results in Figure 1a,b, the inactivation levels of samples with three packages were 1 to 3 logs in lotus root less than in buffer at 150 MPa to 450 MPa. This observation that inactivation of *E. coli* by different treatments was more extensive in the buffer than in the lotus root under all conditions may be because a complex matrix has a protective effect on bacterial inactivation compared with a buffer system [27,28]. However, the reduction of CO_2-packed *E. coli* in lotus root was also more than in vacuum-packed and N_2-packed cells at 250 MPa, 350 MPa, and 450 MPa (Figure 1b).

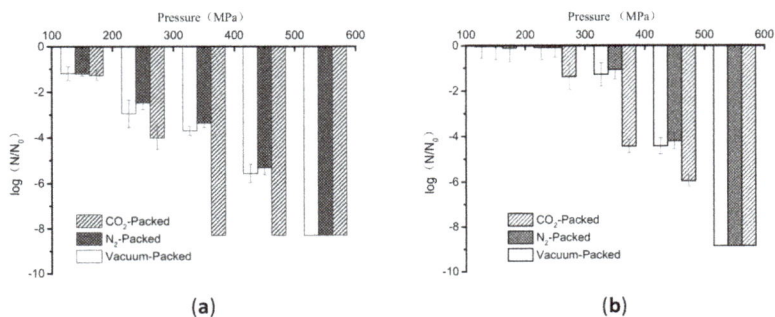

Figure 1. Inactivation of differently packed *E. coli* cells suspended in physiological saline by treatment with different pressures (**a**); Inactivation of differently packed *E. coli* suspended in lotus roots pulps by treatment with different pressures (**b**).

It was reported that the combined treatment with HHP and dissolved CO_2 effectively preserved the low- or medium-acid fruit and vegetable juices compared with HHP treatment [21]. Further, a synergistic effect of the combination of HHP and CO_2 against microorganisms inoculated in poultry sausages was found [20]. This synergistic effect may be due to the increased penetration of CO_2 into the microorganism cells under high pressure. Therefore, we can conclude that the combined application of high pressure and different gases would have different effects on inactivating *E. coli*. The presence of CO_2 could significantly enhance the inactivation of *E. coli* treated with high pressure, which was obvious at a moderately high pressure. Nevertheless, the presence of N_2 did not affect the inactivation at high pressure.

3.2. The Morphology and Intracellular Structure Changes of E. coli Cells Treated with High Pressure Combined with Gas

From the above results, the inactivation of *E. coli* cells by high pressure was indeed affected by gases, which is obvious at 250, 350 MPa, and 450 MPa (Figure 1). Therefore, we used scanning electron microscopy to examine changes in the morphology of CO_2-packed, N_2-packed, and vacuum-packed *E. coli* cells in buffer exposed to these pressures. The untreated *E. coli* cells showed a morphology with a regular rod shape and smooth surface (Figure 2a). After high-pressure treatment at 250 MPa or 350 MPa, both the vacuum- and N_2-packed *E. coli* cells had a similar morphology to the untreated samples (Figure 2b,d,e,g); however, the CO_2-packed cells were collapsed and exhibited holes and wrinkles on the surface (Figure 2c). When the treating pressure increased to 450 MPa, N_2-packed *E. coli* cells were still intact and exhibited similar morphology to that at 250 MPa and 350 MPa (Figure 2d,g,j), while a portion of the vacuum-packed *E. coli* cells was broken and showed cellular debris. Of note, the morphology of CO_2-packed *E. coli* suspensions was further damaged as shown by noticeable hollows, wrinkles, or holes on their surface, and the cellular debris that was caused by cell breakdown (Figure 2f,i).

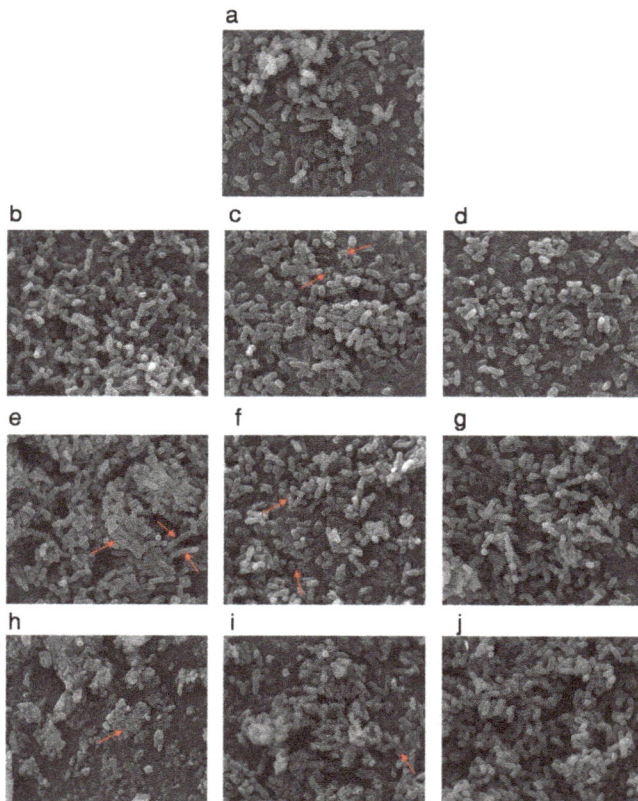

Figure 2. The scanning electron microscopy (SEM) images of differently packed *E. coli* before and after different pressure treatments. Untreated *E. coli* cells (**a**); vacuum-packed (**b**), CO_2-packed (**c**), and N_2-packed (**d**) *E. coli* cells after high hydrostatic pressure (HHP) treatment at 250 MPa: Vacuum-packed (**e**), CO_2-packed (**f**) and N_2-packed (**g**) *E. coli* cells after HHP treatment at 350 MPa; Vacuum-packed (**h**), CO_2-packed (**i**) and N_2-packed (**j**) *E. coli* cells after HHP treatment at 450 MPa. Red arrows represent the remarkable phenotypes.

As shown in the above results, while N_2-packed *E. coli* cells were seriously inactivated by different pressure treatments (2 to 5 log units), their morphology remained stable (Figures 1 and 2d,g,j). This apparent contradiction may be because N_2 is an inert gas and does not dissolve in the water phase, which can induce two phases at high pressure—a water phase and a gas phase [29]. Therefore, destruction of the morphology induced by HHP processing may be decreased in this complex two-phase system. In addition, compressed N_2 may penetrate the cell to balance between the internal and external environment, which has an additional protective effect on morphology.

The effects of HHP treatment combined with gas on the intracellular structure of *E. coli* cells were assessed by transmission electron microscopy. The exposure of vacuum-packed cells to pressures ranging from 250 MPa to 450 MPa induced a slight increase in transparency within the nucleoid areas and the presence of aggregated proteins (Figure 3b,e,h). Although N_2-packed cells showed similar changes in cellular structure to the vacuum-packed cells at 250 MPa (Figure 3b,d), the intracellular damage exhibited by N_2-packed cells was more noticeable than with the vacuum-packed cells at 350 and 450 MPa. This was evidenced by the apparent disorganization of the genome area, the appearance of blank space in the cytoplasm and the condensation of the cytoplasmic material, and the serious intensity and frequency of protein aggregation within the cell cytoplasm (Figure 3g,j). However, the membranes of the N_2-packed cells were organized, whereas the membranes of the vacuum-packed cells displayed winding shapes, and some of them were disrupted or detached from the cytoplasmic content (Figure 3b,d,e,g,h,j). Remarkably, the CO_2-packed cells showed the most damage of the intracellular structures from high-pressure treatment (Figure 3c,f,i). When the treatment pressure increased, the distribution of the amorphous structures became uneven and the genome area was disorganized. Also, there were large clumps of aggregated protein in the cells. The intensity of the damage of the intracellular structures and the weak wrinkling membranes of the CO_2-packed cells increased (Figure 3c,f,i).

Taken together, both the morphology and intracellular structure of CO_2-packed *E. coli* cells were more strongly destroyed by high pressure. Known as the most important gas in MAP, CO_2 can dissolve in the water phase to form carbonic acid (H_2CO_3) to lower the pH or to inhibit the growth of bacteria [18,30,31]. Combined with HHP, this compressed CO_2 dissolved in the liquid state could more easily penetrate the cells [30]. This not only triggered the higher expansion on the sudden release of high pressure to cause more serious cell disruption and membrane damage (morphology damages) but also resulted in the decrease of intracellular pH and disturbance of homeostasis, as well as the extraction of microbial constituents (intracellular structure damage) [32,33]. Also, the major sites of action for CO_2 and HHP treatment were in the cell membrane, and highly-compressed CO_2 could more easily dissolve in and distort these regions [34]. This might explain why CO_2-packed samples by high pressure obtained higher reductions in microbial counts and heavier destruction of the morphology and cellular structures than those treated with HHP alone. However, for N_2-packed *E. coli* cells, the intracellular structure was seriously damaged because it is highly hydrophobic and could dissolve in and distort the cellular core to upset hydrophobic interactions in the proteins [15], but their morphology remained unchanged during HHP treatment. This may explain why the reduction in the N_2-packed *E. coli* cells was less than that of the CO_2-packed cells.

Thus, it seems that HHP treatment combined with CO_2 might inactivate microorganisms by destroying the cellular structure, accompanied by cell rupture. In contrast, the N_2-packed cells were possibly inactivated by HHP through destruction of the cellular structure.

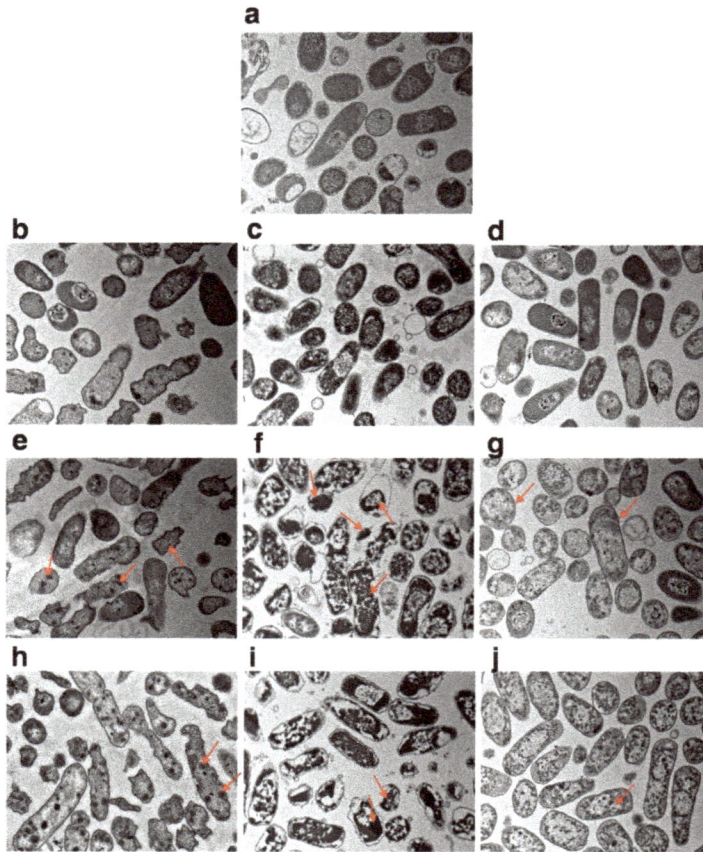

Figure 3. The transmission electron microscopy (TEM) images of differently packed *E. coli* before and after different pressure treatments. Untreated *E. coli* cells (**a**); vacuum-packed (**b**), CO_2-packed (**c**), and N_2-packed (**d**) *E. coli* cells after HHP treatment at 250 MPa: Vacuum-packed (**e**), CO_2-packed (**f**) and N_2-packed (**g**) *E. coli* cells after HHP treatment at 350 MPa; Vacuum-packed (**h**), CO_2-packed (**i**), and N_2-packed (**j**) *E. coli* cells after HHP treatment at 450 MPa. Red arrows represent the remarkable phenotypes.

3.3. Membrane Permeability of E. coli Cells

For an analysis of the membrane damage in *E. coli* cells caused by high pressure, FCM combined with PI and SYTO9 was used. Comparing the profiles of the untreated and HHP treated cells, three groups were distinguished, and then regions were constructed to enumerate events within each group using the CellQuestTM program (Figure 4). Region 1 (R1) corresponded to the living cells with intact membranes (Figure 4). Region 2 (R2) was assigned as *E. coli* cells with unknown cultivability, which were in an intermediate state between dead cells and living cells, having damaged membrane and medium membrane permeability (Figure 4). Region 3 (R3) referred to the inactivated *E. coli* cells with fully damaged membranes, exhibiting high membrane permeability (Figure 4). After a 1-min pressure-holding time at 250 MPa, the majority of the CO_2-packed *E. coli* cells was located in R3, which were higher than that of the N_2-packed cells and vacuum-packed cells located in R3 (Figure 4c–e). For the pressure ranging from 350 MPa to 450 MPa, although there were no significant differences in the proportion of cells in R3 among these treatments, the counts in the other two regions nearly disappeared for the combined treatment of HHP and CO_2, while they still made up a small proportion in the N_2-packed cells and vacuum-packed cells (Figure 4f–k). These results confirmed the synergistic effect of HHP treatment with CO_2 on *E. coli* inactivation.

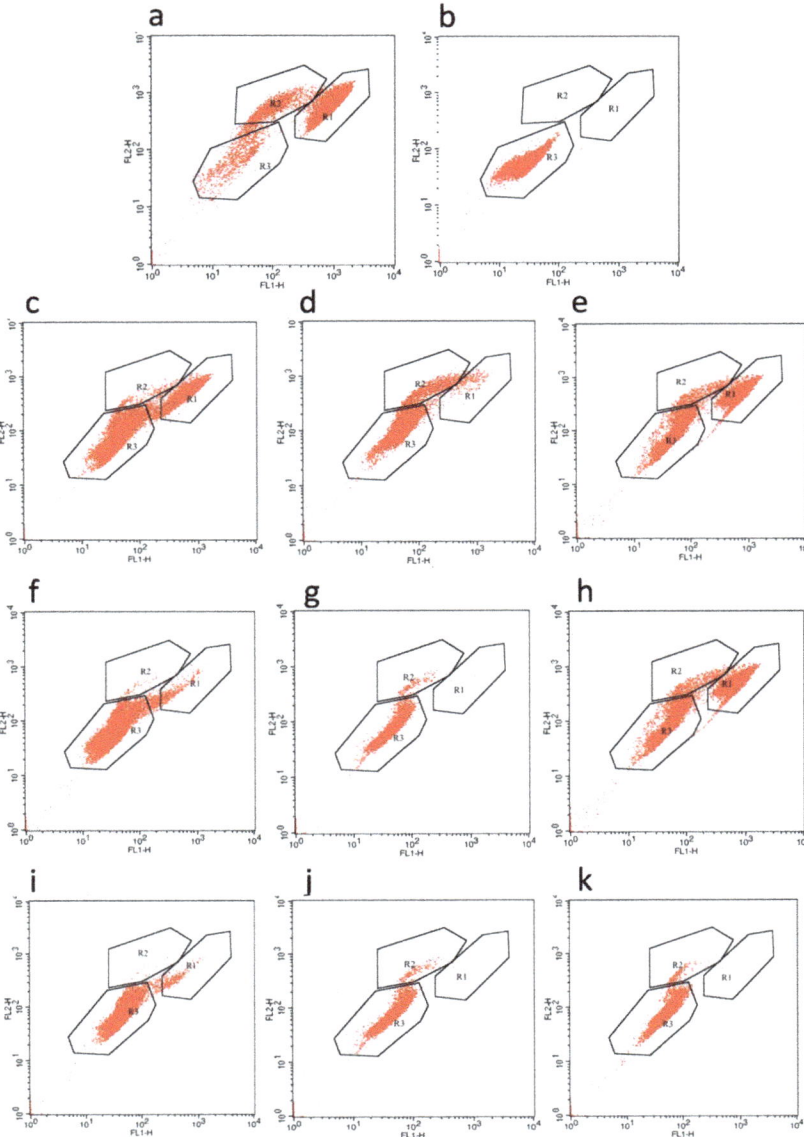

Figure 4. Flow cytometric analysis of *E. coli* differently packaged before and after different pressure treatments. Negative (**a**) and positive (**b**) *E. coli* cells; Vacuum-packed (**c**), CO_2-packed (**d**), and N_2-packed (**e**) *E. coli* cells after HHP treatment at 250 MPa: Vacuum-packed (**f**), CO_2-packed (**g**), and N_2-packed (**h**) *E. coli* cells after HHP treatment at 350 MPa; Vacuum-packed *E. coli* cells (**i**), CO_2-packed (**j**) and N_2-packed (**k**) *E. coli* cells after HHP treatment at 450 MPa.

Surprisingly, the vacuum-packed cells treated with high pressure transferred directly from R1 to R3 and did not go through R2 (Figure 4c,f,i). However, in the presence of gas, the HHP-treated cells transitioned from R1 to R2 and then to R3 as the pressure increased (Figure 4d,e,g,h,j,k). This indicated that the combined use of HHP and gases (CO_2 and N_2) induced intermediate cells. This may be due to the mechanism of inactivation of *E. coli* cells by HHP combined with modified atmosphere packaging, which seems to be different from that of HHP alone.

Here, we show that the morphology of the N_2-packed *E. coli* cells did not change at different pressures but its inner structure was seriously damaged. Also, although both the morphology and intracellular structure of the CO_2-packed *E. coli* cells were strongly destroyed by the high pressure, this damage of the morphology may have been caused by damage to the inner structure. Furthermore, both the CO_2-packed and N_2-packed *E. coli* cells went through an intermediate phase during the high pressure. Based on our results, we propose the following model for the inactivation of cells by combing gas and high pressure. First, the combined use of HHP and gas facilitates the penetration of gas into the *E. coli* cells, which disturbs the intracellular reactions and causes clusters of proteins and the disruption of intracellular enzymes and organelles. Second, a sudden release in pressure ruptures the cells and damages their membranes, leading to the leakage of cytoplasm components. Therefore, this combined treatment might induce a series of cellular damage at first and then act on the membrane components and the extracellular morphology, unlike the independent HHP treatment that directly ruptures the cell membrane and then leads to the loss of internal substances, which would result in bacterial death [29].

4. Conclusions

In this study, the effect of the combination of HHP and gases to inactivate *E. coli* has been studied. The combined use of HHP and CO_2 had a strong synergistic effect on the inactivation of *E. coli* cells, inducing serious destruction in the morphology and the membranes and cellular structure of the cells. In contrast, the combined use of HHP and N_2 showed a similar inactivation effect to HHP alone and destroyed only the cellular structure and the membranes of the cells. Our results provide evidence that the combination of HHP and gases has a different inactivation mechanism than that of HHP treatment. In the presence of gas, the intracellular structure of the cells was damaged at first, and then the membrane and extracellular morphology were destroyed because of the solution of gas and the release of high pressure.

Author Contributions: For research articles with several authors, a short paragraph specifying their individual contributions must be provided. The following statements should be used conceptualization, B.Z., L.Z. and Y.Z.; methodology, B.Z., L.Z. and P.D.; software, P.D. and X.W.; validation, X.W., B.Z. and X.H.; formal analysis, X.W.; investigation, B.Z., L.Z., X.W., P.D., X.H. and Y.Z.; resources, X.H. and Y.Z.; data curation, B.Z.; writing—original draft preparation, B.Z. and Y.Z.; Writing—Review and editing, B.Z., X.W. and Y.Z.

Funding: This work was supported by the National Key Research and Development Plan during the 13th five-year plan period (No. 2017YFD0400504), as well as the key Program of National Natural Science Foundation of China (No. 31530058).

Conflicts of Interest: The authors declare no conflict of interest.

References

1. Ortega-Rivas, E.; Salmeron-Ochoa, I. Nonthermal Food Processing Alternatives and Their Effects on Taste and Flavor Compounds of Beverages. *Crit. Rev. Food Sci.* **2014**, *54*, 190–207. [CrossRef] [PubMed]

2. Fonberg-Broczek, M.; Windyga, B.; Szczawinski, J.; Szczawinska, M.; Pietrzak, D.; Prestamo, G. High pressure processing for food safety. *Acta. Biochim. Pol.* **2005**, *52*, 721–724.

3. Van Impe, J.; Smet, C.; Tiwari, B.; Greiner, R.; Ojha, S.; Stulic, V.; Vukusic, T.; Jambrak, A.R. State of the art of nonthermal and thermal processing for inactivation of micro-organisms. *J. Appl. Microbiol.* **2018**, *125*, 16–35. [CrossRef] [PubMed]

4. Lou, F.F.; Neetoo, H.; Chen, H.Q.; Li, J.R. High Hydrostatic Pressure Processing: A Promising Nonthermal Technology to Inactivate Viruses in High-Risk Foods. *Annu. Rev. Food Sci. Technol.* **2015**, *6*, 389–409. [CrossRef] [PubMed]

5.	Huang, H.W.; Lung, H.M.; Yang, B.B.; Wang, C.Y. Responses of microorganisms to high hydrostatic pressure processing. *Food Control.* **2014**, *40*, 250–259. [CrossRef]

6.	Cruz-Romero, M.; Kelly, A.L.; Kerry, J.P. Influence of packaging strategy on microbiological and biochemical changes in high-pressure-treated oysters (*Crassostrea gigas*). *J. Sci. Food Agric.* **2008**, *88*, 2713–2723. [CrossRef]

7.	Black, E.P.; Setlow, P.; Hocking, A.D.; Stewart, C.M.; Kelly, A.L.; Hoover, D.G. Response of spores to high-pressure processing. *Compr. Rev. Food Sci. Food Saf.* **2007**, *6*, 103–119. [CrossRef]

8.	Sarker, M.R.; Akhtar, S.; Torres, J.A.; Paredes-Sabja, D. High hydrostatic pressure-induced inactivation of bacterial spores. *Crit. Rev. Food Sci.* **2015**, *41*, 18–26. [CrossRef] [PubMed]

9.	San Martin, M.F.; Barbosa-Canovas, G.V.; Swanson, B.G. Food processing by high hydrostatic pressure. *Crit. Rev. Food Sci.* **2002**, *42*, 627–645. [CrossRef]

10.	Queiros, R.P.; Gouveia, S.; Saraiva, J.A.; Lopes-da-Silva, J.A. Impact of pH on the high-pressure inactivation of microbial transglutaminase. *Food Res. Int.* **2019**, *115*, 73–82. [CrossRef]

11.	Pyatkovskyy, T.I.; Shynkaryk, M.V.; Mohamed, H.M.; Yousef, A.E.; Sastry, S.K. Effects of combined high pressure (HPP), pulsed electric field (PEF) and sonication treatments on inactivation of *Listeria innocua*. *J. Food Eng.* **2018**, *233*, 49–56. [CrossRef]

12.	Ross, A.I.V.; Griffiths, M.W.; Mittal, G.S.; Deeth, H.C. Combining nonthermal technologies to control foodborne microorganisms. *Int J. Food Microbiol.* **2003**, *89*, 125–138. [CrossRef]

13.	Roberts, C.M.; Hoover, D.G. Sensitivity of Bacillus coagulans spores to combinations of high hydrostatic pressure, heat, acidity and nisin. *J. Appl. Bacteriol.* **1996**, *81*, 363–368.

14.	Rodriguez-Calleja, J.M.; Cruz-Romero, M.C.; O'Sullivan, M.G.; Garcia-Lopez, M.L.; Kerry, J.P. High-pressure-based hurdle strategy to extend the shelf-life of fresh chicken breast fillets. *Food Control.* **2012**, *25*, 516–524. [CrossRef]

15.	Arao, T.; Hara, Y.; Suzuki, Y.; Tamura, K. Effect of high-pressure gas on yeast growth. *Biosci. Biotech. Bioch.* **2005**, *69*, 1365–1371. [CrossRef] [PubMed]

16.	Debs-Louka, E.; Louka, N.; Abraham, G.; Chabot, V.; Allaf, K. Effect of compressed carbon dioxide on microbial cell viability. *Appl. Environ. Microbiol.* **1999**, *65*, 626–631. [PubMed]

17.	Amanatidou, A.; Slump, R.A.; Gorris, L.G.M.; Smid, E.J. High oxygen and high carbon dioxide modified atmospheres for shelf-life extension of minimally processed carrots. *J. Food Sci.* **2000**, *65*, 61–66. [CrossRef]

18.	Garcia-Gonzalez, L.; Geeraerd, A.H.; Spilimbergo, S.; Elst, K.; Van Ginneken, L.; Debevere, J.; Van Impe, J.F.; Devlieghere, F. High pressure carbon dioxide inactivation of microorganisms in foods: The past, the present and the future. *Int. J. Food Microbiol.* **2007**, *117*, 1–28. [CrossRef]

19.	Lerasle, M.; Federighi, M.; Simonin, H.; Anthoine, V.; Reee, S.; Cheret, R.; Guillou, S. Combined use of modified atmosphere packaging and high pressure to extend the shelf-life of raw poultry sausage. *Innov. Food Sci. Emerg. Technol.* **2014**, *23*, 54–60. [CrossRef]

20.	Al-Nehlawi, A.; Guri, S.; Guamis, B.; Saldo, J. Synergistic effect of carbon dioxide atmospheres and high hydrostatic pressure to reduce spoilage bacteria on poultry sausages. *LWT-Food Sci. Technol.* **2014**, *58*, 404–411. [CrossRef]

21.	Wang, L.; Pan, J.A.; Xie, H.M.; Yang, Y.; Lin, C.M. Inactivation of Staphylococcus aureus and Escherichia coli by the synergistic action of high hydrostatic pressure and dissolved CO_2. *Int. J. Food Microbiol.* **2010**, *144*, 118–125. [CrossRef] [PubMed]

22.	Deng, K.; Serment-Moreno, V.; Welti-Chanes, J.; Paredes-Sabja, D.; Fuentes, C.; Wu, X.; Torres, J.A. Inactivation model and risk-analysis design for apple juice processing by high-pressure CO_2. *J. Food Sci. Technol.* **2018**, *55*, 258–264. [CrossRef] [PubMed]

23.	Li, H.; Xu, Z.Z.; Zhao, F.; Wang, Y.T.; Liao, X.J. Synergetic effects of high-pressure carbon dioxide and nisin on the inactivation of *Escherichia coli* and *Staphylococcus aureus*. *Innov. Food Sci. Emerg. Technol.* **2016**, *33*, 180–186. [CrossRef]

24.	Liao, H.M.; Zhang, F.S.; Liao, X.J.; Hu, X.S.; Chen, Y.; Deng, L. Analysis of Escherichia coli cell damage induced by HPCD using microscopies and fluorescent staining. *Int. J. Food Microbiol.* **2010**, *144*, 169–176. [CrossRef] [PubMed]

25.	Liao, H.M.; Zhang, F.S.; Hu, X.S.; Liao, X.J. Effects of high-pressure carbon dioxide on proteins and DNA in *Escherichia coli*. *Microbiology-Sgm* **2011**, *157*, 709–720. [CrossRef] [PubMed]

26. Amor, K.B.; Breeuwer, P.; Verbaarschot, P.; Rombouts, F.M.; Akkermans, A.D.; De Vos, W.M.; Abee, T. Multiparametric flow cytometry and cell sorting for the assessment of viable, injured, and dead bifidobacterium cells during bile salt stress. *Appl. Environ. Microbiol.* **2002**, *68*, 5209–5216. [CrossRef]

27. Erkmen, O.; Dogan, C. Kinetic analysis of *Escherichia coli* inactivation by high hydrostatic pressure in broth and foods. *Food Microbiol.* **2004**, *21*, 181–185. [CrossRef]

28. Chen, H.Q.; Hoover, D.G. Pressure inactivation kinetics of *Yersinia enterocolitica* ATCC 35669. *Int. J. Food Microbiol.* **2003**, *87*, 161–171. [CrossRef]

29. Abe, F. Exploration of the effects of high hydrostatic pressure on microbial growth, physiology and survival: Perspectives from piezophysiology. *Biosci. Biotech. Bioch.* **2007**, *71*, 2347–2357. [CrossRef]

30. Lo, R.; Xue, T.; Weeks, M.; Turner, M.S.; Bansal, N. Inhibition of bacterial growth in sweet cheese whey by carbon dioxide as determined by culture-independent community profiling. *Int. J. Food Microbiol.* **2016**, *217*, 20–28. [CrossRef]

31. Jakobsen, M.; Bertelsen, G. Solubility of carbon dioxide in fat and muscle tissue. *J. Muscle Food* **2006**, *17*, 9–19. [CrossRef]

32. Zhao, F.; Wang, Y.T.; An, H.R.; Hao, Y.L.; Hu, X.S.; Liao, X.J. New Insights into the Formation of Viable but Nonculturable *Escherichia coli* O157:H7 Induced by High-Pressure CO_2. *Mbio* **2016**, *7*, e00961-16. [CrossRef] [PubMed]

33. Provincial, L.; Guillen, E.; Alonso, V.; Gil, M.; Roncales, P.; Beltran, J.A. Survival of Vibrio parahaemolyticus and Aeromonas hydrophila in sea bream (*Sparus aurata*) fillets packaged under enriched CO_2 modified atmospheres. *Int. J. Food Microbiol.* **2013**, *166*, 141–147. [CrossRef] [PubMed]

34. Hong, S.I.; Pyun, Y.R. Membrane damage and enzyme inactivation of *Lactobacillus plantarum* by high pressure CO2 treatment. *Int. J. Food Microbiol.* **2001**, *63*, 19–28. [CrossRef]

microorganisms

MDPI

Article

Synergism of Mild Heat and High-Pressure Pasteurization Against *Listeria monocytogenes* and Natural Microflora in Phosphate-Buffered Saline and Raw Milk

Abimbola Allison [1], Shahid Chowdhury [1,2] and Aliyar Fouladkhah [1,2,*]

[1] Public Health Microbiology Laboratory, Tennessee State University, Nashville, TN 37209, USA; abimbolaallison20@gmail.com (A.A.); schowdh1@tnstate.edu (S.C.)
[2] Cooperative Extension Program, Tennessee State University, Nashville, TN 37209, USA
* Correspondence: aliyar.fouladkhah@aya.yale.edu; Tel.: +1-970-690-7392

Received: 10 September 2018; Accepted: 1 October 2018; Published: 3 October 2018

Abstract: As many as 99% of illnesses caused by *Listeria monocytogenes* are foodborne in nature, leading to 94% hospitalizations, and are responsible for the collective annual deaths of 266 American adults. The current study is a summary of microbiological hurdle validation studies to investigate synergism of mild heat (up to 55 °C) and elevated hydrostatic pressure (up to 380 MPa) for decontamination of *Listeria monocytogenes* and natural background microflora in raw milk and phosphate-buffered saline. At 380 MPa, for treatments of 0 to 12 min, d-values of 3.47, 3.15, and 2.94 were observed for inactivation of the pathogen at 4, 25, and 50 °C. Up to 3.73 and >4.26 log CFU/mL reductions ($p < 0.05$) of habituated *Listeria monocytogenes* were achieved using pressure at 380 MPa for 3 and 12 min, respectively. Similarly, background microflora counts were reduced ($p < 0.05$) by 1.3 and >2.4 log CFU/mL after treatments at 380 MPa for 3 and 12 min, respectively. Treatments below three min were less efficacious ($p \geq 0.05$) against the pathogen and background microflora, in the vast majority of time and pressure combinations. Results of this study could be incorporated as part of a risk-based food safety management system and risk assessment analyses for mitigating the public health burden of listeriosis.

Keywords: *Listeria monocytogenes*; natural background microflora; raw milk; high-pressure pasteurization; synergism of mild heat and pressure

---

1. Introduction

The epidemiological evidence, derived from the Centers for Disease Control and Prevention (CDC) active surveillance data, indicates that every year in the United States 1591 illness episodes occur due to infections with *Listeria monocytogenes* [1]. These illnesses are almost exclusively associated with contaminated food products (i.e., about 99% of cases are foodborne in nature) and lead to hospitalization in about 94% of cases. Among those hospitalized, 15.9% die annually, one of the highest mortality rates associated with any foodborne pathogen [1,2]. CDC's National Outbreak Reporting System (NORS) also delineated that from 1998 to 2016 there have been at least 66 foodborne outbreaks associated with *Listeria monocytogenes*, leading to 852 illnesses with >72% and >15% hospitalization, and death episodes, respectively [3]. Fluid milk had been associated with recent outbreaks of *Listeria monocytogenes* including a 6-month outbreak in Massachusetts in 2007, a 6-month multistate outbreak in 2014, and an 11-month outbreak in Washington in 2014 [3]. This ubiquitous bacterial pathogen has also created a plethora of consumer insecurity and public health challenges globally, including a 2018 foodborne Listeriosis outbreak in South Africa, which has been categorized as the largest foodborne outbreak in the recorded history of food safety [4]. The elderly, the very young, pregnant women, and immunocompromised are particularly considered as the at-risk population for Listeriosis who comprise approximately 30% of the U.S. population [5]. Raw milk, ready-to-eat products, and dairy products prepared using raw milk are some of the main vehicles for this pathogen in the food chain [3,6].

Consumers of the 21st century have evolving demands and expectations from food products, although consumption of raw milk is a major public health concern, many consumers in developed economies are preferring this product due to their perception for greater healthfulness of raw milk, improved digestion, nutritive value, and preferred organoleptic properties [7]. Although the sale of raw milk had been prohibited in several states, many consumers receive the product through legislative loopholes such as "cow-share" programs [7].

With recent advancements in engineering of high-pressure processing units, this technology is gaining rapid adoption across various sectors of food manufacturing for assuring microbiological safety and extending the shelf-life of various products, providing a clean label, as well as fresh-like organoleptic properties of treated foods [8]. As such, the NACMCF (National Advisory Committee on Microbiological Criteria for Foods) has recently recommended redefinition of pasteurization by including high-pressure processing as a non-thermal pasteurization method [9].

The sales of pressure treated products had been in consistent increase in recent years and are expected to surpass $9 billion annually in the United States [10]. Considering consumers' demand for minimally processed foods, the plethora of foodborne public health episodes associated with *Listeria monocytogenes,* and the increasing momentum in adoption of high-pressure pasteurization in the private food industry, the current study is a microbiological hurdle validation study to investigate synergism of mild heat and elevated hydrostatic pressure for decontamination of raw milk from the pathogen. The study further calculates inactivation indices in raw milk as well as in buffered environment and provides information on decontamination of natural microflora (spoilage organisms) of raw milk as affected by the treatments.

2. Materials and Methods

2.1. Listeria monocytogenes Strains, Preparation of Culture, and Inoculation

Four strains of *Listeria monocytogenes* (ATCC® numbers 51772, 51779, BAA-2657, 13932) were used for inoculation of raw milk and phosphate-buffered saline in separate experiments. The bacterial strains were chosen due to their public health significance, representing diverse ribotypes, PFGE patterns, serotypes (1/2a, 1/2c, and 4b), and lineages [6]. For each strain separately, a loopful from frozen glycerol stock was aseptically transferred into 10 mL Tryptic Soy Broth (Difco, Becton Dickinson, Franklin Lakes, NJ, USA) supplemented with 0.6% yeast extract (TSB + YE) to minimize the acid stress of the cells during preparation of overnight suspension [11]. This bacteriological medium has also been previously used in high-pressure processing treatments to minimize the effect of acid stress during culturing of bacterial inoculum [12]. The inoculated TSB + YE was then incubated at 37 °C for 22–24 h. One loopful of this overnight suspension was then streak plated onto the surface of Tryptic Soy Agar (Difco, Becton Dickinson, Franklin Lakes, NJ, USA), and incubated at 37 °C for 24 h. The plates were then kept for up to a month at 4 °C prior to the experiment.

Five days prior to the experiments, each strain was activated by transferring a single colony from the above-mentioned plates stored at 4 °C, into 10 mL TSB + YE. After incubation at 37 °C for 22–24 h, a 100 µL aliquot was then aseptically sub-cultured into another 10 mL of TSB + YE and re-incubated at 37 °C for 22–24 h. Cells of each strain (2 mL per strain) were then harvested using centrifugal force at 6000 revolutions per min (3548 *g*, for 88 mm rotor) for 15 min (Model 5424, Eppendorf North America, Hauppauge, NY, USA; Rotor FA-45-24-11). After removal of the supernatant and for further removal of sloughed cell components, excreted secondary metabolites, and growth media, the cells were then re-suspended in Phosphate-buffered Saline (PBS, VWR International, Radnor, PA, USA), then re-centrifuged using the above-mentioned time, intensity, and instrumentation. Then, after discarding the supernatant, to improve the external validity of the challenge study, each bacterial strain was then individually habituated in sterilized milk and/or PBS. Strains were habituated at 4 °C for 72 h to allow acclimatization of the pathogen to low temperature and intrinsic factors of the food/medium [13,14]. On the day of the experiments, the individually habituated strains were then composited and used as inoculum for microbiological challenge studies in raw milk and PBS, at target population levels of 5 to 6 log CFU/mL in the raw milk experiments and 7 to 8 log CFU/mL in the PBS experiments. Fresh raw milk was purchased through a cow-share program from the outskirt of Nashville, TN, stored aseptically in a refrigerated cooler during transportation and were utilized for the experiment less than 24 h after the purchase. Levels of inoculation and elements of experimental design were selected (data not shown) based on preliminary trials.

2.2. Mild Heat and High-Pressure Pasteurization

For experiments involving inoculation of raw milk, hydrostatic pressure (Barocycler Hub440, Pressure Bioscience Inc., South Easton, MA) of 380 Megapascal (MPa), i.e., 55,000 pounds per square inch (PSI) and 310 MPa (45,000 PSI) were applied at 4, 25, and 50 °C. Similarly, for experiments involving inoculation of PBS, pressure levels of 380 MPa (55,000 PSI), 310 MPa (45,000 PSI), and 240 MPa (35,000 PSI) were used at 4 °C and 55 °C for time intervals of 0 (untreated control) to 12 min. The pressure intensity levels, as well as temperatures and time combinations, were selected based on preliminary trials [11] and reported based on English and metric units on the graphs due to the popularity of both systems for stakeholders in various regions. The above-mentioned processing unit has chamber size of 16 mL, surrounded with a stainless water jacket connected to a refrigerated circulating water bath (Model refrigerated 1160s, VWR International, Radnor, PA, USA) for precise control of the temperature during the treatments. For monitoring the temperature, two k-type thermocouples (Omega Engineering Inc., Norwalk, CT, USA) were manually inserted inside the wall of the chamber and secured with thermal paste (Model 5 AS5-3.5G, Arctic Silver, Visalia, CA, USA). These thermocouples were connected to the unit's software (HUB PBI 2.3.11 Software, Pressure BioScience Inc., South Easton, MA, USA) that, in addition to chamber pressure, recorded the temperature values every three seconds [11]. The coolant of the circulating water bath and pressure transmission fluid was distilled water (total soluble solids less than 30 parts per million). The chamber of the unit was purged prior to each analysis for removal of residual air, to assure treatments were hydrostatic in nature. All treatments were conducted inside no disk PULSE tubes (Pressure BioScience Inc., South Easton, MA, USA), containing 1.5 mL of habituated inoculum in raw milk or PBS. It is noteworthy that each reported treatment time excludes the time for pressure increase (come-up time of 3 s) and the release time (come-down time of 1 s). These values were recorded and monitored using the Barocycler mode of HUB PBI 2.3.11 Software (Pressure BioScience Inc., South Easton, MA, USA).

2.3. The pH, Neutralization, and Microbiological Analyses

In order to neutralize the intrinsic factors of the food vehicle prior to microbiological enumeration, each sample was neutralized using 3 mL of D/E neutralizing broth (Difco, Becton Dickinson, Franklin Lakes, NJ, USA) per 1 mL of the sample, thus, the detection limit of this study was 0.35 log CFU/mL. After neutralization, for experiments involving the inoculation of raw milk, pressure-treated and untreated controls were 10-fold serially diluted in Maximum Recovery Diluent (Difco, Becton Dickinson, Franklin Lakes, NJ, USA) to maximize the recovery of injured cells. The neutralized diluents were then spread plate onto the surface of PALCAM base agar (Becton, Dickinson and Company, Sparks, MD, USA) supplemented with Ceftazidime (Becton, Dickinson and Company, Sparks, MD, USA) for selective enumeration of *Listeria monocytogenes*, and Tryptic Soy Agar supplemented with yeast extract (TSA ± YE) for enumeration of background microflora. For the experiments using PBS as the vehicle, samples were plated onto TSA ± YE. The addition of 0.6% yeast extract to the medium was based on preliminary trials using concentrations of pyruvic acid and/or yeast extract for maximum recovery of injured cells after pressure treatments [11]. Plates were then incubated at 37 °C for 24–48 h. Incubated plates were then manually counted and converted to log values for further descriptive and inferential analyses. The pH values of substrates were measured twice, once after pressure treatment (prior to neutralization) and once before microbiological analyses (after neutralization) using a digital pH meter (Mettler Toledo, AG, Switzerland) calibrated at pH levels of 4.00, 7.01, and 10.01, prior to measurements.

2.4. Statistical Analyses and Experimental Design

Sample sizes of at least 5 repetitions per time/temperature/pressure were obtained based on an a priori power analysis of existing high-pressure pasteurization data of the public health microbiology laboratory using Proc Power of SAS$_{9.2}$ software (SAS Institute, Cary, NC, USA). The current study is a compilation of two separate experiments using PBS (Figures 1 and 2) and raw milk (Figures 3 and 4) as vehicles. These studies were conducted, analyzed, and reported separately. Each experiment was conducted in two biologically independent repetitions, with each repetition considered as a blocking factor in a randomized complete block design. Each block consisted of three replications, and each replication further consisted of two microbiological repetitions. Thus, each reported value is the mean of 12 independent analyses (i.e., 2 blocks, 3 replications, 2 microbiological repetitions per time/pressure/treatment). Data management, log conversion, and descriptive representation of the data were initially conducted in Microsoft Excel. The raw data were then imported to SAS$_{9.2}$ software (SAS Institute, Cary, NC, USA), for inferential statistics at the type I error level of 5% (α = 0.05). For each experiment, an Analysis of Variance (ANOVA) was conducted using the Generalized Linear Model (Proc GLM) of SAS$_{9.2}$ with two mean separation methods. A Tukey adjustment was utilized for pairwise comparisons of all treated samples and controls and further, a Dunnett adjustment was utilized for comparing treated samples with the untreated controls. Microsoft Excel and GInaFiT version 1.7 [15] software (Katholieke Universiteit, Leuven, Belgium) were further used for calculation of inactivation indices (d-value and K$_{max}$ values).

3. Results and Discussion

3.1. Inactivation of Listeria monocytogenes in Phosphate-Buffered Saline

Investigating the sensitivity of the pathogen in phosphate-buffered saline (PBS) medium would provide the opportunity of exploring the synergism of heat and elevated hydrostatic pressure without the interference of intrinsic and extrinsic factors of a food vehicle. These experiments were conducted by inoculating sterilized PBS with the above-mentioned cocktail of the pathogen. The pH of the media prior to inoculation was 7.54 ± 0.1 and were not different ($p \geq 0.05$) than inoculated samples treated at 4 and 55 °C. Corresponding pH values for inoculated samples at 4 and 55 °C were 7.40 ± 0.1, and 7.44 ± 0.2, respectively. Prior to analyses, samples were kept at refrigeration temperature (4.07 ± 0.2 °C). As further delineated in the Materials and Methods Section 2.2., temperatures before and after treatments were precisely maintained (using a stainless steel jacket connect to refrigerated circulating water bath), monitored (using K-type thermocouples secured inside the chamber wall), and recorded (using HUB PBI 2.3.11 Software). Temperature recordings before and after treatments were similar ($p \geq 0.05$) and was 3.80 ± 0.2 °C prior to treatments at 4 °C and were recorded as 3.81 ± 0.3 °C at the end of the treatments. The corresponding values for samples' temperature treated at 55 °C were 54.83 ± 0.4 °C and 55.11 ± 0.4 °C before and after the treatments, respectively. Temperature of the transmission fluid (distilled water), were precisely controlled and monitored at 4 and/or 55 °C as articulated in Section 2.2.

3.1.1. Sensitivity of Listeria monocytogenes in Phosphate-Buffered Saline at 4 °C

Inactivation of *Listeria monocytogenes* counts were investigated at this temperature under three levels of hydrostatic pressure and for time intervals of 0 min (untreated control) up to 10 min. Counts of the pathogen for untreated controls were 7.86 ± 0.1 log CFU/mL (Figure 1). Treatments for one min resulted in no ($p \geq 0.05$) or only small reductions ($p < 0.05$). As an example, treated samples at 240 MPa, 310 MPa, and 380 MPa after one min had counts of 7.34 ± 0.1, 7.05 ± 0.1, and 6.57 ± 0.1 log CFU/mL (Figure 1D). Longer duration of pressure treatments, predictably resulted in higher inactivation of the pathogen. The counts of *Listeria monocytogenes* were reduced ($p < 0.05$) to 4.88 ± 0.2, 4.38 ± 0.1, and 4.0 ± 0.5 for treatments at 380 MPa after 4, 7, and 10 min, respectively (Figure 1A–C). The corresponding values ($p < 0.05$) at 310 MPa for the above order of treatment times were, 7.05 ± 0.1, 5.99 ± 0.5, and 5.61 ± 0.4, respectively (Figure 1A–C).

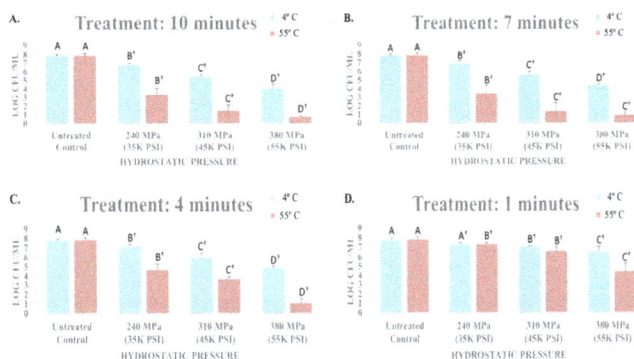

Figure 1. Effects of elevated hydrostatic pressure against four-strain habituated mixture of *Listeria monocytogenes* (ATCC® numbers 51772, 51779, BAA-2657, 13932) in phosphate-buffered saline, treated (Barocycler Hub440, Pressure BioScience Inc., South Easton, MA) at 4 and 55 °C. Within each graph, and for each temperature separately, columns of each pressure intensity level followed by different uppercase letters are representing log CFU/mL values that are statistically ($p < 0.05$) different (Tukey-adjusted ANOVA). Uppercase letters followed by † sign are statistically ($p < 0.05$) different than the untreated control (Dunnett-adjusted ANOVA). (**A**) Treatments for 10 min; (**B**) Treatments for 7 min; (**C**) Treatments for 4 min; (**D**) Treatments for 1 min.

3.1.2. Sensitivity of Listeria monocytogenes in Phosphate-Buffered Saline at 55 °C

In exclusion of intrinsic and extrinsic factors of a food vehicle, *Listeria monocytogenes* exhibited great sensitivity to the combination of mild hydrostatic pressure and heat. Even a one-min treatment at 380 MPa and 55 °C was able to reduce ($p < 0.05$) the pathogen counts to 4.44 ± 0.9 log CFU/mL (Figure 1D). With counts of untreated controls at 7.87 ± 0.2 log CFU/mL, this reduction is equivalent to 3.43 log reductions (e.g., >99.9% of the inoculated pathogen). The pathogen counts were further reduced ($p < 0.05$) to $<1.07 \pm 0.5$, $<0.95 \pm 0.3$, and $<0.75 \pm 0.3$ log CFU/mL, for 4-min, 7-min, and 10-min treatments at 380 MPa and 55 °C. Even milder pressure treatments, coupled with elevated heat resulted in an appreciable reduction ($p < 0.05$) of *Listeria monocytogenes*. As an example, 4-min treatments resulted in pathogen counts of 3.66 ± 0.2 log CFU/mL at 310 MPa and 55 °C, and 4.66 ± 0.6 log CFU/mL at 380 MPa and 4 °C. These counts were lower ($p < 0.05$) at 55 °C relative to those samples treated at 4 °C. Pathogen counts for the above two pressures at 55 °C and for 7 min treatments were $<1.38 \pm 1.0$, and 3.45 ± 0.9 log CFU/mL, respectively (Figure 1A–C).

The current challenge study indicates lower levels of pressure coupled with mild heat could be as efficacious as higher levels of pressure at lower temperatures. As an example, counts of *Listeria monocytogenes* were reduced ($p < 0.05$) from 7.86 ± 0.1 to 4.88 ± 0.2 after a 380 MPa treatment at 4 °C. Similarly, *Listeria monocytogenes* counts were reduced ($p < 0.05$) from 7.87 ± 0.2 to 4.66 ± 0.6 after a 240 MPa treatment at 55 °C (Figure 1A–C). Treatments of 3 to 5 min are current standard procedures in the manufacturing of food products using pressure-based technologies [10].

3.1.3. Inactivation Indices of Listeria monocytogenes in Phosphate-Buffered Saline

Calculation of inactivation indices not only improves the adaptability of a challenge study by private industry but further delineates the synergism of heat and hydrostatic pressure for decontamination of *Listeria monocytogenes* (Figure 2). The index *d*-value is obtained based on a linear model, corresponding to the time (in this study in a unit of min) required at specific conditions (pressure, heat, and other intrinsic, and extrinsic factors) to reduce 90% of the exposed microorganism [16]. Under the condition of this experiment, we observed a *d*-value of 2.77 min for inactivation of *Listeria monocytogenes* at 4 °C and at 380 MPa. The value at the same level of pressure but at 55 °C was reduced to 1.59 min (Figure 2). Similar synergism was observed at lower pressures. The *d*-values were 4.43 and 1.49 for treatments at 310 MPa at 4 and 55 °C, respectively and were 11.61 and 2.06 min for treatments at 240 MPa at 4 and 55 °C, respectively. This indicates that a treatment at lower pressure and higher temperature (310 MPa at 55 °C), could be comparable ($p \geq 0.05$) to a treatment at higher pressure and lower temperature (380 MPa at 4 °C). This synergism could be of great importance to manufacturing facilities using the technology to lower the cost of operation since lower levels of pressure had been associated with lower maintenance cost and higher shelf-life of the pressure vessels [17]. The cost associated with high-pressure processing is currently the main barrier for further adoption of this technology in the private industry [10].

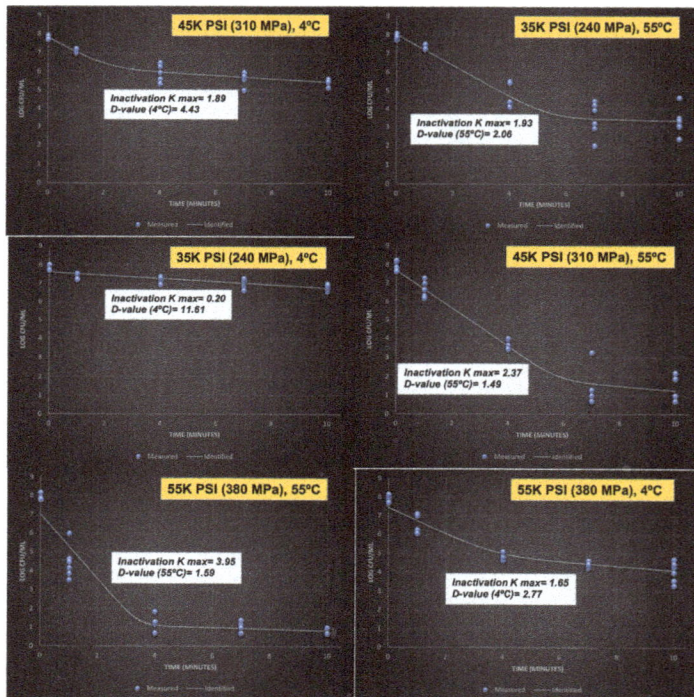

Figure 2. Inactivation rates for four-strain habituated mixture of *Listeria monocytogenes* (ATCC® numbers 51772, 51779, BAA-2657, 13932) exposed to elevated hydrostatic pressure (Barocycler Hub 440, Pressure BioScience Inc., South Easton, MA) in phosphate-buffered saline. K_{max} values are selected from the best-fitted model (goodness-of-fit indicator of R^2 values, $\alpha = 0.05$) using the GInaFiT software. K_{max} values are expressions of number of log cycles of reduction in 1/min unit. The *d*-values provided are calculated based on linear model, exhibiting time required for a log (90%) of microbial cell reduction.

Recent studies also delineate that alternative inactivation indices, particularly those obtained based on non-linear models, could be of great importance for stakeholders since the microbial reduction of many food-pathogen combinations may not follow a linear pattern [11]. As further elaborated in the Materials and Methods Section 2.4., the current study is reporting the non-linear inactivation index K_{max} calculated based on the best-fitted (maximum R^2) model. In contrast to the *d*-value, this index has a unit of 1/min, thus, larger K_{max} values are corresponding to higher/faster microbial inactivation [18]. The K_{max} values at 4 °C were 1.65, 1.89, and 0.20 for treatments at 380 MPa, 310 MPa, and 240 MPa, respectively. The values were increased to 3.95, 2.37, and 1.93 for the above order of the pressure treatments when tested at 55 °C (Figure 2).

3.2. Inactivation of Listeria monocytogenes and Natural Microflora in Raw Milk

Relative to challenge studies conducted in phosphate-buffered saline medium, studying the inactivation of *Listeria monocytogenes* in raw milk could provide a more realistic interpretation, particularly when such studies conducted in presence of natural microflora of the product with the existence of intrinsic and extrinsic factors that had not been altered by any previous treatment.

3.2.1. Sensitivity of Listeria monocytogenes and Natural Microflora in Raw Milk at 4 °C

Under the condition of this experiment, the pH of raw milk samples treated at the various time and pressure intensity levels were similar ($p \geq 0.05$) and ranged from 6.72 ± 0.1 to 6.85 ± 0.1. The pH of untreated raw milk was 6.82 ± 0.1, while the pH of treated milk neutralized in D/E broth prior to microbiological analyses was 7.23 ± 0.1. The temperature recordings of the treatments remained constant ($p \geq 0.05$) before and after treatments, ranging from 3.58 ± 0.3 °C to 3.93 ± 0.3 °C and 3.70 ± 0.4 °C to 3.95 ± 0.1 °C for before and after the pressure treatments, respectively. Counts of selective medium (PALCAM agar), associated with *Listeria monocytogenes* showed the standard deviation of 0.1 to 0.9 (average standard deviation of 0.3) and are summarized in Figure 3A. The pathogen count of untreated controls at 4 °C was 5.30 ± 0.1 log CFU/mL, the count was reduced by 3.35 log for treatment at 380 MPa for 12 min to 1.95 ± 0.4 log CFU/mL. The corresponding log reductions for 9-, 6- and 3-min treatments at 380 MPa were 3.3, 2.8, and 2.1, respectively, exhibiting in excess of a 99% reduction of the pathogen at 380 MPa treatments. At lower pressure intensities of 310 MPa, these reductions showed similar trends: Treatments for 12, 9, 6, and 3 min at this intensity level lead to 2.6, 2.5, 1.8, and 1.4 log reductions ($p < 0.05$), respectively.

Figure 3B, summarizes counts obtained from non-selective media (TSA supplemented with yeast extract) corresponding to inactivation of background microflora. These counts exhibited standard deviations ranging from 0.1 to 0.8 (average standard deviation of 0.3). In the vast majority of the time-pressure combinations at 4 °C, background microflora counts of the raw milk were less sensitive to treatments relative to the inoculated pathogen. As an example, treatments for 12 min, modestly reduced ($p < 0.05$) the background microflora for 2.03 and 1.99 log, for treatment intensity of 380 MPa and 310 MPa, respectively. The higher tolerance of background microflora to elevated hydrostatic pressure had been previously reported in similar products and could be primarily attributed to the presence of spore-forming organisms that are ubiquitous in production and manufacturing environments [19].

3.2.2. Sensitivity of Listeria monocytogenes and Natural Microflora in Raw Milk at 25 °C

Similar to the samples treated at 4 °C, samples treated at 25 °C had comparable ($p \geq 0.05$) pH counts ranging from 6.67 ± 0.1 to 6.77 ± 0.1. Temperature recordings prior and after treatments were controlled at 25 °C ($p \geq 0.05$) and were ranging from 24.10 ± 0.8 to 24.67 ± 0.7 and 23.83 ± 0.5 to 24.63 ± 0.9, prior and after the pressure treatments, respectively. Pathogen load (PALCAM counts) were 5.30 ± 0.1 log CFU/mL (Figure 3A) and were reduced ($p < 0.05$) to 1.58 ± 0.1 log CFU/mL after pressure treatment of 380 MPa for 12 min. Similarly, 3.29 and 2.27 log reductions ($p < 0.05$) were observed for treatments of 9 and 6 min at 380 MPa, respectively (Figure 3A). More than 99% of the inoculated *Listeria monocytogenes* was inactivated at lower pressure of 310 MPa, as result of 12- and

9-min treatments (Figure 3B). Background microflora counts at this temperature were more fastidious to the pressure treatments relative to pathogen counts (Figure 3B). As an example, only 1.86 and 0.81 log reductions were achieved after 12 min of treatment at 380 MPa, and 310 MPa, respectively. Minor differences were observed during inactivation of the pathogen and background microflora comparing treatments of 4 and 25 °C, indicating that these temperatures might be used interchangeably in manufacturing facilities and during validation studies. Reductions obtained in this study are in harmony with previous studies at similar temperatures and pressure intensity levels, where 2.11 log reductions of *Listeria monocytogenes* were reported at 500 MPa after 10 min of treatment of milk at 20 °C [20]. Similar trends were also reported for inactivation of *Listeria monocytogenes* at 400 MPa at temperatures of 20 to 25 °C [21,22]. Studying inactivation of *Listeria monocytogenes* at lower and high temperatures combined with mild elevated hydrostatic pressure, is currently a knowledge gap of hurdle validation data against this pathogen as well as those conducted against the natural background flora of raw milk. A pressure treatment coupled with mild temperature of 50 or 55 °C, could assure microbiological safety of a product while reducing cost associated with pressure vessels shelf-life and high pressure pasteurization maintenance [10].

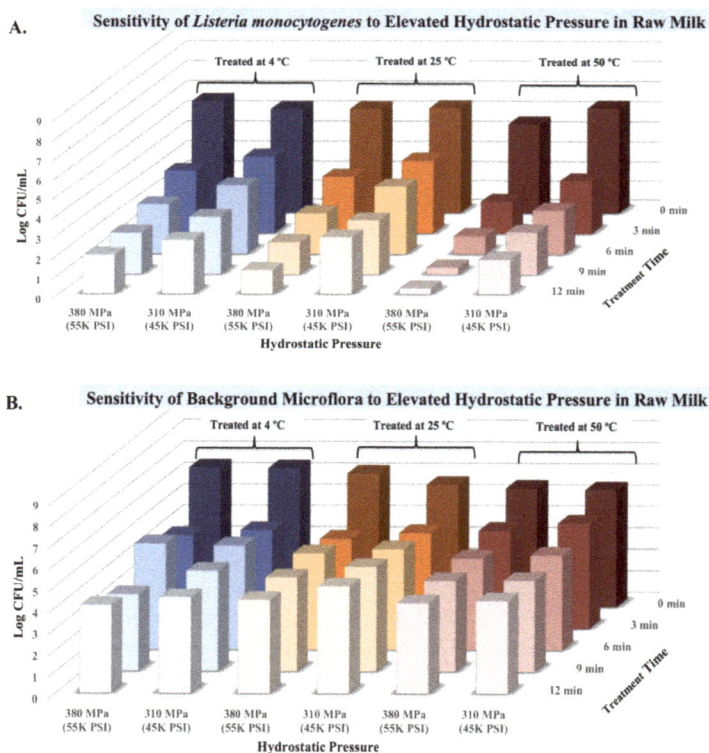

Figure 3. Effects of elevated hydrostatic pressure against four-strain habituated mixtures of *Listeria monocytogenes* (ATCC® numbers 51772, 51779, BAA-2657, 13932) and background microflora in raw milk, treated (Barocycler Hub440, Pressure BioScience Inc., South Easton, MA) at 4, 25, and 50 °C. (**A**) Counts of *Listeria monocytogenes*; (**B**) Counts of background microflora.

3.2.3. Sensitivity of Listeria monocytogenes and Natural Microflora in Raw Milk at 50 °C

Synergism of temperature at 50 °C with elevated hydrostatic pressure resulted in most appreciable reductions ($p < 0.05$) in counts of *Listeria of monocytogenes*. Selective counts, corresponding with the inoculated pathogen (PALCAM counts) were 4.56 ± 0.3 log CFU/mL prior to treatment at 380 MPa (Figure 3A). The counts were reduced ($p < 0.05$) to below the detection limit after 12 min of treatment at 380 MPa/ 50 °C. Similarly, 4.21, 3.67, and 2.08 log reductions ($p < 0.05$) were observed for 9-, 6-, and 3-min treatments at this pressure and temperature combination, respectively (Figure 3A). Treatments of 310 MPa similarly resulted in log reductions ($p < 0.05$) ranging from 2.75 to 3.59 (Figure 3A). Unlike the inoculated pathogen, the natural microflora was affected modestly ($p < 0.05$) even at 50 °C, for 12-min treatments at 380 MPa, exhibiting 1.32 log reductions (Figure 3B). The corresponding log reductions associated with 9-, 6-, and 3-min treatments at 380 MPa/ 50 °C were 1.28, 1.25, and 0.98, respectively, for the natural microflora (Figure 3B). These reductions were also modest at lower pressure intensities, as an example, 1.13 log reductions ($p < 0.05$) were observed as result of pressure treatments at 310 MPa at 50 °C, after 12 min (Figure 3B). The resistance of spoilage microorganisms and natural microflora of the raw milk could be attributed to the presence of spore-forming organisms that are ubiquitous in processing area [11] as previously delineated in Section 3.2.1.

3.3. Synergism of Elevated Hydrostatic Pressure and Heat for Inactivation of Listeria monocytogenes and Natural Microflora

Current data exhibit strong synergism between mild heat and hydrostatic pressure for inactivation of *Listeria monocytogenes* in raw milk. The synergism among extrinsic factors of food against microbial pathogen had been previously discussed as part of "hurdle technology" [23]. With the assumption of a linear relationship between reduction of *Listeria monocytogenes* in raw milk as affected by elevated hydrostatic pressure and mild heat, our study shows 3.47 min are required (Figure 4) at pressure levels of 380 MPa at 4 °C for a 90% reduction of the pathogen (i.e., *d*-value = 3.47). These corresponding *d*-values were reduced to 3.15 and 2.94 at 25 and 50 °C (Figure 4). The assumption of non-linearity and using a best-fitted model obtained by GInaFiT software [15] exhibited similar trends (Figure 4). One of the main challenges for current manufacturers of pressure-treated products is slightly higher production costs relative to conventional products treated solely by thermal processing [10,11]. The current study indicates that pressure treatments at lower intensities, such as 380 MPa and 310 MPa alone or coupled with mild heat could be an alternative to pressures at very high levels of hydrostatic pressure. The lower pressure treatments are typically associated with increased shelf-life of the pressure vessels and reduced cost of pressure processing [17]. Similar synergism was observed for inactivation of natural microflora, although the extent of reductions were less than inactivation rates observed for the pathogen (Figure 3B). As previously discussed, this is due to presence of spore-forming organisms that are ubiquitous in raw milk production and manufacturing area and are considered to be resistant to pressure-based treatments [11].

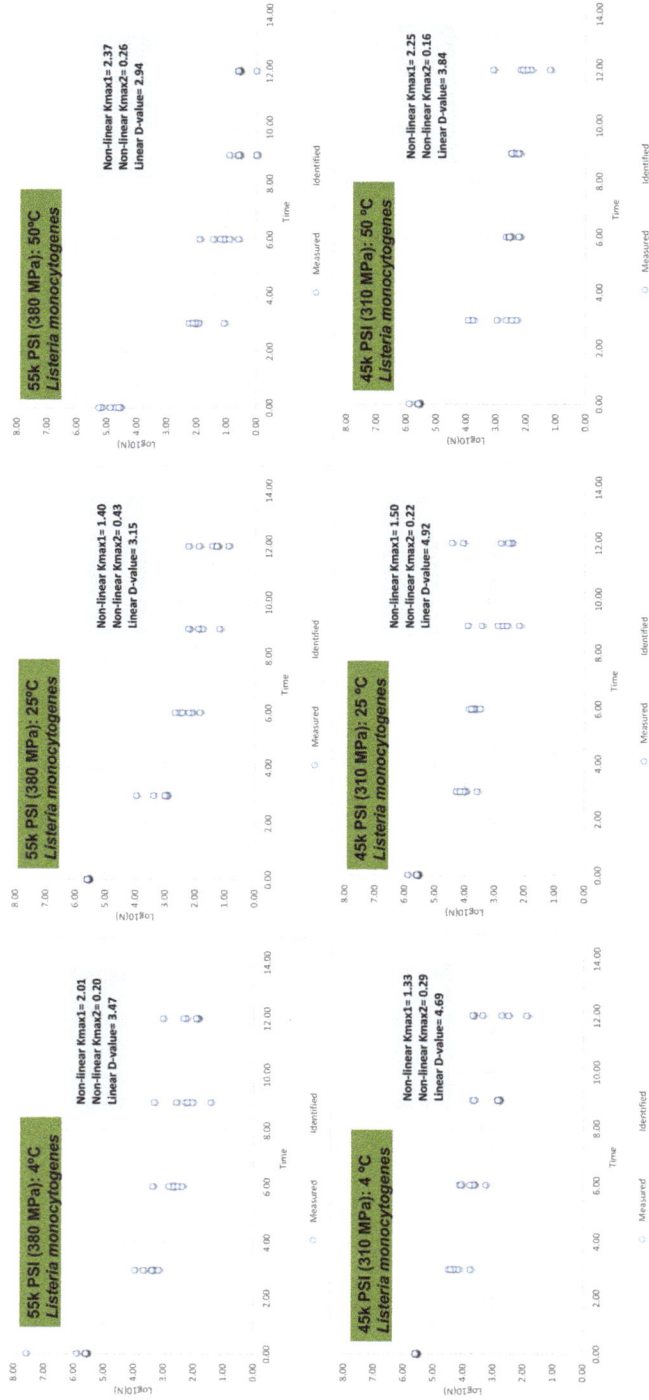

Figure 4. Inactivation indices for four-strain habituated mixture of *Listeria monocytogenes* (ATCC® numbers 51772, 51779, BAA-2657, 13932) exposed to elevated hydrostatic pressure (Barocycler Hub 440, Pressure BioScience Inc., South Easton, MA) in raw milk. K$_{max}$ values are selected from best-fitted biphasic model (goodness-of-fit indicator of R^2 values, $\alpha = 0.05$) using the GInaFiT software. K$_{max}$ values are expression of number(s) of log cycles of reduction in 1/min unit, thus larger values correspond to less time needed for microbial cell reductions for each pressure/temperature combination. The *d*-values provided are calculated based on linear model, exhibiting time required for a log (90%) of microbial cell reductions.

4. Conclusions

Under the conditions of our experiments, we observed *d*-values of 2.77, 4.43, and 11.61 for inactivation of *Listeria monocytogenes* treated at 380 MPa, 310 MPa, and 240 MPa treated in PBS at 4 °C. Corresponding values for pressure treatments in PBS at 55 °C were appreciably lower, delineating strong synergism between heat and hydrostatic pressure for inactivation of *Listeria monocytogenes*. Similar synergistic effects were demonstrated against *Listeria monocytogenes* inoculated in raw milk. Reducing the cost associated with pressure-based technologies are currently the main challenge for further adoption of this emerging technology in various sectors of food manufacturing. The current study delineated the pressure treatments at lower intensity levels, coupled with mild heat could result in an appreciable reduction of *Listeria monocytogenes*. This synergism was demonstrated in laboratory media with the exclusion of intrinsic and extrinsic factors of a food vehicle as well as in studies conducted in raw milk. The synergistic effect of mild heat and pressure-based treatments could be of assistance to manufacturers for avoiding extreme pressures of 600 MPa or above that are typically associated with reduced shelf-life of the pressure vessels, increased manufacturing costs, and increased concern for reduction of the nutritive value of the food.

Author Contributions: A.A.: Doctoral Candidate and Graduate Research Assistant, Public Health Microbiology Laboratory, Tennessee State University. Conducted the laboratory studies, preliminary experiments and co-wrote the first version of the manuscript in partial fulfillment of her doctoral dissertation. S.C.: Research Technician, Public Health Microbiology Laboratory, Tennessee State University. Assisted in the conduct of pressure treatments, laboratory analyses, and data recording. A.F.: Assistant Professor and Founder/Director of Public Health Microbiology Laboratory. Secured extramural funding, prepared the experimental design, statistically analyzed the data, co-wrote, revised and edited the manuscript.

Funding: This research was funded in part by United States Department of Agriculture [grant numbers: 2017-07534; 2017-07975; 2016-07430] and Pressure BioScience Inc. Content of the publication does not necessarily reflect the views of the funding agencies.

Acknowledgments: Financial support in part from the National Institute of Food and Agriculture of the United States Department of Agriculture and Pressure BioScience Inc. is acknowledged gratefully by the authors. Technical contribution and assistance of Monica Henry, Anita Scales, Jayashan Adhikari, Kayla Sampson, Akiliyah Sumlin, Kristin Day, and Braxton Simpson, students and staff of the Public Health Microbiology Laboratory is greatly appreciated. Authors also appreciate the feedback of anonymous reviewers and editorial team of *Microorganisms*.

Conflicts of Interest: The authors declare no conflict of interest. The funding sponsors had no role in the design of the study; in the collection, analyses, or interpretation of data; in the writing of the manuscript, and in the decision to publish the results. The content of the current publication does not necessarily reflect the views of the funding agencies.

References

1. Scallan, E.; Hoekstra, R.M.; Angulo, F.J.; Tauxe, R.V.; Widdowson, M.A.; Roy, S.L.; Jones, J.L.; Griffin, P.M. Foodborne illness acquired in the United States—Major pathogens. *Emerg. Infect. Dis.* **2011**, *17*, 7–15. [CrossRef] [PubMed]
2. Scallan, E.; Hoekstra, R.M.; Mahon, B.E.; Jones, T.F.; Griffin, P.M. An assessment of the human health impact of seven leading foodborne pathogens in the United States using disability adjusted life years. *Epidemiol. Infect.* **2015**, *143*, 2795–2804. [CrossRef] [PubMed]
3. Centers for Disease Control and Prevention (CDC), National Outbreak Reporting System (NORS). 2018. Available online: https://wwwn.cdc.gov/norsdashboard/ (accessed on 3 September 2018).
4. World Health Organization (WHO), Listeriosis—South Africa—Disease Outbreak News. 2018. Available online: http://www.who.int/csr/don/28-march-2018-listeriosis-south-africa/en/ (accessed on 3 September 2018).
5. Fouladkhah, A. The Need for evidence-based outreach in the current food safety regulatory landscape. *J. Ext.* **2017**, *55*, 2COM1.
6. Fouladkhah, A.; Geornaras, I.; Sofos, J.N. Effects of Reheating against *Listeria monocytogenes* inoculated on cooked chicken breast meat stored aerobically at 7 °C. *Food Prot. Trends* **2012**, *32*, 697–704.
7. Markham, L.; Auld, G.; Bunning, M.; Thilmany, D. Attitudes and beliefs of raw milk consumers in Northern Colorado. *J. Hunger Environ. Nutr.* **2014**, *9*, 546–564. [CrossRef]

8. Xu, H.; Lee, H.Y.; Ahn, J. High pressure inactivation kinetics of *Salmonella enterica* and *Listeria monocytogenes* in milk, orange juice, and tomato juice. *Food Sci. Biotechnol.* **2009**, *18*, 861–866.

9. Wang, C.Y.; Huang, H.W.; Hsu, C.P.; Yang, B.B. Recent advances in food processing using high hydrostatic pressure technology. *Crit. Rev. Food Sci. Nutr.* **2016**, *56*, 527–540. [CrossRef] [PubMed]

10. Ting, E. High Pressure Food Processing: Past, Current, and Future. In Session: Industrial Adoption and Validation of High Pressure Based Minimal Processing Technologies (A. Fouladkhah, Session Organizer). **2018**. *Annual Meeting of Institute of Food Technologists*, Chicago, IL. Available online: https://www.pressurebiosciences.com/documents/food (accessed on 31 July 2018).

11. Allison, A.; Daniels, E.; Chowdhury, S.; Fouladkhah, A. Effects of elevated hydrostatic pressure against mesophilic background microflora and habituated *Salmonella* serovars in orange juice. *Microorganisms* **2018**, *6*, 23. [CrossRef] [PubMed]

12. Balasubramaniam, V.M.; Barbosa-Cánovas, G.V.; Lelieveld, H.L. *High Pressure Processing of Food: Principles, Technology and Applications*, 1st ed.; Springer Science & Business Media: New York, NY, USA, 2016; ISBN 978-1-4939-3234-4.

13. Fouladkhah, A.; Geornaras, I.; Nychas, G.J.; Sofos, J.N. Antilisterial properties of marinades during refrigerated storage and microwave oven reheating against post—Cooking inoculated chicken breast meat. *J. Food Sci.* **2013**, *78*, M285–M289. [CrossRef] [PubMed]

14. Koutsoumanis, K.P.; Sofos, J.N. Comparative acid stress response of *Listeria monocytogenes*, *Escherichia coli* O157: H7 and *Salmonella* Typhimurium after habituation at different pH conditions. *Lett. Appl. Microbiol.* **2004**, *38*, 321–326. [CrossRef] [PubMed]

15. Geeraerd, A.H.; Valdramidis, V.P.; Van Impe, J.F. GInaFiT, a freeware tool to assess non-log-linear microbial survivor curves. *Int. J. Food Microbiol.* **2005**, *102*, 95–105. [CrossRef] [PubMed]

16. Jay, J.M.; Loessner, M.J.; Golden, D.A. *Modern Food Microbiology*, 7th ed.; Springer Science & Business Media: New York, NY, USA, 2006; ISBN 978-0-387-23413-7.

17. Manu, D.; Mendoca, A.; Daraba, A.; Dickson, J.; Sebranek, J.; Shaw, A.; Wang, F.; White, S. Antimicrobial efficacy of cinnamaldehyde against *Escherichia coli* O157:H7 and *Salmonella enterica* in carrot juice and mixed berry juice held at 4 °C and 12 °C. *Foodborne Pathog. Dis.* **2017**, *14*, 302–307. [CrossRef] [PubMed]

18. Fouladkhah, A.; Geornaras, I.; Yang, H.; Sofos, J.N. Lactic acid resistance of Shiga toxin-producing *Escherichia coli* and multidrug-resistant and susceptible *Salmonella* Typhimurium and *Salmonella* Newport in meat homogenate. *Food Microbiol.* **2013**, *36*, 260–266. [CrossRef] [PubMed]

19. Daryaei, H.; Balasubramaniam, V.M. Kinetics of *Bacillus coagulans* spore inactivation in tomato juice by combined pressure–heat treatment. *Food Control* **2013**, *30*, 168–175. [CrossRef]

20. Linton, M.; Mackle, A.B.; Upadhyay, V.K.; Kelly, A.L.; Patterson, M.F. The fate of *Listeria monocytogenes* during the manufacture of Camembert-type cheese: A comparison between raw milk and milk treated with high hydrostatic pressure. *Innov. Food Sci. Emerg. Technol.* **2008**, *9*, 423–428. [CrossRef]

21. Erkmen, O.; Dogan, C. Effects of ultra high hydrostatic pressure on *Listeria monocytogenes* and natural flora in broth, milk and fruit juices. *Int. J. Food Sci. Technol.* **2004**, *39*, 91–97. [CrossRef]

22. Hayman, M.M.; Anantheswaran, R.C.; Knabel, S.J. The effects of growth temperature and growth phase on the inactivation of *Listeria monocytogenes* in whole milk subject to high pressure processing. *Int. J. Food Microbiol.* **2007**, *115*, 220–226. [CrossRef] [PubMed]

23. Leistner, L.; Gould, G.W. *Hurdle Technologies: Combination Treatments for Food Stability, Safety and Quality*; Springer Science & Business Media: New York, NY, USA, 2012; ISBN 9781461507437.

microorganisms

MDPI

Article

Biocidal Effectiveness of Selected Disinfectants Solutions Based on Water and Ozonated Water against *Listeria monocytogenes* Strains

Krzysztof Skowron [1,*], Ewa Wałecka-Zacharska [2], Katarzyna Grudlewska [1], Agata Białucha [1], Natalia Wiktorczyk [1], Agata Bartkowska [1], Maria Kowalska [3], Stefan Kruszewski [4] and Eugenia Gospodarek-Komkowska [1]

[1]   Department of Microbiology, Nicolaus Copernicus University in Toruń, L. Rydygier Collegium Medicum in Bydgoszcz, 9 M. Skłodowska-Curie St., 85-094 Bydgoszcz, Poland; katinkag@gazeta.pl (K.G.); agatabialucha@wp.pl (A.B.); natalia12127@gmail.com (N.W.); agata.bartkowska@o2.pl (A.B.); gospodareke@cm.umk.pl (E.G.-K.)
[2]   Department of Food Hygiene and Consumer Health, Wrocław University of Environmental and Life Sciences, 31 C.K. Norwida St., 50-375 Wrocław, Poland; ewa.walecka@upwr.edu.pl
[3]   Department of Food Analytics and Environmental Protection, Faculty of Chemical Technology and Engineering, UTP University of Science and Technology, Seminaryjna 3, 85-326 Bydgoszcz, Poland; maria.kowalska@utp.edu.pl
[4]   Biophysics Department, Faculty of Pharmacy, Collegium Medicum of Nicolaus Copernicus University, Jagiellońska 13-15 St., 85–067 Bydgoszcz, Poland; skrusz@cm.umk.pl
*   Correspondence: skowron238@wp.pl; Tel.: +48-52-585-3838

Received: 30 March 2019; Accepted: 9 May 2019; Published: 10 May 2019

Abstract: The aim of this study was to compare the biocidal effectiveness of disinfectants solutions prepared with ozonated and non-ozonated water against *Listeria monocytogenes*. Six *L. monocytogenes* strains were the research material (four isolates from food: meat (LMO-M), dairy products (LMO-N), vegetables (LMO-W), and fish (LMO-R); one clinical strain (LMO-K) and reference strain ATCC 19111). The evaluation of the biocidal effectiveness of disinfectant solutions (QAC—quaternary ammonium compounds; OA—oxidizing agents; ChC—chlorine compounds; IC—iodine compounds; NANO—nanoparticles) was carried out, marking the MBC values. Based on the obtained results, the effectiveness coefficient (A) were calculated. The smaller the A value, the greater the efficiency of disinfection solutions prepared on the basis of ozonated versus non-ozonated water. Ozonated water showed biocidal efficacy against *L. monocytogenes*. Among tested disinfectants, independent on type of water used for preparation, the most effective against *L. monocytogenes* were: QAC 1 (benzyl-C12-18-alkydimethyl ammonium chlorides) (1.00×10^{-5}–1.00×10^{-4} g/mL) in quaternary ammonium compounds, OA 3 (peracetic acid, hydrogen peroxide, bis (sulphate) bis (peroxymonosulfate)) (3.08×10^{-4}–3.70×10^{-3} g/mL) in oxidizing agents, ChC 1 (chlorine dioxide) (5.00×10^{-8}–7.00×10^{-7} g/mL) in chlorine compounds, IC 1 (iodine) (1.05–2.15 g/mL) in iodine compounds, and NANO 1 (nanocopper) (1.08×10^{-4}–1.47×10^{-4} g/mL) in nanoparticles. The values of the activity coefficient for QAC ranged from 0.10 to 0.40, for OA—0.15–0.84, for ChC—0.25–0.83, for IC—0.45–0.60, and for NANO—0.70–0.84. The preparation of disinfectants solution on the basis of ozonated water, improved the microbicidal efficiency of the tested disinfectant, especially the quaternary ammonium compounds. An innovative element of our work is the use of ozonated water for the preparation of working solutions of the disinfection agents. Use ozonated water can help to reduce the use of disinfectant concentrations and limit the increasing of microbial resistance to disinfectants. This paper provides many new information to optimize hygiene plans in food processing plants and limit the spread of microorganisms such as *L. monocytogenes*.

Keywords: *Listeria monocytogenes*; ozon; ozonated water; non-ozonated water; disinfectants; biocidal effectiveness

1. Introduction

Listeria monocytogenes causes listeriosis. This intracellular pathogen is widespread in the environment, from where it can enter the digestive tract of animals and humans [1]. The main source of *L. monocytogenes* is food, especially smoked fish, cheese, delicatessen meat products, milk, seafood, eggs, and vegetables [2].

Since *L. monocytogenes* is able to survive under food processing conditions it constitutes a serious threat in food processing plants. To prevent the spread of infection caused by this pathogen, chemical disinfection processes using compounds such as chlorine, iodine, oxidizing, phenolic, quaternary ammonium compounds, alcohols, aldehydes, or metal nanoparticles are carried out [3].

Recently ozone has become an alternative disinfectant. Because of its antibacterial properties, ozone is widely used for the disinfection of drinking water and sewage as well as in the food industry. In the disinfection processes, ozone is used in gaseous or aqueous form depending on the type of decontaminated surfaces. Low concentrations of ozone and short duration of action are sufficient to eliminate microorganisms [4]. It is active against bacteria (such as *Listeria* spp., *Escherichia* spp., *Salmonella* spp.), viruses, fungus, fungal spores, and protozoa [5]. Also, the constant ozonation of water with low ozone concentration (0.5 mg/L) intended for washing vegetables has resulted in a reduction in the number of mesophilic and coliform bacteria on the surface of lettuce and peppers. This method was less dense in relation to the elimination of mold and fungi. Also the type of vegetable plays a role in the effectiveness of the method (more effective for peppers) [6]. Ozone disturbs the integrity of the bacterial cell membrane by oxidizing phospholipids and lipoproteins. In the case of fungi, ozone inhibits microbial growth in a certain phase. In the case of viruses, ozone damages the viral capsid and interferes with the viral replication cycle [7]. Thanomsub et al. [8], using scanning electron microscopy (SEM), showed deformation of Gram-negative cells exposed to ozone at a concentration of 0.167/mg/min/L. After 60 min exposure, the cells were sunken and concave, while after 90 min of exposure, they showed lysis. The use of ozone does not require high temperatures and carries many economic benefits [9]. However, its short half-life period is associated with the need to produce it at the place of use. The half-life of ozone in an aqueous solution at 20 °C is approximately 20–30 min [10]. The use of ozonated water in the food industry determines the inclusion of organic matter and pH values. Arayan et al. [11] showed that organic pollution affects the ozone's water half-life, and temperature is also an important factor. An increase in the temperature of ozone water caused a decrease in biocidal effectiveness [11]. Moreover ozone may be corrosive for the treated surface [12]. On the other hand its usage reduces the amount of other disinfectants, and herby the amount of their toxic by-products [4].

The aim of the study was to compare the biocidal effectiveness against *L. monocytogenes* strains of thirteen selected disinfectants, for which solutions were prepared using sterile hard water and ozone water. The aim was also to assess the stability of solutions of three disinfectants which effectiveness was significantly higher in ozonated water compared to hard water.

2. Materials and Methods

2.1. Bacterial Strains

The study was conducted on 6 *L. monocytogenes* strains isolated in 2015 from the territory of the Kuyavian-Pomeranian Voivodeship (Poland), including four isolates from food: meat (LMO-M), dairy products (LMO-N), vegetables (LMO-W), and fish (LMO-R), one clinical strain (LMO-K) from the collection of the Department of Microbiology, Nicolaus Copernicus University in Toruń, L. Rydygier Collegium Medicum in Bydgoszcz. These strains were susceptible to all antibiotics tested (penicillin, ampicillin, meropenem, erythromycin, cotrimoxazole) in accordance with the EUCAST v.8.00 recommendations [13]. The study material also included the reference strain *L. monocytogenes* ATCC 19111.

2.2. Ozonated Water

In the experiment, the hard water was used, prepared according to the Polish Standard PN-EN 1276: 2010 [14]. A solution A (19.84 g $MgCl_2$ (Avantor, Gliwice, Poland) and 46.24 g $CaCl_2$ (Avantor, Gliwice, Poland) was dissolved in 1000 mL H_2O) and B (35.02 g $NaHCO_3$ (Avantor, Gliwice, Poland) dissolved in 1000 mL H_2O) was prepared to obtain a hard water. Both solutions were sterilized. Then, 6 mL of solution A and 8 mL of solution B were added to 700 mL of sterile water, mixed thoroughly and made up with sterile water to 1000 mL. The ozonation process of 100 mL of sterile hard water (temperature 20 °C, pH = 7.0) was carried out using a 20W ZE-H103 Orientee ozonator (ELTOM, Warsaw, Poland) with a diffuser and with the function of water and air ionization for 45 min. Ozone was generated in a laminar chamber at constant humidity and air temperature. Ozone concentration was determined in the reaction mixtures using DPD (*N,N*-diethyl-1,4-phenylene diammonium sulfate) Method, according ISO 7393-2:2017 [15], cuvette test and DR3900 Benchtop VIS Spectrophotometer provided by Hach (Frederick, Maryland, USA). This procedure of ozone determination was designed for water samples by Hach company. According to the procedure, the samples are treated with oxidizing agent *N,N*-diethyl-1,4-phenylene diammonium sulfate (DPD) to form a red dye. It was determined by visible spectrophotometry (λmax = 552 nm).

Due to the short typical half-life time of ozone in water (15–25 min for pH = 7–10) [16], ozonated water was immediately used to prepare solutions.

2.3. Disinfectants

The study used 13 disinfectants. The composition and concentration (in accordance with the manufacturer's instructions) of working solutions needed to prepare the dilution series are presented in Table 1.

Taking into account the information provided by the manufacturer of a particular disinfector for the preparation of a commercial working solution (100%), the following solutions were prepared: 200%, 180%, 160%, 140%, 120%, 100%, 80%, 60%, 40%, 20%, 10%, 2%, and 1% of working solutions concentrations. The concentrations were selected in order to carry out the procedure for assessing the minimum bactericidal concentrations described in the further part of the methodology. The specific concentrations for particular tested disinfectants were presented in Table 2. Two independent dilution series were prepared—one using sterile hard water [14], the other—sterile ozonated water, immediately after its preparation.

Table 1. Characteristics of disinfectants.

Group of Disinfectants	Name	Active Substance	Working Solution Concentration (g/mL)
Quaternary ammonium compounds	QAC 1	benzyl-C12-18-alkydimethyl ammonium chlorides	2.0×10^{-3}
	QAC 2	benzyl-C12-16 alkyldimethyl chlorides	2.55×10^{0}
	QAC 3	didecyldimethylammonium chloride, benzyl-C12-16-alkyldimethyl chlorides	2.97×10^{0}
Oxidizing agents	OA 1	hydrogen peroxide, silver nitrate	1.20×10^{1}
	OA 2	perlactic acid	4.90×10^{0}
	OA 3	peracetic acid, hydrogen peroxide	6.15×10^{-3}
	OA 4	bis (sulphate) bis (peroxymonosulfate) pentapotassium, benzenesulfonic acid, C10-13 alkyl derivatives, sodium salts, malic acid, sulfamic acid	2.00×10^{-2}
Chlorine compounds	ChC 1	chlorine dioxide	1.00×10^{-5}
	ChC 2	hypochlorous acid calcium salt	2.00×10^{-3}
Iodine compounds	IC 1	iodine	6.15×10^{0}
	IC 2	iodine	1.17×10^{1}
Nanoparticles	NANO 1	nanocopper	1.50×10^{-4}
	NANO 2	nanosilver	1.50×10^{-4}

Table 2. The specific concentrations for particular tested disinfectants.

Group of Disinfectants	Name	Initial Concentration (g/mL)	Final Concentration (g/mL)
Quaternary ammonium compounds	QAC 1	4.00×10^{-3}; 3.60×10^{-3}; 3.20×10^{-3}; 2.80×10^{-3}; 2.40×10^{-3}; 2.00×10^{-3}; 1.60×10^{-3}; 1.20×10^{-3}; 8.00×10^{-4}; 4.00×10^{-4}; 2.00×10^{-4}; 4.00×10^{-5}; 2.00×10^{-5}	2.00×10^{-3}; 1.80×10^{-3}; 1.60×10^{-3}; 1.40×10^{-3}; 1.20×10^{-3}; 1.00×10^{-3}; 8.00×10^{-4}; 6.00×10^{-4}; 4.00×10^{-4}; 2.00×10^{-4}; 1.00×10^{-4}; 2.00×10^{-5}; 1.00×10^{-5}
	QAC 2	5.1×10^{0}; 4.59×10^{0}; 4.08×10^{0}; 3.57×10^{0}; 3.06×10^{0}; 2.55×10^{0}; 2.04×10^{0}; 1.53×10^{0}; 1.02×10^{0}; 5.1×10^{-1}; 2.6×10^{-1}; 5.00×10^{-2}; 2.6×10^{-2}	2.55×10^{0}; 2.30×10^{0}; 2.04×10^{0}; 1.79×10^{0}; 1.53×10^{0}; 1.28×10^{0}; 1.02×10^{0}; 7.70×10^{-1}; 5.10×10^{-1}; 2.60×10^{-1}; 1.30×10^{-1}; 2.60×10^{-2}; 1.00×10^{-2}
	QAC 3	5.94×10^{0}; 5.35×10^{0}; 4.75×10^{0}; 4.16×10^{0}; 3.56×10^{0}; 2.97×10^{0}; 2.38×10^{0}; 1.78×10^{0}; 1.19×10^{0}; 5.90×10^{-1}; 3.00×10^{-1}; 5.90×10^{-2}; 3.00×10^{-2}	2.97×10^{0}; 2.67×10^{0}; 2.38×10^{0}; 2.08×10^{0}; 1.78×10^{0}; 1.49×10^{0}; 1.19×10^{0}; 8.90×10^{-1}; 5.90×10^{-1}; 2.30×10^{-1}; 1.50×10^{-1}; 3.00×10^{-2}; 1.50×10^{-2}
Oxidizing agents	OA 1	2.40×10^{1}; 2.16×10^{1}; 1.92×10^{1}; 1.68×10^{1}; 1.44×10^{1}; 1.20×10^{1}; 9.60×10^{0}; 7.20×10^{0}; 4.80×10^{0}; 2.40×10^{0}; 1.20×10^{0}; 2.40×10^{-1}; 1.20×10^{-1}	1.20×10^{1}; 1.08×10^{1}; 9.60×10^{0}; 8.40×10^{0}; 7.20×10^{0}; 6.00×10^{0}; 4.80×10^{0}; 3.60×10^{0}; 2.40×10^{0}; 1.20×10^{0}; 6.00×10^{-1}; 1.20×10^{-1}; 6.00×10^{-2}
	OA 2	9.80×10^{0}; 8.82×10^{0}; 7.84×10^{0}; 6.86×10^{0}; 5.88×10^{0}, 4.90×10^{0}; 3.92×10^{0}; 2.94×10^{0}; 1.96×10^{0}; 9.80×10^{-1}; 4.90×10^{-1}; 9.80×10^{-2}; 4.90×10^{-2}	4.90×10^{0}; 4.41×10^{0}; 3.92×10^{0}; 3.43×10^{0}; 2.45×10^{0}; 2.40×10^{0}; 1.96×10^{0}; 1.47×10^{0}; 9.80×10^{-1}; 4.90×10^{-1}; 2.50×10^{-1}; 5.00×10^{-2}; 2.50×10^{-2}
	OA 3	1.20×10^{-2}; 1.00×10^{-2}; 9.80×10^{-3}; 8.60×10^{-3}; 7.40×10^{-3}; 6.15×10^{-3}; 4.92×10^{-3}; 3.69×10^{-3}; 2.46×10^{-3}; 1.23×10^{-3}; 6.15×10^{-4}; 1.23×10^{-4}; 6.15×10^{-5}	6.15×10^{-3}; 5.50×10^{-3}; 4.90×10^{-3}; 4.30×10^{-3}; 3.70×10^{-3}; 3.10×10^{-3}; 2.46×10^{-3}; 1.90×10^{-3}; 1.23×10^{-3}; 6.15×10^{-4}; 3.10×10^{-4}; 6.15×10^{-5}; 3.10×10^{-5}
	OA 4	4.00×10^{-2}; 3.60×10^{-2}; 3.20×10^{-2}; 2.80×10^{-2}; 2.40×10^{-2}; 2.00×10^{-2}; 1.60×10^{-2}; 1.20×10^{-2}; 8.00×10^{-3}; 4.00×10^{-3}; 2.00×10^{-3}; 4.00×10^{-4}; 2.00×10^{-4}	2.00×10^{-2}; 1.80×10^{-2}; 1.60×10^{-2}; 1.40×10^{-2}; 1.20×10^{-2}; 1.00×10^{-2}; 8.00×10^{-3}; 6.00×10^{-3}; 4.00×10^{-3}; 2.00×10^{-3}; 1.00×10^{-3}; 2.00×10^{-4}; 1.00×10^{-4}
Chlorine compounds	ChC 1	2.00×10^{-5}; 1.8×10^{-5}; 1.6×10^{-5}; 1.4×10^{-5}; 1.20×10^{-5}; 1.00×10^{-5}; 8.00×10^{-6}; 6.00×10^{-6}; 4.00×10^{-6}; 2.00×10^{-6}; 1.00×10^{-6}; 2.00×10^{-7}; 1.00×10^{-7}	1.00×10^{-5}; 9.00×10^{-6}; 8.00×10^{-6}; 7.00×10^{-6}; 6.00×10^{-6}; 5.00×10^{-6}; 4.00×10^{-6}; 3.00×10^{-6}; 2.00×10^{-6}; 1.00×10^{-6}; 2.00×10^{-7}; 1.00×10^{-7}
	ChC 2	4.00×10^{-3}; 3.60×10^{-3}; 3.20×10^{-3}; 2.80×10^{-3}; 2.40×10^{-3}; 2.00×10^{-3}; 1.60×10^{-3}; 1.20×10^{-3}; 8.00×10^{-4}; 4.00×10^{-4}; 2.00×10^{-4}; 4.00×10^{-5}; 2.00×10^{-5}	2.00×10^{-3}; 1.80×10^{-3}; 1.60×10^{-3}; 1.40×10^{-3}; 1.20×10^{-3}; 1.00×10^{-3}; 8.00×10^{-4}; 6.00×10^{-4}; 4.00×10^{-4}; 2.00×10^{-4}; 1.00×10^{-4}; 2.00×10^{-5}; 1.00×10^{-5}
Iodine compounds	IC 1	1.23×10^{1}; 1.11×10^{1}; 9.84×10^{0}; 8.61×10^{0}; 7.38×10^{0}; 6.15×10^{0}; 4.92×10^{0}; 3.69×10^{0}; 2.46×10^{0}; 1.23×10^{0}; 6.20×10^{-1}; 1.20×10^{-1}; 6.20×10^{-2}	6.15×10^{0}; 5.54×10^{0}; 4.92×10^{0}; 4.31×10^{0}; 3.69×10^{0}; 3.08×10^{0}; 2.46×10^{0}; 1.85×10^{0}; 1.23×10^{0}; 6.20×10^{-1}; 3.10×10^{-1}; 6.20×10^{-2}; 3.10×10^{-2}
	IC 2	2.34×10^{1}; 2.11×10^{1}; 1.87×10^{1}; 1.64×10^{1}; 1.40×10^{1}; 1.17×10^{1}; 9.36×10^{0}; 7.02×10^{0}; 4.68×10^{0}; 2.34×10^{0}; 1.17×10^{0}; 2.30×10^{-1}; 1.20×10^{-1}	1.17×10^{1}; 1.05×10^{1}; 9.36×10^{0}; 8.19×10^{0}; 7.02×10^{0}; 5.85×10^{0}; 4.68×10^{0}; 3.51×10^{0}; 2.34×10^{0}; 1.17×10^{0}; 5.90×10^{-1}; 1.20×10^{-1}; 5.90×10^{-2}
Nanoparticles	NANO 1	3.00×10^{-3}; 2.70×10^{-3}; 2.40×10^{-3}; 2.10×10^{-3}; 1.80×10^{-3}; 1.50×10^{-3}; 1.20×10^{-3}; 9.00×10^{-4}; 6.00×10^{-4}; 3.00×10^{-4}; 1.50×10^{-4}; 3.00×10^{-5}; 1.50×10^{-5}	1.50×10^{-4}; 1.35×10^{-4}; 1.20×10^{-4}; 1.05×10^{-4}; 9.00×10^{-5}; 7.50×10^{-5}; 6.00×10^{-5}; 4.50×10^{-5}; 3.00×10^{-5}; 1.50×10^{-5}; 7.50×10^{-6}; $1,50 \times 10^{-6}$; 7.50×10^{-7}
	NANO 2	3.00×10^{-3}; 2.70×10^{-3}; 2.40×10^{-3}; 2.10×10^{-3}; 1.80×10^{-3}; 1.50×10^{-3}; 1.20×10^{-3}; 9.00×10^{-4}; 6.00×10^{-4}; 3.00×10^{-4}; 1.50×10^{-4}; 3.00×10^{-5}; 1.50×10^{-5}	1.50×10^{-4}; 1.35×10^{-4}; 1.20×10^{-4}; 1.05×10^{-4}; 9.00×10^{-5}; 7.50×10^{-5}; 6.00×10^{-5}; 4.50×10^{-5}; 3.00×10^{-5}; 1.50×10^{-5}; 7.50×10^{-6}; $1,50 \times 10^{-6}$; 7.50×10^{-7}

QAC 1—benzyl-C12-18-alkyldimethyl ammonium chlorides; QAC 2—benzyl-C12-16 alkyldimethyl chlorides; QAC 3—didecyldimethylammonium chloride, benzyl-C12-16-alkyldimethyl chlorides; OA 1—hydrogen peroxide, silver nitrate; OA 2—perlactic acid; OA 3—peracetic acid, hydrogen peroxide bis (sulphate) bis (peroxymonosulfate); OA 4—pentapotassium, benzenesulfonic acid, C10-13 alkyl derivatives, sodium salts, malic acid, sulfamic acid; ChC 1—chlorine dioxide; ChC 2—hypochlorous acid calcium salt; IC 1—iodine, IC 2—iodine; NANO 1—nanocopper; NANO 2—nanosilver.

2.4. Preparation of Bacterial Suspensions

From cultures of *L. monocytogenes* strains obtained on Columbia Agar with 5% sheep blood (CAB, bioMérieux, Marcy-l'Étoile, France) suspensions of a density of 0.5 MacFarland standard (5.80×10^8 CFU $\times$ mL^{-1}) were prepared in 3 mL of Mueller Hinton Broth (MHB, Becton Dickinson, Franklin Lakes, New Jersey, USA). For this purpose, the optical density for the sterile MHB (Mueller Hinton Broth, Becton Dickinson, Franklin Lakes, New Jersey, USA) medium was first established. A sterile swab was then collected from a single colony grown on Columbia Agar with 5% sheep blood (CAB, bioMérieux, Marcy-l'Étoile, France) and loaded into the MHB (Mueller Hinton Broth, Becton Dickinson, Franklin Lakes, New Jersey, USA) medium, followed by measurement of the optical density of the suspension and subsequent colonization of *L. monocytogenes* added if necessary. The optical density of the suspension was set at 0.5 + the optical density of the sterile MHB (Mueller Hinton Broth, Becton Dickinson, Franklin Lakes, New Jersey, USA). The measurements were made with a DEN-1B denitometer from Biogenet (Józefów, Poland).

2.5. Assessment of Biocidal Effectiveness of Ozonated and Non-Ozonated Water

The suspensions of the tested *L. monocytogenes* strains (100 μL) were pipetted into Eppendorf (1.5 mL, Genoplast, Poland) tubes. After centrifugation (3000 rpm per 5 min) of the bacterial suspensions, 150 μL of an ozone solution of appropriate concentration (the mean value determined according to point 2.2–2.32 μg O$_3$/mL) was added to the sediments. Immediately after preparation of the suspension in ozonated/non-ozonated water, a row of decimal dilutions were made in sterile PBS (Phosphate-buffered saline, Avantor, Gliwice, Poland). Each dilution (100 μL) was seeded into a Columbia Agar with 5% sheep blood (CAB, bioMérieux, Marcy-l'Étoile, France) and incubated for 24 h at 37°C. In this way, the initial number of *L. monocytogenes* was determined. The number of bacteria in the obtained suspension was 3.60–4.20 $\times$ 10^8 CFU $\times$ mL^{-1}. The negative control was 150 μL of ozon water, and the positive control—150 μL of bacterial suspension of a given strain suspended in sterile hard water. The study was carried out in triplicate for each strain and each tested concentration.

After 5 min of treatment of the suspensions with ozonated water, sample were transferred to 900 μL neutralizer (10 g Tween 80 (Sigma Aldrich, Saint Louis, Missouri, USA), 1 g lecithin (Sigma Aldrich, Saint Louis, Missouri, USA), 0.5 g histidine L (Sigma Aldrich, Saint Louis, Missouri, USA), 2.5 g Na$_2$S$_2$O$_3$ (Avantor, Gliwice, Poland), 3.5 g C$_3$H$_3$NaO$_3$ (Avantor, Gliwice, Poland), 1000 mL sterile water). Lecithin neutralizes quaternary ammonia compounds while phenolic disinfectants and hexachlorophene are neutralized by Tween. Together, lecithin and Tween neutralize ethanol. Histidine inactivates aldehydes, especially formaldehyde and gluteraldehyde. Sodium thiosulfate neutralizes iodine and chlorine, whereas sodium pyruvate neutralizes active oxygen and peroxides [17,18]. After 5 min of exposure, linear cultures were made on the CAB (Columbia Agar with 5% sheep blood, bioMérieux, Marcy-l'Étoile, France) substrate plate sectors that were incubated for 24 h at 37 °C. After incubation, the concentration of ozonated water was analyzed, which enabled the inactivation of the tested strains of *L. monocytogenes*. The effect of sterile hard water [15] on the number of recovered bacteria was also assessed.

After determining the concentration range in which the value of the minimum bactericidal concentration (MBC) of ozonated water was located, the procedure was repeated, preparing solutions with a concentration varying by 1% in this range, in order to accurately determine the MBC (minimum bactericidal concentration).

To check the durability of ozonated water, the same test cycle was carried out one and two h after the ozonation process was completed. All plates with banded cultures were incubated under the conditions described above, and then the results were read.

2.6. Evaluation of the Effectiveness of Disinfectants

The suspensions of the tested strains of *L. monocytogenes* (100 µL) and 100 µL of the appropriate concentration of disinfectant were introduced into the wells of a multi-well polystyrene plate. The target concentration of the disinfectant in the well plate was respectively 100%, 90%, 80%, 70%, 60%, 50%, 40%, 30%, 20%, 10%, 5%, 1%, and 0.5% concentration working solution of a particular disinfectant. The specific concentrations for particular tested disinfectants were presented in Table 2. The negative control consisted of 200 µL of sterile MHB (Mueller Hinton Broth, Becton Dickinson, Franklin Lakes, New Jersey, USA) medium, and a positive control—200 µL of bacterial suspension. After 5 min of the agent's action on bacterial suspensions, 100 µL of liquid was transferred from each well to 900 µL of neutralizer. After 5 min of neutralization, band-cultures were made on the CAB (Columbia Agar with 5% sheep blood, bioMérieux, Marcy-l'Étoile, France) segments, which were incubated for 24 h at 37°C. After incubation, the minimum concentration allowing inactivation of the tested strains of *L. monocytogenes* was read for solutions based on non-ozonated and ozonated water.

After determining the concentration range in which the value of the minimum bactericidal concentration (MBC) of a given disinfectant was located, the procedure was repeated by preparing solutions with a concentration varying by 1% in this range, to accurately determine the MBC.

Based on the obtained results, for all strains and disinfectants used in the studies, the effectiveness coefficient (A) were calculated [19]. The smaller the value of the coefficient A, the greater the efficiency of disinfecting solutions based on ozonated water compared to non-ozonated disinfectants.

The effectiveness factor was calculated from the formula:

$$A = b/c$$

where:

A—effectiveness coefficient,

b—effective concentration of disinfectant in active solution (ozonated water),

c—effective concentration of disinfectant in non-ozonated water solution.

In order to assess the decrease in the number of *L. monocytogenes* under the influence of disinfectants prepared on the basis of ozonated and non-ozonated water, the initial number of tested bacteria and the number of bacteria isolated at preMBC disinfectant concentration were determined. For this purpose, a series of decimal dilutions was made for the prepared suspension of a given strain and then plated on CAB (Columbia Agar with 5% sheep blood, bioMérieux, Marcy-l'Étoile, France) agar. The cultures were incubated at 37°C for 24 h. After this time, the grown colonies were counted and converted into logarithmic units. Similarly, the suspension subjected to action of the disinfectant concentration directly preceding MBC (preMBC) was treated. To determine preMBC, the MBC value from Table 3 was checked and the directly preceding concentration was selected from Table 2. The decreases in the number of bacteria were calculated from the formula:

$$R = \log_i - \log_{preMBC}$$

where:

R—reduction in bacteria number (log CFU)

$\log_i$—initial bacteria number

$\log_{preMBC}$—nuber of bacteria recovered from suspension trated with preMBC concentration of disinfectant

Table 3. Minimal bactericidal concentration of tested disinfectant depending of water type (g/mL).

Group of Disinfectants	Disinfectant	Water Type	Minimal Bactericidal Concentration of Disinfectant (g/mL) Strain					
			LMO-ATCC	LMO-W	LMO-M	LMO-N	LMO-R	LMO-K
Quaternary ammonium compounds	QAC 1	Nonozonated	1.00×10^{-4} a	1.00×10^{-4} a	4.00×10^{-5} b	4.00×10^{-5} b	4.00×10^{-5} b	1.00×10^{-4} a
		Ozonated	4.00×10^{-5} b	2.00×10^{-5} b	1.00×10^{-5} b	1.00×10^{-5} b	1.00×10^{-5} b	4.00×10^{-5} b
	QAC 2	Nonozonated	1.28×10^{-1} c	1.28×10^{-1} c	1.28×10^{-1} c	1.28×10^{-1} c	1.28×10^{-1} c	1.28×10^{-1} c
		Ozonated	1.28×10^{-2} d	1.28×10^{-2} d	1.28×10^{-2} d	1.28×10^{-2} d	1.28×10^{-2} d	1.28×10^{-2} d
	QAC 3	Nonozonated	1.49×10^{-1} c	1.49×10^{-1} c	1.49×10^{-1} c	1.49×10^{-1} c	1.49×10^{-1} c	1.49×10^{-1} c
		Ozonated	1.49×10^{-2} d	2.97×10^{-2} d	1.49×10^{-2} d	1.49×10^{-2} d	1.49×10^{-2} d	1.49×10^{-2} d
	OA 1	Nonozonated	1.20×10^{1} e	1.20×10^{1} e	1.20×10^{1} e	1.20×10^{1} e	1.20×10^{1} e	1.20×10^{1} e
		Ozonated	1.01×10^{1} e	1.01×10^{1} e	1.01×10^{1} e	1.01×10^{1} e	1.01×10^{1} e	1.01×10^{1} e
Oxidizing agents	OA 2	Nonozonated	4.90×10^{0} k	Ineffective	Ineffective	4.90×10^{0} k	Ineffective	4.90×10^{0} k
		Ozonated	2.70×10^{0} f	2.79×10^{0} f	3.38×10^{0} f,k	2.70×10^{0} f	3.68×10^{0} f,k	2.70×10^{0} f
	OA 3	Nonozonated	1.42×10^{-3} g	1.42×10^{-3} g	1.85×10^{-3} g	1.85×10^{-3} g	3.69×10^{-3} g	1.42×10^{-3} g
		Ozonated	3.08×10^{-4} a	3.08×10^{-4} a	4.31×10^{-4} a	3.08×10^{-4} a	5.54×10^{-4} a	3.08×10^{-4} a
	OA 4	Nonozonated	4.20×10^{-3} h	4.20×10^{-3} h	4.20×10^{-3} h	4.20×10^{-3} h	4.20×10^{-3} h	4.20×10^{-3} h
		Ozonated	1.60×10^{-3} g	1.60×10^{-3} g	1.60×10^{-3} g	1.60×10^{-3} g	1.60×10^{-3} g	1.60×10^{-3} g
Chlorine compounds	ChC 1	Nonozonated	2.00×10^{-7} i	4.00×10^{-7} i	7.00×10^{-7} i	4.00×10^{-7} i	6.00×10^{-7} i	4.00×10^{-7} i
		Ozonated	5.00×10^{-8} j	1.00×10^{-7} i	3.00×10^{-7} i	1.00×10^{-7} i	2.00×10^{-7} i	1.00×10^{-7} i
	ChC 2	Nonozonated	2.40×10^{-1} c	3.20×10^{-1} c	3.60×10^{-1} c	3.20×10^{-1} c	2.80×10^{-1} c	4.0×10^{-1} c
		Ozonated	2.00×10^{-1} c	2.00×10^{-1} c	2.00×10^{-1} c	2.00×10^{-1} c	2.00×10^{-1} c	2.80×10^{-1} c
Iodine compounds	IC 1	Nonozonated	2.15×10^{0} f	2.15×10^{0} f	2.15×10^{0} f	2.15×10^{0} f	2.15×10^{0} f	2.15×10^{0} f
		Ozonated	1.05×10^{0} l	1.29×10^{0} l	1.29×10^{0} l	1.29×10^{0} l	1.29×10^{0} l	1.05×10^{0} l
	IC 2	Nonozonated	3.63×10^{0} f,k	3.63×10^{0} f,k	3.86×10^{0} f,k	3.63×10^{0} f,k	3.63×10^{0} f,k	3.86×10^{0} f,k
		Ozonated	1.64×10^{0} l	1.64×10^{0} l	1.76×10^{0} l	1.64×10^{0} l	1.69×10^{0} l	1.76×10^{0} l
Nanoparticles	NANO 1	Nonozonated	1.37×10^{-4} a	1.37×10^{-4} a	1.41×10^{-4} a	1.37×10^{-4} a	1.47×10^{-4} a	1.37×10^{-4} a
		Ozonated	1.08×10^{-4} a	1.14×10^{-4} a	1.14×10^{-4} a	1.08×10^{-4} a	1.19×10^{-4} a	1.08×10^{-4} a
	NANO 2	Nonozonated	Ineffective	Ineffective	Ineffective	Ineffective	Ineffective	Ineffective
		Ozonated	1.20×10^{-4} a	1.20×10^{-4} a	1.20×10^{-4} a	1.20×10^{-4} a	1.20×10^{-4} a	1.20×10^{-4} a

QAC 1—benzyl-C12-18-alkyldimethyl ammonium chlorides; QAC 2—benzyl-C12-16 alkyldimethyl chlorides; QAC 3—didecyldimethylammonium chloride, benzyl-C12-16-alkyldimethyl chlorides; OA 1—hydrogen peroxide, silver nitrate; OA 2—perlactic acid; OA 3—peracetic acid, hydrogen peroxide bis (sulphate) bis (peroxymonosulfate); OA 4—pentapotassium, benzenesulfonic acid, C10-13 alkyl derivatives, sodium salts, malic acid, sulfamic acid; ChC 1—chlorine dioxide; ChC 2—hypochlorous acid calcium salt; IC 1—iodine, IC 2—iodine; NANO 1—nanocopper; NANO 2—nanosilver; LMO-ATCC—*L. monocytogenes* ATCC 19111, LMO-W—strain isolated from vegetables, LMO-M—strain isolated from meat, LMO-N—strain isolated from fish, LMO-R—clinical strain, LMO-K—strain isolated from dairy products, LMO-R—strain isolated from fish, a—l—values marked with different letters differ statistically significantly ($p \le 0.05$).

2.7. Assessment of the Stability of QAC 2, OA 3, and ChC1 Solutions

Three disinfectants, for which the effectiveness coefficient was the lowest, were evaluated for the stability (QAC 2, OA 3, and ChC 1). The prepared working solutions, both in non-ozonated and ozonated water, were stored at room temperature. The effectiveness of these agents on *L. monocytogenes* strains was evaluated immediately after preparation of the solutions, after 12 and 24 h, determining the MBC values in individual time intervals. This part of the experiment was designed to assess whether the possible increased effectiveness of disinfectants is not prolonged due to some reactions between ozone and active substances. On the basis of MBC values, the coefficient A was calculated for particular disinfectant, strains and time of storage.

2.8. Statistical Analysis

Statistical analysis was performed in the STATISTICA 13.1 PL program (StatSoft). Significance of differences between the ozonated water effectiveness, MBC values of disinfectants, reduction in *L. monocytogenes* number and maximal effectiveness coefficients for disinfectants were assessed.

2.8.1. Biocidal Effectiveness of Ozonated and Non-Ozonated Water

It was checked how the MBC value of the ozonated water varied depending on the time (0, 1, and 2 h) and the *L. monocytogenes* strain. Experiment was made in 3 replications. As independent variables, the strain as well as the time elapsed from the ozonation of water were treated, and as a dependent variable the determined value of MBC was recognized. Significance of differences between mean MBC values for the combination of both independent variables was checked. For this purpose, the general line models (GLM) were used. The multi-way ANOVA was conducted. The Tukey post-hoc test was used for significance of $\alpha = 0.05$.

2.8.2. Evaluation of the Effectiveness of Disinfectants

It was checked how the MBC value of tested disinfectants varied depending on the type of water (ozonated/non-ozonated) used for solutions preparation. Experiment was made in 3 replications. As independent variables, the disinfectant, type of water, as well as the strain were treated, and as a dependent variable the determined value of disinfectants MBC was recognized. Significance of differences between mean MBC values for the combination of all independent variables was checked. For this purpose, the general line models (GLM) were used. The multi-way ANOVA was conducted. The Tukey post-hoc test was used for significance of $\alpha = 0.05$.

2.8.3. Reduction in Bacteria Number

It was checked how the reduction in bacteria number varied depending on the type of water (ozonated/non-ozonated) used for disinfecting solutions preparation. Experiment was made in 3 replications. As independent variables, the type of water was treated, and as a dependent variable the determined reduction in bacteria number was recognized. Significance of differences between mean reduction obtained for particular disinfectant depending on type of water was checked. For this purpose, the general line models (GLM) were used. The one-way ANOVA was conducted. The Tukey post-hoc test was used for significance of $\alpha = 0.05$.

2.8.4. Maximal Effectiveness Coefficients for Disinfectants

It was checked how the maximal effectiveness coefficients for disinfectants varied depending on the type of water (ozonated/non-ozonated) used for disinfecting solutions preparation. Experiment was made in 3 replications. As independent variables, the type of water was treated, and as a dependent variable the determined maximal effectiveness coefficients for disinfectants was recognized. Significance of differences between mean coefficient value obtained for particular disinfectant depending on type of water was checked. For this purpose, the general line models (GLM) were used. The one-way ANOVA was conducted. The Tukey post-hoc test was used for significance of $\alpha = 0.05$.

3. Results

3.1. Assessment of Biocidal Effectiveness of Ozonated and Non-Ozonated Water

The non-ozonated water did not show biocidal efficacy against the tested strains of *L. monocytogenes*.

The determined concentration of ozone in ozonated water was 2.32 µg/mL ($\pm$ 0.022 µg/mL). The ozonated water, used immediately after preparation, inhibited bacterial growth at ozone concentrations 1.86–1.96 µg/mL, with the lowest resistance characterized for the reference strain (1.86 µg/mL) and the highest—for strains derived from meat and fish (1.96 µg/mL). For the ozonated water after two h from the end of the ozonation process, for all tested *L. monocytogenes* strains, the total lack of efficacy was demonstrated (Figure 1).

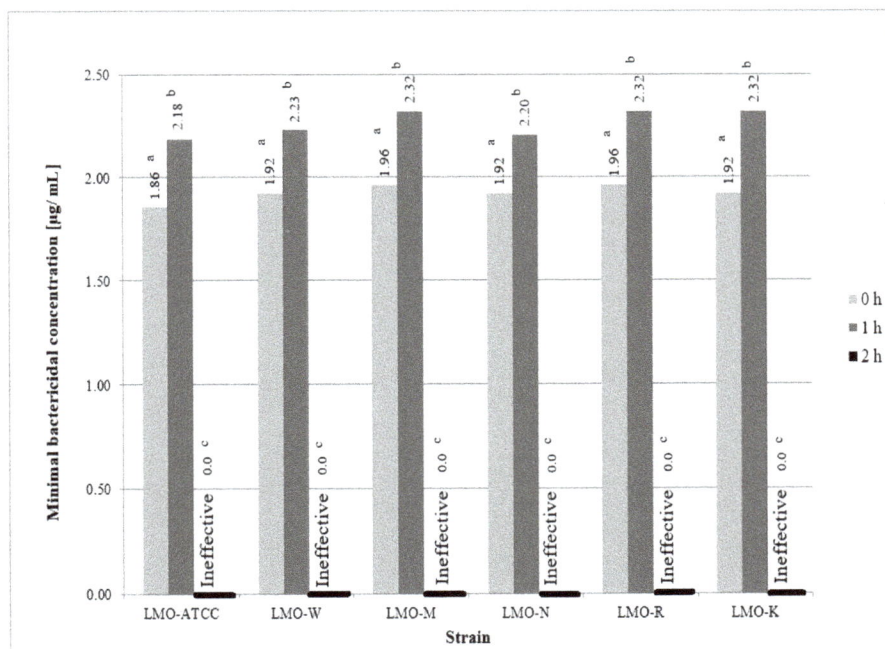

Figure 1. Effectiveness of ozonated water against the tested strains of *L. monocytogenes* (LMO-ATCC—*L. monocytogenes* ATCC 19111, LMO-W—strain isolated from vegetables, LMO-M—strain isolated from meat, LMO-N—strain isolated from dairy products, LMO-R—strain isolated from fish, LMO-K—clinical strain.; a,b,c—variables with different letters are statistically different ($p \leq 0.05$).

3.2. Evaluation of Effectiveness of Disinfectant

Disinfectants based on ozonated water were characterized by a higher biocidal efficiency than solutions based on non-ozonated water (Table 3). In most cases these differences were statistically significant ($p \leq 0.05$) (Table 3).

Among the tested disinfectants, regardless of type of water used for preparation, the most effective against *L. monocytogenes* were: QAC 1 (1.00×10^{-5}–1.00×10^{-4} g/mL) in quaternary ammonium compounds, OA 3 (3.08×10^{-4}–3.70×10^{-3} g/mL) in oxidizing agents, ChC 1 (5.00×10^{-8}–7.00×10^{-7} g/mL) in chlorine compounds, IC 1 (1.05–2.15 g/mL) in iodine compounds and NANO 1

$(1.08 \times 10^{-4}$–1.47×10^{-4} g/mL) in nanoparticles (Table 3). In case of NANO 2, the solution based on non-ozonated water, was totally ineffective against all tested *L. monocytogenes* strains. Moreover, the OA 2 solution based on non-ozonated water was also ineffective against LMO-W, LMO-M and LMO-R strains (Table 3).

The MBC value of tested disinfectants determined for particular examined strains of *L. monocytogenes* were very similar, so the effect was not strain-dependent. None strain-dependent differences in MBC were stated in case of QAC 2, QAC 3, OA 1, OA 4 and NANO 2 (Table 3). In all cases the concentration of disinfectants equal MBC value cause decrease in bacteria number below the detection limit. In Table 4, the logarithmic decreases in bacteria number after using the disinfectant concentration directly preceding MBC (preMBC), for solution based on ozonated and non-ozonated water, are presented. For solutions of disinfectants prepared on the basis of ozonated water, the determined preMBC values were lower, and despite this the logarithmic decreases in the number of *L. monocytogenes* were found to be higher. The observed differences in decreases were, in most cases, statistically significant ($p \leq 0.05$) (Table 4).

Table 4. Logarithmic decreases in bacteria number after using the disinfectant concentration directly preceding MBC (preMBC).

Group of Disinfectants	Disinfectant	Water Type	Reduction in Bacteria Number (log CFU)					
			Strain					
			LMO-ATCC	LMO-W	LMO-M	LMO-N	LMO-R	LMO-K
Quaternary ammonium compounds	QAC 1	Initial no.	8.81	8.69	8.71	8.76	8.66	8.73
		Nonozonated	7.23 [a]	6.96 [a]	6.84 [a]	6.92 [a]	6.76 [a]	7.07 [a]
		Ozonated	8.28 [b]	8.00 [b]	7.88 [b]	7.97 [b]	7.80 [b]	8.12 [b]
	QAC 2	Initial no.	8.81	8.69	8.71	8.76	8.66	8.73
		Nonozonated	7.05 [a]	6.78 [a]	6.62 [a]	6.75 [a]	6.50 [a]	6.90 [a]
		Ozonated	7.93 [b]	7.65 [b]	7.62 [b]	7.49 [a]	7.36 [b]	7.77 [b]
	QAC 3	Initial no.	8.81	8.69	8.71	8.76	8.66	8.73
		Nonozonated	6.70 [a]	6.43 [a]	6.27 [a]	6.40 [a]	6.15 [a]	6.55 [a]
		Ozonated	7.49 [a]	7.22 [a]	7.06 [a]	7.19 [a]	6.93 [a]	7.34 [a]
Oxidizing agents	OA 1	Initial no.	8.81	8.69	8.71	8.76	8.66	8.73
		Nonozonated	6.52 [a]	6.26 [a]	6.10 [a]	6.22 [a]	5.98 [a]	6.37 [a]
		Ozonated	7.23 [b]	6.96 [a]	6.79 [a]	6.93 [a]	6.67 [a]	7.07 [a]
	OA 2	Initial no.	8.81	8.69	8.71	8.76	8.66	8.73
		Nonozonated	2.64 [a]	2.43 [a]	2.37 [a]	2.17 [a]	2.26 [a]	2.53 [a]
		Ozonated	5.11 [b]	4.87 [b]	4.59 [b]	4.82 [b]	4.70 [b]	4.98 [b]
	OA 3	Initial no.	8.81	8.69	8.71	8.76	8.66	8.73
		Nonozonated	6.61 [a]	6.35 [a]	6.18 [a]	6.31 [a]	6.06 [a]	6.46 [a]
		Ozonated	7.40 [a]	7.13 [a]	6.97 [a]	7.08 [a]	6.80 [a]	7.25 [a]
	OA 4	Initial no.	8.81	8.69	8.71	8.76	8.66	8.73
		Nonozonated	6.43 [a]	6.18 [a]	6.02 [a]	6.12 [a]	5.83 [a]	6.29 [a]
		Ozonated	7.35 [b]	7.04 [b]	6.90 [b]	7.00 [b]	6.75 [b]	7.10 [b]
Chlorine compounds	ChC 1	Initial no.	8.81	8.69	8.71	8.76	8.66	8.73
		Nonozonated	7.67 [a]	7.39 [a]	7.23 [a]	7.33 [a]	7.06 [a]	7.52 [a]
		Ozonated	8.55 [b]	8.30 [b]	8.24 [b]	8.10 [a]	7.91 [b]	8.33 [b]
	ChC 2	Initial no.	8.81	8.69	8.71	8.76	8.66	8.73
		Nonozonated	7.31 [a]	6.95 [a]	6.71 [a]	6.93 [a]	6.58 [a]	7.08 [a]
		Ozonated	7.85 [a]	7.56 [a]	7.30 [a]	7.55 [a]	7.41 [b]	7.65 [a]

Table 4. *Cont.*

Group of Disinfectants	Disinfectant	Water Type	Reduction in Bacteria Number (log CFU)					
			Strain					
			LMO-ATCC	LMO-W	LMO-M	LMO-N	LMO-R	LMO-K
Iodine compounds	IC 1	Initial no.	8.81	8.69	8.71	8.76	8.66	8.73
		Nonozonated	6.48 [a]	6.22 [a]	6.05 [a]	6.16 [a]	5.93 [a]	6.37 [a]
		Ozonated	7.54 [b]	7.26 [b]	7.17 [b]	7.31 [b]	7.03 [b]	7.38 [b]
	IC 2	Initial no.	8.81	8.69	8.71	8.76	8.66	8.73
		Nonozonated	6.49 [a]	6.21 [a]	6.03 [a]	6.15 [a]	5.90 [a]	6.33 [a]
		Ozonated	7.63 [b]	7.34 [b]	7.16 [b]	7.30 [b]	7.04 [b]	7.47 [b]
Nanoparticles	NANO 1	Initial no.	8.81	8.69	8.71	8.76	8.66	8.73
		Nonozonated	6.17 [a]	5.91 [a]	5.75 [a]	5.87 [a]	5.69 [a]	6.03 [a]
		Ozonated	7.05 [b]	6.79 [b]	6.62 [b]	6.75 [b]	6.50 [b]	6.90 [b]
	NANO 2	Initial no.	8.81	8.69	8.71	8.76	8.66	8.73
		Nonozonated	1.76 [a]	1.57 [a]	1.30 [a]	1.49 [a]	1.37 [a]	1.66 [a]
		Ozonated	6.03 [b]	5.75 [b]	5.87 [b]	5.91 [b]	5.63 [b]	6.17 [b]

QAC 1—benzyl-C12-18-alkydimethyl ammonium chlorides; QAC 2—benzyl-C12-16 alkyldimethyl chlorides; QAC 3—didecyldimethylammonium chloride, benzyl-C12-16-alkyldimethyl chlorides; OA 1—hydrogen peroxide, silver nitrate; OA 2—perlactic acid; OA 3—peracetic acid, hydrogen peroxide bis (sulphate) bis (peroxymonosulfate); OA 4—pentapotassium, benzenesulfonic acid, C10-13 alkyl derivatives, sodium salts, malic acid, sulfamic acid; ChC 1—chlorine dioxide; ChC 2—hypochlorous acid calcium salt; IC 1—iodine, IC 2—iodine; NANO 1—nanocopper; NANO 2—nanosilver; LMO-ATCC—*L. monocytogenes* ATCC 19111, LMO-W—strain isolated from vegetables, LMO-M—strain isolated from meat, LMO-N—strain isolated from dairy products, LMO-R—strain isolated from fish, LMO-K—clinical strain, a,b—values marked with different letters differ statistically significantly ($p \leq 0.05$) (Tested separately for each disinfectant and each strain depending on water type—ozonated/non-ozonated).

3.3. Coefficients of Effectiveness of Disinfectants

The values of the activity coefficient for all strains and disinfectants are shown in Table 5. These values for quaternary ammonium compounds ranged from 0.10 to 0.40, for oxidizing agents—from 0.15 to 0.84, for chlorine compounds—from 0.25 to 0.83, for iodine compounds—from 0.45 to 0.60 and for nanoparticles—from 0.70 to 0.84 (Table 5). It was shown, ozonated water had the greatest impact on the efficiency of the quaternary ammonium compounds whereas did not significantly improve the effectiveness of nanoparticles (Figure 2).

Table 5. Efficiency coefficient values for tested strains and disinfectants.

Disinfectant	Efficiency Coefficient (A)					
	Strain					
	LMO-ATCC	LMO-W	LMO-M	LMO-N	LMO-R	LMO-K
QAC 1	0.40 [a]	0.20 [b]	0.25 [b,g]	0.25 [b,g]	0.25 [b,g]	0.40 [a]
QAC 2	0.10 [c,f]	0.10 [c,f]	0.10 [c,f]	0.10 [c,f]	0.10 [c,f]	0.10 [c,f]
QAC 3	0.10 [c,f]	0.20 [b]	0.10 [c,f]	0.10 [c,f]	0.10 [c,f]	0.10 [c,f]
OA 1	0.84 [d]	0.84 [d]	0.84 [d]	0.84 [d]	0.84 [d]	0.84 [d]
OA 2	0.55 [e,i]	Ineffective	Ineffective	0.55 [e,i]	Ineffective	0.55 [e,i]
OA 3	0.22 [b]	0.22 [b]	0.23 [b]	0.17 [b,f]	0.15 [b,f]	0.22 [b]
OA 4	0.38 [a]	0.38 [a]	0.38 [a]	0.38 [a]	0.38 [a]	0.38 [a]
ChC 1	0.25 [b,g]	0.25 [b,g]	0.43 [a]	0.25 [b,g]	0.33 [a,g]	0.25 [b,g]
ChC 2	0.83 [d]	0.63 [e,h]	0.56 [e]	0.63 [e,h]	0.71 [h,j]	0.70 [h,j]
IC 1	0.49 [a,i]	0.60 [e,h]	0.60 [e,h]	0.60 [e,h]	0.60 [e,h]	0.49 [a,i]
IC 2	0.45 [a,i]	0.45 [a,i]	0.45 [a,i]	0.45 [a,i]	0.45 [a,i]	0.45 [a,i]
NANO 1	0.70 [h,j]	0.84 [d]	0.81 [d]	0.79 [d,j]	0.81 [d]	0.79 [d,j]
NANO 2	Ineffective	Ineffective	Ineffective	Ineffective	Ineffective	Ineffective

QAC 1—benzyl-C12-18-alkydimethyl ammonium chlorides; QAC 2—benzyl-C12-16 alkyldimethyl chlorides; QAC 3—didecyldimethylammonium chloride, benzyl-C12-16-alkyldimethyl chlorides; OA 1—hydrogen peroxide, silver nitrate; OA 2—perlactic acid; OA 3—peracetic acid, hydrogen peroxide bis (sulphate) bis (peroxymonosulfate); OA 4—pentapotassium, benzenesulfonic acid, C10-13 alkyl derivatives, sodium salts, malic acid, sulfamic acid; ChC 1—chlorine dioxide; ChC 2—hypochlorous acid calcium salt; IC 1—iodine, IC 2—iodine; NANO 1—nanocopper; NANO 2—nanosilver; LMO-ATCC—*L. monocytogenes* ATCC 19111, LMO-W—strain isolated from vegetables, LMO-M—strain isolated from meat, LMO-N—strain isolated from dairy products, LMO-R—strain isolated from fish, LMO-K—clinical strain, a–j—values marked with different letters differ statistically significantly ($p \leq 0.05$).

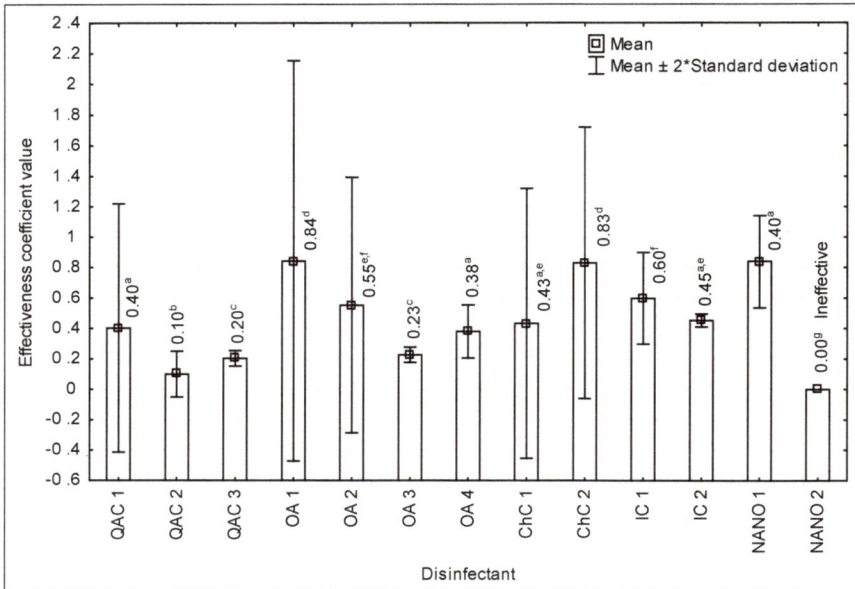

Figure 2. Values of efficacy coefficients for the tested disinfectants, including their belonging to distinguished groups (QAC—quaternary ammonium compounds, OA—oxidizing agents, ChC—chlorine compounds, IC—iodine compounds, NANO—nanoparticles); a–g—variables with different letters are statistically different ($p \leq 0.05$)

It was not possible to select the *L. monocytogenes* strain, for which the increase in the effectiveness of all tested disinfectants would be the greatest or the smallest after preparation of solutions with ozonated water (Table 5).

The lowest maximal value of the efficacy coefficient, and therefore the highest increase in microbicidal effectiveness after using ozonated water, was demonstrated for QAC 2 (0.10) and QAC 3 (0.20) and OA 3 (0.23). The values of the efficacy coefficient differ significantly ($p > 0.05$) between QAC 2 versus QAC 3 and OA 3 (Figure 2). The highest maximal efficacy coefficients, indicating a small improvement in the bactericidal effectiveness of solutions based on ozonated water in relation to solutions prepared with non-ozonated water, were found for OA 1 (0.84), ChC 2 (0.83) and NANO 1 (0.84). The above values of coefficients differed statistically significantly ($p \leq 0.05$) with the majority of values calculated for the remaining disinfectants (Figure 2).

3.4. Assessment of the Stability of QAC 2, OA 3, and ChC 1 Solutions

QAC 2, OA 3, and ChC 1 were chosen as the agents from different groups with the lowest maximal value of efficacy to assess the stability of the solutions. The stability results of the tested solutions are shown in Table 6.

It has been shown that during the storage of disinfectant solutions, their biocidal activity decreases against *L. monocytogenes* strains. Decrease in the biocidal activity of disinfectant solutions, which were prepared using both ozonated and non-ozonated water, was observed. The greater decrease in biocidal activity was visible for solutions prepared with the use of ozonated water (Table 6).

The activity of the QAC 2 solutions decreased over all tested strains after 24 h of storage, both for solutions with non-ozonated and ozonated water. In the case of OA 3 and ChC 1 disinfectants, a decrease in activity was observed for all tested strains of *L. monocytogenes* after 12 and 24 h of storage, for both type of water used for solutions preparation (Table 6).

Table 6. Stability assessment of stored solutions of QAC 2, OA 3, and ChC 1 after 0. 12 and 24 h.

Disinfectant	Water Type	Storage Time (h)	Minimal Bactericidal Concentration of Disinfectant [g/cm^3] Strain					
			LMO-ATCC	LMO-W	LMO-M	LMO-N	LMO-R	LMO-K
QAC 2 (working solution: 2.55 × 10^0 g/mL)	Nonozonated	0	1.28×10^{-4} a	1.28×10^{-4} a	1.28×10^{-4} a	1.28×10^{-4} a	1.28×10^{-4} a	1.28×10^{-4} a
		12	1.28×10^{-4} a	1.28×10^{-4} a	1.28×10^{-4} a	1.28×10^{-4} a	1.28×10^{-4} a	1.28×10^{-4} a
		24	1.53×10^{-4} a	1.53×10^{-4} a	1.53×10^{-4} a	1.53×10^{-4} a	1.53×10^{-4} a	1.53×10^{-4} a
	Ozonated	0	1.28×10^{-5} b	1.28×10^{-5} b	1.28×10^{-5} b	1.28×10^{-5} b	1.28×10^{-5} b	1.28×10^{-5} b
		12	1.02×10^{-4} a	1.28×10^{-5} b	1.28×10^{-5} b	1.28×10^{-5} b	1.28×10^{-5} b	1.02×10^{-4} a
		24	1.53×10^{-4} a	1.53×10^{-4} a	1.53×10^{-4} a	1.53×10^{-4} a	1.79×10^{-4} a	1.53×10^{-4} a
OA 3 (working solution: 6.15 × 10^{-3} g/mL)	Nonozonated	0	1.42×10^{-3} c	1.42×10^{-3} c	1.42×10^{-3} c	1.80×10^{-3} c	3.81×10^{-3} d,f	1.42×10^{-3} c
		12	1.66×10^{-3} c	1.72×10^{-3} c	2.40×10^{-3} c,d	2.20×10^{-3} c,d	4.30×10^{-3} e,f	1.60×10^{-3} c
		24	2.40×10^{-3} c,d	2.58×10^{-3} c,d	3.40×10^{-3} d	3.20×10^{-3} d	5.23×10^{-3} e	2.50×10^{-3} c,d
	Ozonated	0	3.10×10^{-4} g	3.10×10^{-4} g	4.30×10^{-4} g	3.10×10^{-4} g	6.15×10^{-4} h	2.46×10^{-4} g
		12	1.60×10^{-3} c	1.66×10^{-3} c	2.40×10^{-3} c,d	2.15×10^{-3} c,d	4.12×10^{-3} e,f	1.60×10^{-3} c
		24	2.40×10^{-3} c,d	2.65×10^{-3} c,d	3.40×10^{-3} d	3.9×10^{-3} d,f	5.10×10^{-3} e	2.50×10^{-3} c,d
ChC 1 (working solution: 1.00 × 10^{-5} g/mL)	Nonozonated	0	2.00×10^{-7} i	4.00×10^{-7} i,k	7.00×10^{-7} i	4.00×10^{-7} i,k	6.00×10^{-7} i,k	4.00×10^{-7} i,k
		12	3.00×10^{-7} i	6.00×10^{-7} i,k	8.00×10^{-7} i	5.00×10^{-7} i,k	8.00×10^{-7} i	5.00×10^{-7} i,k
		24	5.00×10^{-7} i,k	8.00×10^{-7} i	1.00×10^{-6} i	7.00×10^{-7} i	1.10×10^{-6} i	7.00×10^{-7} i
	Ozonated	0	5.00×10^{-8} m	1.00×10^{-7} i	3.00×10^{-7} i	1.00×10^{-7} i	2.00×10^{-7} i	1.00×10^{-7} i
		12	2.00×10^{-7} i	5.00×10^{-7} i,k	8.00×10^{-7} i	5.00×10^{-7} i,k	8.00×10^{-7} i	4.00×10^{-7} i,k
		24	5.00×10^{-7} i,k	8.00×10^{-7} i	1.00×10^{-6} i	8.00×10^{-7} i	1.10×10^{-6} i	7.00×10^{-7} i

QAC 2—benzyl-C12-16 alkyldimethyl chlorides; OA 3—peracetic acid, hydrogen peroxide bis (sulphate) bis (peroxymonosulfate); ChC 1—chlorine dioxide; LMO-ATCC—*L. monocytogenes* ATCC 19111, LMO-W—strain isolated from vegetables, LMO-M—strain isolated from meat, LMO-N—strain isolated from dairy products, LMO-R—strain isolated from fish, LMO-K—clinical strain, a–m—values marked with different letters differ statistically significantly ($p \leq 0.05$).

For all the disinfectants included in this part of the study, a gradual increase in the efficiency coefficient was observed during storage (Table 7). For OA 3 and ChC 1, after 12 h from preparation, the antilisterial effectiveness of the solution based on ozonated and non-ozonated water was almost identical. For QAC 2, this was observed after 24 h (Table 7). After 24 h, in the case of QAC 2 for LMO-R strain, OA 3 strains for LMO-W and LMO-N strains and ChC 1 for LMO-N strain, the effectiveness of solution based on ozonated water was lower than for those prepared on non-ozonated water (Table 7).

Table 7. Efficiency coefficient values for tested strains and disinfectants after storage of solutions.

Disinfectant	Storage Time (h)	Efficiency Coefficient (A)					
		Strain					
		LMO-ATCC	LMO-W	LMO-M	LMO-N	LMO-R	LMO-K
QAC 2 (working solution: 2.55×10^0 g/mL)	0	0.10 [a]	0.10 [a]	0.10 [a]	0.10 [a]	0.10 [a]	0.10 [a]
	12	0.80 [b]	0.10 [a]	0.10 [a]	0.10 [a]	0.10 [a]	0.80 [b]
	24	1.00 [b,c]	1.00 [b,c]	1.00 [b,c]	1.00 [b,c]	1.17 [c]	1.00 [b,c]
OA 3 (working solution: 6.15×10^{-3} g/mL)	0	0.22 [a]	0.22 [a]	0.23 [a]	0.17 [a]	0.16 [a]	0.17 [a]
	12	0.96 [b]	0.96 [b]	1.00 [b,c]	0.97 [b]	0.96 [b]	1.00 [b,c]
	24	1.00 [b,c]	1.02 [b,c]	1.00 [b,c]	1.06 [b,c]	0.98 [b,c]	1.00 [b,c]
ChC 1 (working solution: 1.00×10^{-5} g/mL)	0	0.25 [a,e]	0.25 [a]	0.43 [d]	0.25 [a,e]	0.33 [d,e]	0.25 [a,e]
	12	0.67 [f]	0.83 [b]	1.00 [b,c]	1.00 [b,c]	1.00 [b,c]	0.80 [b]
	24	1.00 [b,c]	1.00 [b,c]	1.00 [b,c]	1.14 [b,c]	1.00 [b,c]	1.00 [b,c]

QAC 2—benzyl-C12-16 alkyldimethyl chlorides; OA 3—peracetic acid, hydrogen peroxide bis (sulphate) bis (peroxymonosulfate); ChC 1—chlorine dioxide; LMO-ATCC—*L. monocytogenes* ATCC 19111, LMO-W—strain isolated from vegetables, LMO-M—strain isolated from meat, LMO-N—strain isolated from dairy products, LMO-R—strain isolated from fish, LMO-K—clinical strain, a–f—values marked with different letters differ statistically significantly ($p \leq 0.05$).

4. Discussion

Ozone is one of the strongest disinfectants, which are active after a short contact time and in low concentration. It has biocidal effect on many types of microorganisms [4]. Muthukumar and Muthuchama's [20] studies confirmed a decrease in the number of microorganisms by 2.00×10^6 CFU in 1g of raw chicken samples. While Sheelamary and Muthukumar [21] showed complete inactivation of microorganisms isolated from samples of milk and its products after only 15 min with emission of 0.2 g O_3/h.

In the literature, research work on the evaluation of the effectiveness of gas ozone in combination with various physical methods has been found [22–25]. Sung et al. [23] examined the action of gas ozone and high temperature on inactivation of *L. monocytogenes* present in apple juice. The synergistic effect was obtained when the temperature was 50 °C. Kumar et al. [25] during a 10-min exposure to ozone gas and UV showed a decrease in the number of *L. monocytogenes* in fresh brine by more than 9 log CFU/mL, and hourly ozonation combined with ten-minute UV irradiation resulted in a reduction of microbes over 5 log CFU/mL in used brines. The effect of low concentrations of ozone and metal ions in the reduction of *L. monocytogenes* was studied by Kang et al. [26], who showed that the use of ozone in concentrations of 0.2 and 0.4 ppm in combination with 1 mM $CuCl_2$ and 0.1 mM $AgNO_3$ by 30 min is much more effective than using ozonated water ($p < 0.05$) only. Marino et al. [27] evaluated the effect of gas ozone and ozone in water on the survival of *Pseudomonas fluorescens*, *Staphylococcus aureus*, and *L. monocytogenes* cells in the biofilm structure on the surface of stainless steel. They showed that the use of ozone in the aqueous solution affected the reduction of the number of bacteria by 1.61–2.14 log CFU/cm² after 20 min exposure, while the reduction values were higher (3.26–5.23 log CFU/cm²) in the case of biofilms treated with ozone under dynamic flow conditions. They also showed that *S. aureus* was the most sensitive species for ozone dissolved in water [27]. Korany et al. [28] showed that the use of ozonated water (1 min) against the structure of *L. monocytogenes* biofilm on the surface of polystyrene at the concentration of 1.0, 2.0, and 4.0 ppm resulted in a bacterial number reduction of 0.9, 3.4, and 4.1 log CFU/cm², respectively. Moreover Korany et al. [28] found that quaternary ammonium compounds (QAC) (100/400 ppm), chlorine (100/200 ppm), chlorine dioxide (2.5/5.0 ppm) and peracetic acid (PAA)

(80/160 ppm) resulted in a reduction of 2.4/3.6, 2.0/3.1, 2.4/3.8, and 3.6/4.8 log CFU/cm^2, respectively. The antimicrobial efficacy of all tested disinfectants against the 7-day *L. monocytogenes* biofilm was significantly lower compared to 2-day biofilms, and the biofilm age having the minor impact on the effectiveness of PAA [28].

To date, information on the synergistic effect of ozonated water and disinfectants was not found. However, it should be expected that the implementation of a solution of disinfectants based on ozonated water will increase the effectiveness of the tested agents, at least up to the level of additive action.

In this study, the biocidal efficacy of ozonated water and non-ozonated water was evaluated. Non-ozonated water was characterized by a complete lack of biocidal activity, against all tested strains of *L. monocytogenes*. The ozonated water, used immediately after preparation, containing 1.86–1.96 µg O$_3$/cm^3, was effective for all tested isolates. The LMO-M and LMO-R strains were characterized by the highest resistance (MBC: 1.96 µg/cm^3), and the LMO-ATCC strain (MBC: 1.86 µg/cm^3) was the most sensitive. Fishburn et al. [29] showed that the use of ozonated water in vegetable washing can cause a decrease in the number of *L. monocytogenes* strains by about 0.5 log CFU/g (broccoli, lettuce) to 1.5 log CFU/g (green onion), in relation to washing in non-ozonated water. Larivière-Gauthier et al. [30] have shown that the use of ozonated water containing 3.5 ppm ozone improves the efficiency of cleaning and disinfection in a pork cutting plant, increasing the proportion of free surfaces from *L. monocytogenes* by 12.5%. Arayan et al. [11] showed that the lowest biocidal concentration against *Staphylococcus aureus* (strains isolated from food products) was 0.5 ppm of ozone with an exposure time of 0.1 min. The addition of organic pollutants such as fetal bovine serum (FBS) resulted in a decrease in the biocidal effectiveness of the ozonated water [11]. Also Korany et al. [28] showed lower efficacy of disinfectants, including ozonated water, in case of the presence of organic contamination (diluted milk and apple juice) in the environment.

The results of our study showed that after 1 and 2 h of storage of ozonated water its activity decreased, as evidenced by the need to use higher concentrations of ozone to eliminate *L. monocytogenes* strains after longer storage period of ozonated water. This was probably related to the decomposition of ozone, which is unstable and has a water stability of 20–40 min. This is confirmed by the studies of Białoszewski et al. [12] who used ozonated water after 30 min of its preparation and observed a decrease in ozone concentration from the initial 2.5–3.0 µg/mL to 1.3–1.5 µg/mL. Despite the decrease in ozone content, they did not show a decrease in the biocidal effectiveness of water. Research carried out by Seki et al. [10] showed that storage of ozonated water at 4 °C and 25 °C significantly affected its durability. The highest decrease (90.0%) in ozone concentration was noticed after one week of storage at 25 °C. However, tests carried out at 4 °C by Seki et al. [10] showed that after one week the concentration of ozone in water maintained at the level of above 90%, after a month about 65%, and after one year of storage ozone was not found in water. The effect of biocidal ozonated water stored at 4°C against *Escherichia coli* has been confirmed. Seki et al. [10] also assessed the effect of freezing and thawing on the durability of ozone in water. They showed that the first freeze-thaw cycle did not affect the ozone concentration in water, but after four cycles it was found that the ozone concentration decreased to about 90% [10].

This study allowed to assess and compare the biocidal efficacy of solutions of thirteen selected disinfectants made using non-ozonated water and ozonated water against *L. monocytogenes* strains. It was shown that disinfectants based on ozonated water were characterized by higher microbicidal effectiveness, compared to solutions based on non-ozonated water and were more effective than ozonated water alone. This shows the synergistic effect of ozonated water and disinfectants. This may be related to the same target site of the test compound and ozone in bacterial cells. For example, QACs or peracetic acid act similarly to ozone, destabilizing the bacterial cell membrane, making it more permeable [31,32]. The synergistic mechanism of action can also be the result of influence of ozone and the active substance of the disinfectants on other structures of the bacterial cell, which will result in an increase of microbicidal effect. Oxidizing compounds (e.g., hydrogen peroxide, hypochlorous acid, and peracetic acid) cause the oxidation of thiol groups of cysteine residues, which are often found in the active sites of many bacterial enzymes, such as, for example, dehydrogenases.

It was found that the use of ozonated water for solution preparation, improves the microbicidal properties of all tested disinfectants, with the exception of nanosilver. The greatest impact of ozonated water on disinfectant effectiveness was noted for quaternary ammonium compounds. This type of disinfectants displayed the highest biocidal efficacy against tested isolates. The strong biocidal activity of quaternary ammonium compounds against *L. monocytogenes* strains is confirmed by the results of Chavant et al. [33], which showed 98% effectiveness in eliminating planktonic forms. In contrast, Aarnisalo et al. [34] have shown that quaternary ammonium compounds are characterized by lower biocidal efficacy against *L. monocytogenes* strains than disinfectants based on chlorine, ethanol, isopropanol and peracetic acid. In our study, the LMO-R strain was the most resistant to the effect of NANO 1, OA 2, and OA 3 preparations belonging to oxidizing compounds, and the LMO-M strain was the most resistant to the ChC 1 compound from the group of chlorine derivatives. The LMO-ATCC strain was characterized by the highest sensitivity against ChC 1. Heir et al. [35] and Popowska et al. [36] showed different sensitivity of *L. monocytogenes* to disinfectants depending on the origin of the strain.

The conducted studies also evaluated the durability of solutions with the lowest efficiency index for QAC 2, OA 3, and ChC 1. The decrease in the biocidal activity of disinfectant solutions prepared using non-ozonated water may suggest that they are gradually inactivated during storage.

The results of our study show that ozonated water have microbicidal properties. Since ozonated water and disinfectants show a synergistic effect of a biocidal action its combination can be used to more effectively eradicate microorganisms. Moreover ozonated water can reduce the working concentration of disinfectants. However, it is important to remember about the short half-life of ozone and to use it directly or in a short time after the preparation of appropriate disinfecting solutions.

To elucidate the mechanism of joint action of disinfectants and ozonated water on the disruption of bacterial cell structure further research are needed. In our study, the Columbia Agar with 5% sheep blood was used, which is a non-selective medium. The use of such a medium provided suitable conditions for the regeneration of sub-lethal injured cells. Presence of selective and differential agents in a medium could inhibit the growth of such cells relative to non-selective medium and thus counts obtained from treated samples enumerated on selective media are typically lower than those obtained from non-selective media. This tendency is confirmed by studies by Fouladkhah et al. [37] who showed that when assessing the growth of bacteria on selective medium on day 0 the number of colonies ranged from 1.5 ± 0.8 to 2.0 ± 0.8 log CFU/cm^2 and increased to values from 2.9 ± 0.5 to 4.3 ± 0.4 log CFU/cm^2 on 7 day. In turn, the results obtained after cultivation on non-selective media ranged from 2.0 ± 0.5 to 2.3 ± 0.4 log CFU/cm^2 on day 0 and from 6.4 ± 0.6 to 7.1 ± 0.4 log CFU/cm^2 on day 7 [37].

5. Conclusions

We can state that the ozonated water, in contrast to non-ozonated water, showed biocidal efficacy against examined *L. monocytogenes* strains though its activity decreases over time. Moreover, it was demonstrated that ozonated water improved the biocidal effectiveness of disinfecting agents. Among the tested solutions of disinfectant, the most effective group were quaternary ammonium compounds and chlorine compounds. The lowest biocidal activity against the tested strains of *L. monocytogenes* were characterized by nanoparticles. The biocidal activity of disinfectants against tested *L. monocytogenes* strains decreases during storage regardless of the disinfectant type.

Author Contributions: Conceptualization: K.S. and E.W.-Z.; Methodology: K.S., E.W.-Z., and M.K.; Validation: K.S., E.G.-K., and K.G.; Formal Analysis: S.K..; Investigation: K.G., A.B., N.W., A.B., and M.K.; Resources: K.S., E.W.-Z., and E.G.-K.; Data Curation: S.K., K.G., N.W., and A.B.; Writing—Original Draft Preparation: K.G., N.W., A.B., A.B., and M.K.; Writing—Review and Editing: K.S., E.W.-Z., and S.K.; Visualization: N.W., K.G., and A.B.; Supervision: K.S. and E.W.-Z.; Project Administration: K.S. and E.G.-K.; Funding Acquisition: E.G.-K.

Funding: This research was financially supported by the Nicolaus Copernicus University with funds from the maintenance of the research potential of the Department of Microbiology DS-UPB.

Conflicts of Interest: The authors declare no conflict of interest.

References

1. Orsi, R.H.; Wiedmann, M. Characteristics and distribution of *Listeria* spp., including *Listeria* species newly described since 2009. *Appl. Microbiol. Biotechnol.* **2016**, *100*, 5273–5287. [CrossRef] [PubMed]
2. European Food Safety Authority (EFSA). The European Union summary report on trends and sources of zoonoses, zoonotic agents and food-borne outbreaks in 2017. *EFSA J.* **2018**, *16*, 5500.
3. Książczyk, M.; Krzyżewska, E.; Futoma-Kołoch, B.; Bugla-Płoskońska, G. Oddziaływanie związków dezynfekcyjnych na komórki bakteryjne w kontekście bezpieczeństwa higieny i zdrowia publicznego. *Postępy Hig. Med. Dośw.* **2015**, *69*, 1042–1055.
4. Demir, F.; Atguden, A. Experimental Investigation on the Microbial Inactivation of Domestic Well Drinking Water using Ozone under Different Treatment Conditions. *Ozone-Sci. Eng.* **2016**, *38*, 25–35. [CrossRef]
5. Wysok, B.; Uradziński, J.; Gomółka-Pawlicka, M. Ozone as an alternative disinfectant—A review. *Pol. J. Food Nutr. Sci.* **2006**, *15*, 3–8.
6. Alexopoulos, A.; Plessas, S.; Ceciu, S.; Lazar, V.; Mantzourani, I.; Voidarou, C.; Stavropoulou, E.; Bezirtzoglou, E. Evaluation of ozone efficacy on the reduction of microbial population of fresh cut lettuce (*Lactuca sativa*) and green bell pepper (*Capsicum annuum*). *Food Control* **2013**, *30*, 491–496. [CrossRef]
7. Elvis, A.M.; Ekta, J.S. Ozone therapy: A clinical review. *J. Nat. Sci. Biol. Med.* **2011**, *2*, 66–70. [CrossRef]
8. Thanomsub, B.; Anupunpisit, V.; Chanphetch, S.; Watcharachaipong, T.; Poonkhum, R.; Srisukonth, C. Effects of ozone treatment on cell growth and ultrastructural changes in bacteria. *J. Gen. Appl. Microbiol.* **2002**, *48*, 193–199. [CrossRef]
9. Khadere, M.A.; Yousef, A.E.; Kim, J.-G. Microbiological aspects of ozone applications in food: A review. *J. Food Sci.* **2001**, *66*, 1242–1252. [CrossRef]
10. Seki, M.; Ishikawa, T.; Terada, H.; Nashimoto, M. Microbicidal Effects of Stored Aqueous Ozone Solution Generated by Nano-bubble Technology. *In Vivo* **2017**, *31*, 579–583. [PubMed]
11. Arayan, L.T.; Reyes, A.W.B.; Hop, H.T.; Xuan, H.T.; Yang, H.S.; Chang, H.H.; Kim, S. Optimized applications of ozonated water as an effective disinfectant for *Staphylococcus aureus* clearance in an abattoir setting. *J. Prev. Vet. Med.* **2017**, *41*, 71–74. [CrossRef]
12. Białoszewski, D.; Bocian, E.; Bukowska, B.; Czajkowska, M.; Sokół-Leszczyńska, B.; Tyski, S. Antimicrobial activity of ozonated water. *Med. Sci. Monit.* **2010**, *16*, 71–75.
13. European Committee on Antimicrobial Susceptibility Testing (2018) Breakpoints Tables for Interpretation of MICs and Zones Diameters. Version 8.0. Available online: http://www.eucast.org (accessed on 1 March 2019).
14. PN-EN 1276:2010. Chemiczne środki dezynfekcyjne i antyseptyczne—Ilościowa zawiesinowa metoda określania działania bakteriobójczego chemicznych środków dezynfekcyjnych i antyseptycznych stosowanych w sektorze żywnościowym, warunkach przemysłowych i domowych oraz zakładach użyteczności publicznej - metoda badania i wymagania (faza 2, etap 1). (In Polish). Available online: http://sklep.pkn.pl/pn-en-1276-2010e.html (accessed on 1 March 2019).
15. International Organization for Standardization (ISO). *Water Quality—Determination of Free Chlorine and Total Chlorine—Part 2: Colorimetric Method Using N,N-dialkyl-1,4-phenylenediamine, for Routine Control Purposes*; ISO: Geneva, Switzerland, 2017.
16. Kasprzyk-Hordern, B.; Ziółek, M.; Nawrocki, J. Catalytic ozonation and methods of enhancing molecular ozone reactions in water treatment. *Appl. Catal. B Environ.* **2003**, *46*, 639–669. [CrossRef]
17. American Public Health Association (APHA). Technical Committee on Microbiological Methods for Foods. In *Compendium of Methods for the Microbiological Examination of Foods*; APHA: Washington, DC, USA, 1984.
18. U.S. Food and Drug Administration (FDA). *Bacteriological Analytical Manual*; AOAC: Arlington, VA, USA, 2017.
19. Skowron, K.; Grudlewska, K.; Krawczyk, A.; Gospodarek-Komkowska, E. The effectiveness of radiant catalytic ionization in inactivation of *Listeria monocytogenes* planktonic and biofilm cells from food and food contact surfaces as a method of food preservation. *J. Appl. Microbiol.* **2018**, *124*, 1493–1505. [CrossRef]
20. Muthukumar, A.; Muthuchamy, M. Optimization of ozone in gaseous phase to inactive *Listeria monocytogenes* on raw chicken samples. *Food Res. Int.* **2013**, *54*, 1128–1130. [CrossRef]

21. Sheelamary, M.; Muthukumar, M. Effectiveness of Ozone in Inactivating *Listeria monocytogenes* from Milk Samples. *World J. Young Res.* **2011**, *1*, 40–44.

22. Patil, S.; Valdramidis, V.; Frias, J.M.; Cullen, P.; Bourke, P. Ozone inactivation of acid stressed *Listeria monocytogenes* and *Listeria innocua* in orange juice using a bubble column. *Food Control* **2010**, *21*, 1723–1730. [CrossRef]

23. Sung, H.J.; Song, W.J.; Kim, K.W.; Ryu, S.; Kang, D.H. Combination effect of ozone and heat treatments for the inactivation of *Escherichia coli* O157:H7, *Salmonella Typhimurium*, and *Listeria monocytogenes* in apple juice. *Int. J. Food Microbiol.* **2013**, *171*, 147–153. [CrossRef] [PubMed]

24. Song, W.J.; Shin, J.Y.; Ryu, S.; Kang, D.H. Inactivation of *Escherichia coli* O157:H7, *Salmonella Typhimurium* and *Listeria monocytogenes* in apple juice at different pH levels by gaseous ozone treatment. *J. Appl. Microbiol.* **2015**, *119*, 465–474. [CrossRef] [PubMed]

25. Kumar, G.D.; Williams, R.C.; Sumner, S.S.; Eifert, J.D. Effect of ozone and ultraviolet light on *Listeria monocytogenes* populations in fresh and spent chill brines. *Food Control* **2016**, *59*, 172–177. [CrossRef]

26. Kang, S.N.; Kim, K.J.; Park, J.H.; Lee, O.H. Effect of a combination of low level ozone and metal ions on reducing *Escherichia coli* O157:H7 and *Listeria monocytogenes*. *Molecules* **2013**, *18*, 4018–4025. [CrossRef]

27. Marino, M.; Maifreni, M.; Baggio, A.; Innocente, N. Inactivation of Foodborne Bacteria Biofilms by Aqueous and Gaseous Ozone. *Front. Microbiol.* **2018**, *9*, 2024. [CrossRef]

28. Korany, A.M.; Hua1, Z.; Green, T.; Hanrahan, I.; El-Shinawy, S.H.; El-Kholy, A.; Hassan, G.; Zhu, M.J. Efficacy of Ozonated Water, Chlorine, Chlorine Dioxide, Quaternary Ammonium Compounds and Peroxyacetic Acid Against *Listeria monocytogenes* Biofilm on Polystyrene Surfaces. *Front. Microbiol.* **2018**, *9*, 2296. [CrossRef]

29. Fishburn, J.D.; Tang, Y.; Frank, J.F. Efficacy of Various Consumer-Friendly Produce Washing Technologies in Reducing Pathogens on Fresh Produce. *Food Protect. Trends* **2011**, *32*, 456–466.

30. Lariviere-Gauthier, G.; Letellier, A.; Quessy, S.; Fournaise, S.; Fravalo, P. Assessment of the efficiency of ozonated water as bacterial contamination reduction tool in a pork cutting plant. *SafePork* **2013**, *42*, 143–146.

31. Ioannou, C.J.; Hanlon, G.W.; Denyer, S.P. Action of disinfectant quaternary ammonium compounds against *Staphylococcus aureus*. *Antimicrob. Agents Chemother.* **2007**, *51*, 296–306. [CrossRef]

32. Leggett, M.J.; Schwarz, J.S.; Burke, P.A.; McDonnell, G.; Denyer, S.P.; Maillard, J.-Y. Mechanism of sporicidal activity for the synergistic combination of peracetic acid and hydrogen peroxide. *Appl. Environ. Microbiol.* **2016**, *82*, 1035–1039. [CrossRef]

33. Chavant, P.; Gaillard- Martine, B.; Hébraud, M. Antimicrobial effects of sanitizers against planktonic and sessile *Listeria monocytogenes* cells according to the growth phase. *FEMS Microbiol. Lett.* **2004**, *236*, 241–248. [CrossRef]

34. Aarnisalo, K.; Lundén, J.; Korkeala, H.; Wirtanen, G. Susceptibility of *Listeria monocytogenes* strains to disinfectants and chlorinated alkaline cleaners at cold temperatures. *LWT* **2008**, *40*, 1041–1048. [CrossRef]

35. Heir, E.; Lindstedt, B.A.; Røtterud, O.J.; Vardund, T.; Kapperud, G.; Nesbakken, T. Molecular epidemiology and disinfectant susceptibility of *Listeria monocytogenes* from meat processing plants and human infections. *Int. J. Food Microbiol.* **2004**, *96*, 85–96. [CrossRef]

36. Popowska, M.; Olszak, M.; Markiewicz, Z. Susceptibility of *Listeria monocytogenes* strains isolated from dairy products and frozen vegetables to antibiotics inhibiting murein synthesis and to disinfectants. *Pol. J. Microbiol.* **2006**, *55*, 279–288. [PubMed]

37. Fouladkhah, A.; Geornaras, I.; Sofos, J.N. Biofilm formation of O157 and Non-O157 Shiga toxin-producing *Escherichia coli* and multidrug-resistant and susceptible *Salmonella* Typhimurium and Newport and their inactivation by sanitizers. *J. Food Sci.* **2013**, *78*, 880–886. [CrossRef] [PubMed]

microorganisms

MDPI

Review

Development of Salmonellosis as Affected by Bioactive Food Compounds

Ajay Kumar [1,*], Abimbola Allison [2], Monica Henry [2], Anita Scales [2] and
Aliyar Cyrus Fouladkhah [2,3,*]

[1] Division of Gastroenterology, Hepatology and Nutrition, Department of Pediatrics, University of Virginia School of Medicine, Charlottesville, VA 22908, USA
[2] Public Health Microbiology Laboratory, Tennessee State University, Nashville, TN 37209, USA; abimbolaallison20@gmail.com (A.A.); mhenry3@my.tnstate.edu (M.H.); ascales3@my.tnstate.edu (A.S.)
[3] Cooperative Extension Program, Tennessee State University, Nashville, TN 37209, USA
* Correspondence: KumarAjay.Healthsciences@gmail.com (A.K.); afouladk@tnstate.edu or aliyar.fouladkhah@aya.yale.edu (A.C.F.); Tel.: +1-434-924-8960 (A.K.); +1-970-690-7392 (A.C.F.)

Received: 30 August 2019; Accepted: 17 September 2019; Published: 18 September 2019

Abstract: Infections caused by *Salmonella* serovars are the leading cause of foodborne hospitalizations and deaths in Americans, extensively prevalent worldwide, and pose a considerable financial burden on public health infrastructure and private manufacturing. While a comprehensive review is lacking for delineating the role of dietary components on prevention of Salmonellosis, evidence for the role of diet for preventing the infection and management of Salmonellosis symptoms is increasing. The current study is an evaluation of preclinical and clinical studies and their underlying mechanisms to elaborate the efficacy of bioactive dietary components for augmenting the prevention of *Salmonella* infection. Studies investigating dietary components such as fibers, fatty acids, amino acids, vitamins, minerals, phenolic compounds, and probiotics exhibited efficacy of dietary compounds against Salmonellosis through manipulation of host bile acids, mucin, epithelial barrier, innate and adaptive immunity and gut microbiota as well as impacting the cellular signaling cascades of the pathogen. Pre-clinical studies investigating synergism and/or antagonistic activities of various bioactive compounds, additional randomized clinical trials, if not curtailed by lack of equipoise and ethical concerns, and well-planned epidemiological studies could augment the development of a validated and evidence-based guideline for mitigating the public health burden of human Salmonellosis through dietary compounds.

Keywords: dietary bioactive components; salmonellosis; bile acids; epithelial barrier; gut microbiota

1. Introduction

Despite increased awareness and development of treatments such as antimicrobial interventions in manufacturing and antibiotic therapies in healthcare facilities for over a hundred years, *Salmonella* serovars are still a major concern in infectious diseases related premature morbidity and mortality [1–4]. Various serovars of *Salmonella* are the leading cause of foodborne hospitalizations and deaths in Americans causing over one million, about 20,000, and 378 annual illness, hospitalization, and deaths episodes, respectively [5]. Non-typhoidal *Salmonella* serovars are also the leading agent among most common foodborne infectious diseases, responsible for highest number (32,900 years) of disability adjusted life year (DALY), annually [6]. National Antimicrobial Resistance Monitoring System (NARMS) and other epidemiological sampling also reveal a widespread presence of multiple drug resistance (MDR) phenotypes of the pathogen in various facilities—as an example, 0.6% of ground meat samples may harbor MDR *Salmonella* [7] with approximately 7% of them displaying MDR-AmpC phenotype [8].

As such, the U.S. Department of Health and Human Services had categorized non-typhoidal *Salmonella* as a "serious threat" to the public health [9].

Data from world population also indicate that the pathogen is one of the leading causes of deaths associated with diarrheal diseases globally, with estimated 3.4 million cases (invasive non-typhoidal *Salmonella* serovars) and over 600,000 deaths annually [10,11]. The bacterium is a Gram-negative organism with a complicated and evolving nomenclature, currently consist of two species, at least six sub-species, and over 2500 serovars [12].

Changes in production and manufacturing practices, increased international commerce and travel, increased proportion of at-risk populations for infectious diseases, and changes in population's eating habits during last few decades had contributed to increased incidences of *Salmonella* infections [13,14]. *Salmonella* serovars induce acute inflammation in the intestinal track after infection and utilizes the environment to further proliferate and colonize [15–17]. Colonization resistance against *Salmonella* is modulated by gut microflora, intestinal immunity, epithelium, and quality and quantity of digestive fluids. Various food components have been shown to modulate these factors and could be a potential intervention for reducing the likelihood of enteric infections.

Over the past 20 years, role of the dietary agents in shaping immunity against enteric infections has becoming increasingly evident [18–22] and piqued the interest in nutritional interventions for enteric infections. Several dietary components ranging from polyphenolic compounds, fibers, micronutrients, fatty acids, peptides, and carbohydrates of plant and animal origin had been shown efficacious against *Salmonella* serovars in various experimental models [22–32]. These associations are the result of an array of potential biochemical pathways, very complex and dynamic in nature, including interactions among dietary components, gut epithelium, digestive system, immune system and gut microbiota as affected by various seasons [33–37]. Better understanding of these underlying mechanisms could reduce *Salmonella* prevalence in the food chain though modifications in food animal diets. It could further reduce the public health burden of non-typhoidal *Salmonella* serovars by mitigating severe symptoms and reducing the pathogen DALY and mortality rate in healthcare facilities for Salmonellosis patients. Hence, the current work is a review of *Salmonella* infection studies as affected by various dietary components with discussions of the mechanisms of action and types of preclinical, animal models, and clinical studies employed.

2. Current Status of Knowledge

2.1. Effect of Dietary Components against Salmonella: In-Vitro Models

Dietary components may prevent infection outcome by directly affecting the pathogen multiplication and virulence [7,38,39] or by modulating host response to the pathogens [26,37]. To test the direct effects of dietary components on pathogens, researchers have used food extracts or dietary bioactive components on *Salmonella* cultures. Summary of potential relationships between dietary components and *Salmonella* infections are presented in Figure 1. Following treatment of dietary components, multiplication, and gene expression for virulence and motility could be measured. These models are comparatively less expensive and less cumbersome to assess the efficacy of dietary components for *Salmonella* infection. For instance, several essential oils were added to *Salmonella* growth media at various doses and *Salmonella* multiplication was compared with untreated controls [38]. Among 28 tested essential oils, *Origanum heracleoticum*, *Cinnamomum cassia*, *Corydothymus capitatus*, *Satureja montana*, and *Cinnamomum verum* were particularly effective against *Salmonella* Typhimurium [38]. Citrus flavonoids were similarly evaluated on *Salmonella* virulence gene expression [39]. In the study, Naringenin, a flavanone present in grapefruit, repressed 24 genes in pathogenicity island of *Salmonella* Typhimurium LT2 and further down-regulated 17 genes associated with the pathogen motility [39]. Most recent studies also reveal similar trends, for example, various essential oils extracted from *Aloysia triphylla*, *Cinnamomum zeylanicum*, *Cymbopogon citratus*, *Litsea cubeba, Mentha piperita*, and *Syzygium aromaticum* had been shown to be efficacious against *Salmonella* serovars during in vitro challenge studies [40].

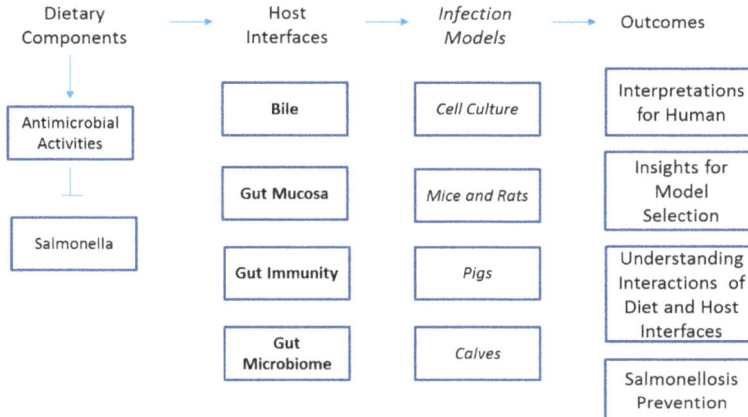

Figure 1. Relationships between dietary bioactive components and *Salmonella* infection. Dietary bioactive components such as fiber, amino acids, vitamins and minerals, fatty acids, and polyphenols improve the gut epithelium, microbiota, and immunity that may eventually lead to increased resistance to *Salmonella* infection.

It is noteworthy that aforementioned studies are conducted without host interaction and interpretation and generalization of the results should be drawn with caution and after further investigations in presence of host cells.

Orally infected *Salmonella* can enter circulation through various routes. It can invade several phagocytic and non-phagocytic cells depending upon serotype. In the murine model, *Salmonella* invades both phagocytic and epithelial non-phagocytic cell types. Hence, in vitro models of *Salmonella* entry have been developed to assess the effect of a test compound on a host. The *Salmonella* entry model could reveal the mechanism of action of a test compound on an organism. Several human and mouse cell lines such as Caco-2 [41] and RAW264.7 have been used in the literature to test efficacy of the compounds against *Salmonella* entry. For example, secretory immunoglobulin A (SIgA) was demonstrated to be a potent inhibitor for *Salmonella* Typhimurium entry into polarized monolayers of HeLa cells [42].

Salmonella contains the pathogenicity islands for the secretion of effector molecules to infect the target cells [43]. The molecules released by these secretory systems change the host cell cytoskeleton to facilitate *Salmonella* entry. The in vitro *Salmonella* entry models are impactful in studying the effects of dietary components on *Salmonella* as well as on host cells. However, these studies do not represent involvement of all host cell types that are simultaneously present in gastrointestinal area of humans. Dietary components could affect *Salmonella* virulence by affecting secretory systems or by competing with *Salmonella* for the receptors on host cells [39]. Host cells can also release the cytokines in response to the dietary components that can affect *Salmonella* virulence or motility. Therefore, in vitro models of *Salmonella* infection can have great implications for assessing mechanisms of actions by the dietary components.

2.2. Summary of Effect of Dietary Components on Salmonella Infection in Rodent Models

The fecal shedding of *Salmonella*, tissue colonization, local and systemic inflammatory changes, survival and weight reduction are the major observable changes associated with *Salmonella* infections in rodents. Bovee-Oudenhoven et al. showed reduced *Salmonella* fecal shedding when fructooligosaccharides were fed to the male Wistar rats as compared to the cellulose-fed group after 2 weeks of dietary intervention [44]. Furthermore, dietary fructooligosaccharides increased fecal *Lactobacilli* count and increased the translocation of *Salmonella* to the liver and spleen with an increase in fecal mucin as compared to cellulose fed rats [19]. The author concluded that dietary fructooligosaccharides decreased *Salmonella* colonization but increased the translocation potentially due to irritation of mucosal membrane. Some of the mice strains succumb easily to *Salmonella* infection and hence survival rate is the primary indicator of the dietary efficacy against infection. Hitchins et al. showed that feeding of freeze dried yoghurt to male weanling Sprague-Dawley rats increased overall survival rate and weight of the animals after intraperitoneal *Salmonella* challenge as compared to rats fed on milk diet for 1 week [45]. Similarly, dietary feeding of Herba Pogostemonis extract to Balb/c mice increased the overall survival rate as compared to control diet fed animals after intraperitoneal *Salmonella* challenge [46]. Feeding of Herba Pogostemonis (*Pogostemon cablin* Bantham extract) also reduced *Salmonella* liver damage as compared to control diet fed animals [46]. Recent studies similarly show association among various bioactive food compounds and prevention of Salmonellosis. Supplementing the diet of albino rats with olive oil, as an example, had been shown to have efficacy against *Salmonella* Typhi as a natural antimicrobial and non-toxic immune modulator [47]. These studies show that there are measurable markers for *Salmonella* infections in rodents and they can be used as a model to mimic *Salmonella* infections in human host.

Both foodborne pathogens and dietary components pass through the stomach acid, when ingested. Hence, gastric acidity is one of the important factors in determining stability of enteric pathogens. In a randomized controlled clinical trial, gastric hypochlorhydria (low hydrochloric acid) was found to be associated with increased *Salmonella* infections [48]. This hypothesis was also confirmed in the rodent model of *Salmonella* infection. Tennant et al. [49] showed that treatment of mice with antacids resulted in the decreased infectious dose of *Salmonella* as compared to normal mice.

Similar results were also observed in a constitutively hypochlorhydric mice (proton pump mutation) as compared to the normal mice [49]. Additionally, gastric pH not only affects the survival of pathogens but also affects digestion and absorption of foods. Lucas et al. showed that an increase in pH from 1.5 to 2.5 reduced digestion of the kiwifruit peptides [50]. Gastric pH also modulates absorption of micronutrients such as zinc. Henderson et al. observed higher plasma zinc levels in the young healthy volunteers at low pH as compared to plasma level in higher gastric pH volunteers [51]. The gastric pH is considerably different across species. For instance, mean gastric pH in mice is 3.1–4.5 and in rats ranges from 3.2 to 3.9, whereas in the humans it is 1.5–3.5. In addition to gastric pH, intestinal pH is also different in rodents as compared to humans. Mice and rats have a mean intestinal pH of 5.2 and 6.6, respectively, as compared to 7.2 in humans [52]. These studies show that gastric and intestinal pH could potentially affect bioactivity of dietary components and should be considered as one of the important factors in selecting a study model.

In rodents and humans, several disease symptoms can be confounding due to the differences in their anatomy and physiology. For example, in the non-typhoidal salmonellosis, vomiting and diarrhea are the main symptoms in humans. However, anatomically mice cannot vomit and due to this reason, the assessment of diarrhea could be very difficult in mice. In these cases, it becomes harder to translate the finding into clinical applications. Hence, these limitations of rodent models should be taken into consideration while interpreting the results from the dietary intervention studies for *Salmonella* infections in the rodent models for application in human clinical trials.

2.3. Summary of Effect of Dietary Components on Salmonella Infection in Pig Models

Pigs have been used in several studies involving dietary interventions [53]. Pigs have many more similarities to the human gastrointestinal tract as compared to rodents. Humans and pigs are similar in the body composition, cardiovascular, renal, nutritional, immunological, metabolic, and gastrointestinal aspects [53]. As such, several studies have been conducted in pig models of *Salmonella* infection interactions with dietary interventions. Michiels et al. demonstrated that supplementation of a mixture of formic, sorbic, and benzoic acid to the piglets for 35 days, significantly reduced the *Salmonella* fecal shedding as compared to the control group after oral challenge [54]. Dietary organic acids increase fecal cytotoxicity to *Salmonella*, but the effect can be dependent upon the environmental temperature. Rajtak et al. exhibited that supplementation of a pig diet with organic acid (Potassium-diformate) reduced the survival of *Salmonella* in pig feces when incubated at 22 °C but not at 4 °C [55]. Boyen et al. fed the supplemented diet with the coated butyric acid (2 g/kg of diet) to the pigs for 12 days and orally challenged the animals with *Salmonella* [56]. Fecal shedding of *Salmonella* was decreased in the coated butyric acids fed animals as compared to the un-coated group. It was hypothesized that coating prevents the degradation of fatty acids in the intestinal tract [56]. Dietary supplements also reduced inflammation after *Salmonella* infection in pigs. Chen et al. supplemented the pig diet with arginine (0.5%) for 1 week and infected the pigs intramuscularly with *Salmonella* [57]. Effects of various essential oils have been similarly reviewed by Omonijo et al. as effective antimicrobials in Swine production [58].

Fecal *Salmonella* shedding is one of the distinctive biomarkers of *Salmonella* infection in pig models. However, *Salmonella* colonization patterns are different in pigs as compared to humans. For instance, *Salmonella* Typhimurium has been observed to colonize in tonsils and respiratory tissues of infected pigs [59], whereas in humans, it does not colonize at those sites. The pig stomach is 2–3 times larger compared to humans [52], this anatomical difference may have impacts on *Salmonella* survival and digestibility of dietary components. Pig cecum is also several folds larger than the human cecum and may have implications in the *Salmonella* colonization [52]. In humans, stomach pH before eating is around 5, however, in pigs it is below 2. Consequently, pigs release a much greater extent of bile in the duodenum as compared to humans. Due to antimicrobial activities, bile could impact colonization of *Salmonella* in the proximal small intestine. Additionally, it can modulate digestion and absorption of the dietary components. Besides these differences, pigs are different in gastrointestinal thickness of mucus, and gastrointestinal motility and transit, as compared to humans. The distal small intestine of pigs contains a larger number of microbes as compared to humans and can degrade some carbohydrates with low digestibility compared to humans [60]. Hence, similar dietary interventions in pigs and humans may exhibit different potential. The pig immune system also differs from humans, however, implications of this difference have not been studied in regard to enteric infections. For instance, the gut of neonate piglets completely lacks leukocytes whereas human infants have a few leukocytes at birth [61]. Pig intestine contains a larger number of Peyer's patches as compared to humans throughout the intestine [61].

2.4. Summary of Effect of Dietary Components on Salmonella Infection in Calf Models

Although there are appreciable differences between monogasters and ruminants, calves develop very similar clinical and pathological features such as diarrhea and enteritis to human, hence, calves are considered one of most reliable models to mimic the human non-typhoidal salmonellosis [36]. These similarities have been also discussed by Higginson et al. [62].

After *Salmonella* infection, the calves show similar clinical symptoms as humans such as fever, diarrhea, anorexia and dehydration and the intestinal pathological changes [63]. Hill et al. revealed that feeding of a commercially available blend of butyric acid, coconut oil, and flax oil to the male Holstein calves for 28 days altered the inflammatory response to intraperitoneal *Salmonella* toxoid as compared to the control group [64]. The dietary blend reduced hyperthermia, hypophagia, and serum TNF-α but increased the IL-4 as compared to the control group [64].

Despite above-mentioned similarities, calves also exhibit significant anatomical and physiological differences in the digestive system relative to humans. A ruminant's stomach is four chambered and contains a large number of microflora that digests fibers, especially cellulose which remain undigested in humans. Sugars are fermented in ruminant stomach and as a result several volatile fatty acids are produced [65]. Most of the carbohydrates are converted into volatile fatty acids and a very small proportion of carbohydrates are absorbed as glucose. Additionally, the ruminant microflora differs from the human gut microflora to a great extent [66]. Hence, the same dietary components may produce different metabolites and physiological effects as compared to humans. Logistically, calves need a large amount of food and it is very expensive to conduct the dietary experimental studies in this model.

2.5. Summary of Dietary Interventions for Salmonella Infection in Humans

A variety of *Salmonella* serovars infect humans. Epidemiological studies have shown that typhoidal and non-typhoidal salmonellosis are the predominant types of infections [67]. Salmonellosis is clinically prognosed by headache, diarrhea, constipation, abdominal pain, chills, loss of appetite, and fever with an incubation time varying from hours to several days [68,69]. Typhoidal salmonellosis is less prevalent in the United States as compared to other developing countries [67]. In contrast, non-typhoidal salmonellosis presents a major and persisting public health challenge in North America. From 1998 to 2017, over 2600 single or multi-state non-typhoidal *Salmonella* outbreaks have occurred in the United States associated with animal and plant based foods [70]. Symptoms could be self-limiting, lasting for 1 week without treatment but could also lead to serious complications if left untreated, especially in immunocompromised subjects and those in at-risk populations [69]. Antimicrobial therapy is the first choice of treatment in persistent human salmonellosis. However, as discussed in the introduction section, the problem of drug resistance has become more prevalent due to extensive therapeutic use of antibiotics in healthcare facilities and subtherapeutic doses during animal food production [71]. Hence, dietary prophylactic interventions could be further utilized for prevention and alleviating symptoms of *Salmonella* infections. A few dietary prophylactic studies have been conducted in children for prevention of *Salmonella* infections. Stool frequency, vomiting, and *Salmonella* fecal shedding are the parameters measured in these clinical trials. Several other disease conditions also affect the incidence of *Salmonella* infections. Di Cagno et al. revealed that administration of gluten free diets in children with celiac disease did not reduce *Salmonella* shedding from stool as compared to healthy children [72]. Other dietary interventions are effective in reducing *Salmonella* infection. Lara et al. showed that feeding of dairy products containing probiotic mixtures of various strains of *Lactobacillus* to healthy children for 6 weeks decreased *Salmonella* serovars adhesion to the intestinal mucin [73]. Dietary interventions can also reduce frequency of stool and vomiting in *Salmonella* infected children. Rabbani et al. revealed that feeding of cooked banana for 1 week in children having persistent diarrhea, reduced frequency of the stool and vomiting as compared to children fed only with rice diet [74]. In another clinical trial in children, fermented food (lactic-acid fermented cereal gruel) was fed to healthy children three times a day for 2 weeks. After 2 weeks of feeding, stool swabs were taken from the treated and non-treated groups and analyzed for the presence of enteropathogenic bacteria including *Salmonella*. The fermented food reduced the presence of enteropathogenic bacteria as compared to the control diet [75]. These studies show that dietary interventions can be effective in the management of diarrheal diseases. However, there are several constraints in conducting dietary studies in *Salmonella* infections in humans that are prophylactic in nature. In addition to clinical equipoise, the major issues in conducting human clinical trials are time, cost, availability of appropriate stool and serum biomarkers and overall patient compliance and ethics. In presence of these curtailments, a dietary intervention could be pre-clinically evaluated in a relevant animal model to predict the safety and efficacy of the compound prior to administration in clinical trials [76].

It is noteworthy that bioactive compounds and probiotic diet might have a positive effect on colonization of *Salmonella* serovars in gastrointestinal area. As an example, a probiotic diet containing *Enterococcus* spp. could lead to increased fecal excreting and colonization of *Salmonella* in organs of piglets [77]. The current study is limited to discussing the literature that demonstrates antagonistic efficacy against colonization of *Salmonella* serovars, rather than those enhancing proliferation of the pathogen. Table 1 summarizes the pros and cons of *Salmonella* models discussed in the current study.

Table 1. Strengths and weaknesses of *Salmonella* models discussed in the current study.

Infection Model	Strength	Weaknesses
Salmonella Culture (Non-Host)	Direct interaction with the pathogen without confounders	Does not represent the interaction of dietary components with the host
Co-culture of *Salmonella* with host cell	Increased complexity of interaction compared to only pathogen culture, represents effects of intervention on the pathogen as well as the host	Does not represent involvement of all the host cell types that simultaneously happen together in human
Rodent Models	Represent a complex living system, very economical and convenient, ease in genetic manipulation to know mechanistic pathways	No diarrhea and vomiting, different intestinal immunity, different gastric environment, and anatomical structures
Pig Models	Similar to humans in body composition, cardiovascular, renal, nutritional, immunological, metabolic, and gastrointestinal aspects	Different than humans in *Salmonella* colonization pattern, gastric acidity, bile quantities, mucus thickness, immune system, not economical, not convenient
Calf Models	Develop similar clinical and pathological features such as diarrhea and enteritis	Stomach structure is different, not economical
Clinical Trials	The ideal model	Difficult to study the preventive effects of interventions due to ethical considerations

3. Potential Mechanisms of Protection against *Salmonella* Infections

3.1. Alteration in Bile Quality and Quantity

Bile is an important digestive fluid synthesized by the liver of many vertebrates. Bile plays a role in digestion of fats in small intestine by emulsification, micelle formation. As a result, absorption of fat-soluble vitamins such as vitamin A, D, E, and K is also increased in the presence of bile. Bile is stored in the gall bladder and released into the duodenum after receiving stimuli in the form of semi-digested fats and proteins from stomach. After digestion of fats, the majority of the bile is reabsorbed in terminal ileum. Cholycystokinin and secritin hormones in the gut control this process. Bile is alkaline and composed of phospholipids, bile acids, and surfactants. In the duodenum alkaline pH neutralizes stomach acid [78].

In addition to digestive role of bile, it exhibits an antimicrobial role against gastrointestinal pathogens [79]. Both bile quality and bile quantity may determine the multiplication of enteric pathogen [80]. Bile salts have been shown to act as antimicrobials especially on *Salmonella* and other enteric infections [81,82]. Different dietary fibers have been shown to affect bile composition to different extent and to improve colonization resistance against enteric pathogens [83]. Inagaki et al. showed that bile acids induces genes involved in enteroprotection by inhibiting pathogenic overgrowth and mucosal injury in the ileum in a mouse model of infection [84]. Diet consists of several compounds of plant and animal origin and hence considered as a multi-targeting intervention for prevention of enteric infection. Xu et al. [85] showed that consumption of dietary medium chain fatty acids increased fecal bile acids (cholic acid) significantly as compared to control group in C57BL/6J Mice. Kollanoor et al. demonstrated that feeding of Caprylic acid (a medium chain fatty acid) to poultry significantly reduced *Salmonella* infection in the intestine as well in organs therapeutically [83]. Further, in vitro study in hepatocytes showed that addition of medium chain fatty acids in culture media enhances cell surface expression and transport capacity of bile salt export pump (BSEP/ABCB11) [86]. Costarelli et al. compared diets containing different fatty acids in healthy premenopausal women and found that dietary linoleate increased postparandial plasma bile acid and cholycytokinin as compared to low fat diet [87]. Dietary fish oil increased fecal bile acids in a rodent model without increased gene expression for bile synthesis in the liver [88]. This study suggests although not all fatty acids increase bile acid synthesis in the liver, some could reduce bile absorption in the ileum. Studies have further exhibited that the change in bile acid release alters pH of the intestine and affects *Salmonella* adherence and survival. Several *Salmonella* genes are affected in the presence or absence of bile. Both bile quality and quantity have been shown repress *Salmonella* virulence in gut environment in in vivo models [89–91]. Antunes et al. [92] showed that *Salmonella* could multiply in the gall bladder of susceptible mice and causes typhoid. Bile acids exert antimicrobial actions on pathogens by virtue of their detergent properties. Cholic and deoxycholic acids in bile can damage bacterial DNA [79].

Dietary factors such as fiber may bind to bile acids and reduce reabsorption in colon [93]. Oat bran, pectin, and guar gum have been shown to increase bile acids in fecal matter [94–96]. Reduction in reabsorption of bile acids in the large intestine modulates the gut hormone feedback system and stimulates the liver to synthesize more bile acids [78]. This process could reduce alkalinity of the small intestine, and may increase gut motility, making the gut environment unfit for *Salmonella* infection [97].

3.2. Gut Mucosa

In order to reach epithelium, *Salmonella* needs to cross luminal barriers. Intestinal mucous is the first line of defense to *Salmonella* in the small intestine of rodents and humans [98]. Mucus in the small intestine is single layered and loosely attached to epithelium as compared to double-layered mucus of colon. Mucous is made up of secretory proteins called mucins and the predominant mucin in small intestine is Muc2 [99]. Abnormalities in mucous layers, underproduction of Muc2 by goblet cells and mutated Muc2 results in elevated risk for bacterial infection [100]. A study shows that during *Salmonella* infection, the mucin layer is disrupted and *Salmonella* obtain access to epithelium [101].

Various components of diet have been shown to upregulate expression of Muc2 in intestinal cells. Willemsen et al. showed that treatment of intestinal epithelial and fibroblast co-culture with short chain fatty acids significantly increased expression of Muc2 [102]. Ingestion of dietary fibers (soluble and insoluble) has been shown to increases proliferation of goblet cells and sialylated mucin in the small intestine of rats [103]. In another study, feeding of inulin/fructans in a rodent trial significantly increases mucous layer thickness in the colon and increases the number of goblet cells in crypts of distal jejunum as compared to control diet [104]. Morita et al. similarly exhibited that intake of dietary resistant starch in rodents reduces endotoxin influx from intestinal tissue and hypothesized that it could be partially due to alterations in mucosal barrier functions [105].

3.3. Antimicrobial Activities

After crossing the mucin layer in the gut, enteric pathogens need to penetrate epithelial layer in order to infect the organism. Human gut epithelia consist of a monolayer of epithelial cells. It separates the gut lumen from the lamina propria. Intestinal epithelial cellular junctions affect intestinal permeability as well as transcytosis capacity of individual cells. Strong cellular junctions are necessary to avoid the invasion of pathogens through epithelium. *Salmonella* can breach the epithelial barrier by employing para-cellular and trans-cellular mechanisms, including actin cytoskeleton of the epithelial cells and the secretion of the effector molecules [30].

Dietary components have been discussed in the past as factors to modulate the epithelial barrier [106]. Diet can have both positive and negative impacts on epithelial integrity. Liu et al. showed that when a high grain diet was fed to male goats, it resulted in the disruption of the ruminal epithelium as measured by the presence of systemic lipopolysacharide (LPS) [107]. However, diet can also impact epithelial integrity positively. In a study by Nofrarias et al., pigs were fed resistant starch for 97 days and consequently increased hypertrophy, reduced apoptosis in the crypts, lymphoid nodules in the colon, and increased mucin sulfuration were observed. These changes promoted epithelial protection compared to the control dietary group containing digestible starch [108]. Dietary components can also modulate the epithelial proteins such as occludins that secure junctions between the adjacent cells in the gut epithelium. Enteric pathogenic bacteria secrete LPS that causes inflammation and escalates loss of protein occludin that decreases the barrier function of epithelium. Park et al. showed in a rodent trial that dietary administration of gangliosides (a lipid) prevents LPS induced degradation of the occludin and reduces the total nitric oxide in the gut mucosa [109]. An in vitro study with Caco-2 cells demonstrated that addition of quercetin (a flavonoid) induces expression of zonula occludens-2, occludin, and claudin-1 and claudin-4 as compared to the control group [110]. All of these proteins play an important part in maintaining epithelial integrity. *Salmonella* entry into epithelial cells can result in epithelial necrosis and apoptosis. Int-407 cell line (human intestinal cell line) showed a significantly lesser extent of necrosis and apoptosis during *Salmonella* infection when treated with sterols and fatty acids found in the root extract of *Hemidusmus indicus* as compared to an untreated cell line [111]. Hence, protection of the epithelium can be considered an important target of dietary interventions in *Salmonella* infections.

3.4. Gut Microbiome

The gut contains more than a trillion symbiotic bacteria that play a major role in developing immunity as well as resistance against enteric infections. Initially it was hypothesized that the genetic factors were responsible for susceptible and resistant mouse strains against the enteric infections. However, currently literature delineates that the genetic factors are only one of the determinants of composition and structure of the gut microflora. As an example, Willing et al. successfully transferred the microbiota from resistant to susceptible mice and observed a delayed colonization of *Citrobacter rodentium* and mortality in susceptible strain [112]. In the same study, native gut microbiota of resistant mice was depleted by oral streptomycin (20 mg) 24 h prior to transplantation and replaced by the microbiota from susceptible mice. As a result, the oral antibiotic treatment reduced the innate defenses and a severe infection pathology was observed as compared to mice in control group. This experiment demonstrates that gut microbiota plays an important role in fighting the infection [112]. Similarly, mice were given a combination of antibiotics (Streptomycin, Vancomycin, Ampicillin, Neomycin, and Metronidazole) for 1 week in drinking water and later orally challenged with *Salmonella* Typhimurium 14028. The mice on the antibiotics showed a significantly higher number of *Salmonella* DNA in the cecum and large intestine as compared to control mice group [113]. The gut microbiota may affect enteric infections by modulating the intestinal immunity or by the direct competition. Symbiotic gut microbiota competes with pathogens for the nutrients such as iron and carbon sources [114]. Stelter et al. showed that *Salmonella*-induced mucosal lactins kills symbiotic gut microflora and then *Salmonella* takes advantage of this process for survival in gastrointestinal tract [115]. *Salmonella* induces acute inflammation in mice and neutrophils are recruited at the site of infection. Gill et al. showed that neutrophil elastases can shift mice gut microbiota and increase *Salmonella* colonization, while neutralization of neutrophil elastases decrease colonization of *Salmonella* [116]. These studies show that gut microbiota play an important role in protection from *Salmonella* infections and modulation of gut microflora for prevention of enteric infections warrants further studies.

Given the role of gut microbiota in protection against *Salmonella*, several studies have been conducted to test effects of dairy and native gut probiotics on *Salmonella* colonization. Probiotics are the microorganisms that induce health benefits when consumed in effective doses. *Lactobacillus* and *Streptococcus* are two widely studied categories of probiotics and their effectiveness against *Salmonella* is articulated by Castillo et al. [117]. *Lactobacillus rhamnosus* has been shown to reduce *Salmonella* adhesion to epithelial cells in in vitro model of *Salmonella* infection [118]. Probiotics not only compete with *Salmonella* for nutrients but also enhance protective immunity against the pathogen. Castillo et al. showed that oral administration of *Lactobacillus* in mice changes cytokine production and Toll Like Receptor (TLR) expression that is protective for mice against *Salmonella* infection [119]. Moreover, probiotics such as *Bifidobacterium* can directly affect virulence of *Salmonella* by releasing the molecules that down-regulate the expression of pathogenicity islands 1 and 2 [120]. Hence, *Lactobacillus* and *Bifidobacterium* have emerged as potential contributors for protection against enteric infections such as *Salmonella* serovars.

Diet is a major factor in the establishment of gut microbiome. As an example, previous studies exhibit that a change of diet from low-fat, high plant-based polysaccharide to the high-fat, and high simple sugar diet, could change structure of the gut microbiota very rapidly [121]. A shift of low-fat diet to the Western diet also changes metabolic pathways and modulates gene expression in gut microbiome [122]. Humanized mice (mice transplanted with human gut microflora) when fed a Western-type diet, showed an increased adiposity and this trait was transmissible through the transplantation of the gut microbiota in other mice [122]. Diet could also modulate gut microbiota directly by providing prebiotics—many studies have exhibited the efficacy of the prebiotics such as dietary fiber, fatty acids, and polyphenols for a shift in gut microflora [123,124].

3.5. Gut Immunity

The immune system of the gastrointestinal tract is the largest segment of the mammalian immune system. The gut encounters massive amounts of pathogens and dietary antigens that need to be neutralized. These functions emphasize the importance of gut immune system. The mucosal immune system is equipped with innate and adaptive immune defense mechanisms. Innate immunity provides the first line of defense against pathogens. The major players of the innate immune defense are macrophages, monocytes, neutrophils, epithelial cells, natural killer (NK) cells, and dendritic cells (DCs) [125]. Dendritic cells, macrophages, and epithelial cells are also termed as antigen presenting cells (APCs) because of their capacity of processing and presenting foreign antigens to other cells. APCs have a series of receptors called Pattern Recognition Receptors (PRRs) on their surfaces such as TLRs and Nod Like Receptors (NODs) to recognize the pathogens [126]. These receptors recognize motifs on pathogens known as the Pathogen Associated Molecular Patterns (PAMPs) [127]. The innate immune cells release inflammatory cytokines and mediators after sensing the PAMPs [128]. However, if innate immunity fails to resolve the inflammation and eliminate pathogen, adaptive immunity enters this process. In the gut adaptive immune system, the predominant response is antibody mediated and is represented by the Immunoglobulin A (IgA) [129]. The IgA is chiefly produced by the B cells in the intestinal mucosa triggered by anti-inflammatory cytokines such as TGF-β and IL-10 [130]. Hence, both innate and adaptive immune responses are required in the protection against infection and depends upon type of pathogen.

The role of the gut immune system in protection from enteric infections has been studied intensely [131–133]. Primary *Salmonella* infection increases interferon gamma (IFN-γ), tumor necrosis factor alpha (TNF-α), and interleukin 12 (IL-12) in circulation and in local tissues [133–135]. Major sources of IFN-γ and TNF-α are neutrophils and macrophages [136]. IL-12 is a cytokine induced in response to several bacteria and mediates onset of the Th1 protective response. Natural killer T (NKT) cells produce IFN-γ in response to IL-12 [137]. Infected macrophages also interact with NK cells in order to produce IFN-γ in humans [138]. Even though initial innate immune response restricts infection to a certain extent, it fails to inhibit multiplication of pathogens in deeper tissues. Hence, immune response is switched to adaptive response after some time and is achieved mainly by induction of CD4+ T cells, CD8+ T cells, and B cells [139]. In experimental models, depletion of CD4+ T cells had a more pronounced effect on protection from *Salmonella* as compared to CD8+ T cells. However, underlying mechanisms are not clear. The second major adaptive response to *Salmonella* is induction of the B cells to produce antibodies such as IgA. The antibodies bind *Salmonella* and prevent entry into deeper tissues. Administration of B cell hybridoma producing *Salmonella* specific IgA has been shown to prevent oral *Salmonella* infection in the mice [140]. These studies exhibited the potentially appreciable role of bioactive compounds for augmenting host immunity against *Salmonella* infections.

Dietary components such as dietary fiber and prebiotics manipulate both the innate and adaptive immunity [141]. Galdeano et al. demonstrated that feeding of probiotic fermented milk to the rats increases the number of macrophages and DCs with an increase in IFN-γ, TNF-α, and IL-12 after 5 days of nutrition [142]. Nutrients such as glutamine, arginine, vitamin A, and zinc have protective impacts against enteric infections [143]. Macrophages play an important role in clearance of *Salmonella* in primary infections. Modified arabinoxylan rice bran improves the phagocytic function of macrophages in the in vitro models of RAW264.7 cells [144]. Treatment of macrophages with the modified arabinoxylan rice bran increased the attachment and phagocytosis of yeast cells with an increase in TNF-α and IL-6 [144]. Wang et al. showed an enhanced *Salmonella* specific immune response in the orally vaccinated mice with attenuated *Salmonella* and fed with white button mushroom powder as compared to the only vaccinated mice [145]. The white button mushroom fed mice had higher number of *Salmonella* specific fecal IgA, IFN-γ, and TNF-α in splenocytes. These mice also showed an increased number of DCs and activation marker CD40 in splenocytes as compared to the control mice [145]. These studies show that dietary interventions could modulate pro-inflammatory responses and manipulate the innate and adaptive immunity [141].

4. Conclusions

Various dietary components could have considerable efficacy on prevention of *Salmonella* serovars infections. These effects may involve various mechanisms through impacting the gastrointestinal microbiota, immune system, and epithelium. The efficacy of various bioactive compounds for inhibiting the proliferation of *Salmonella* serovars in various in vitro, in vivo, animal models, and randomized studies reviewed creates the opportunity of mitigating the burden of Salmonellosis through dietary intervention. Despite striking similarities, animal models have major differences with human anatomy, as such delineated differences should be considered diligently for interpretation of these studies. Clinical equipoise, cost, time, and other ethical issues are also major curtailments for further conduct of randomized clinical trials with human subjects. The vast majority of the discussed literature demonstrate efficacy and mechanism of action of a sole bioactive compound. Pre-clinical studies investigating synergism and/or antagonistic activities of an array of bioactive compounds, additional randomized clinical trials, and well-planned epidemiological studies with comprehensive plans for control of confounders could augment the development of a validated and evidence-based guideline for mitigating the public health burden of human Salmonellosis through dietary compounds.

Author Contributions: A.K. co-wrote the first version of the manuscript. Authors A.A., M.H., and A.S. assisted in completion of the manuscript and reviewing the references for accuracy and formatting. A.F. co-wrote, revised, and edited the manuscript.

Acknowledgments: All authors read and approved the final manuscript. Author A.K. would like to express gratitude to former committee members of his doctoral degree for their support and encouragements. Financial support in part from the National Institute of Food and Agriculture of USDA (2017-07534; 2017-07975; 2017-06088) is acknowledged gratefully by the corresponding authors.

Conflicts of Interest: The authors declare no conflict of interest.

References

1. Herrick, R.L.; Buchberger, S.G.; Clark, R.M.; Kupferle, M.; Murray, R.; Succop, P. A Markov model to estimate Salmonella morbidity, mortality, illness duration, and cost. *Health Econ.* **2012**, *21*, 1169–1182. [CrossRef] [PubMed]
2. Fouladkhah, A.; Geornaras, I.; Yang, H.; Belk, K.; Nightingale, K.K.; Woerner, D.; Smith, G.C.; Sofos, J.N. Sensitivity of Shiga Toxin-Producing *Escherichia coli*, Multidrug Resistant *Salmonella*, and Antibiotic Susceptible *Salmonella* to Lactic Acid on Inoculated Beef Trimmings. *J. Food Prot.* **2012**, *75*, 1751–1758. [CrossRef] [PubMed]
3. Fouladkhah, A.; Geornaras, I.; Sofos, J. Biofilm Formation of O157 and Non-O157 Shiga Toxin-Producing *Escherichia coli* and Multidrug-Resistant and Susceptible *Salmonella* Typhimurium and Newport and Their Inactivation by Sanitizers. *J. Food Sci.* **2013**, *78*, M880–M886. [CrossRef] [PubMed]
4. Fouladkhah, A.; Geornaras, I.; Yang, H.; Sofos, J. Lactic Acid Resistance of Shiga Toxin-Producing *Escherichia coli* and Multidrug-Resistant and Susceptible *Salmonella* Typhimurium and *Salmonella* Newport in Meat Homogenate. *Food Microbiol.* **2013**, *36*, 260–266. [CrossRef] [PubMed]
5. Scallan, E.; Hoekstra, R.M.; Angulo, F.J.; Tauxe, R.V.; Widdowson, M.A.; Roy, S.L.; Jones, J.L.; Griffin, P.M. Foodborne Illness Acquired in the United States—major Pathogens. *J. Emerg. Infect. Dis.* **2011**, *17*, 7–15. [CrossRef] [PubMed]
6. Scallan, E.; Hoekstra, R.M.; Mahon, B.E.; Jones, T.F.; Griffin, P.M. An Assessment of the Human Health Impact of Seven Leading Foodborne Pathogens in the United States Using Disability Adjusted Life Years. *Epidemiol. Infect.* **2015**, *143*, 2795–2804. [CrossRef] [PubMed]

7. Bosilevac, J.M.; Arthur, T.M.; Bono, J.L.; Brichta-Harhay, D.M.; Kalchayanad, N.; King, D.A.; Shackelford, S.D.; Wheeler, M.L.; Koohmaraie, M. Prevalence and Enumeration of *Escherichia coli* O157: H7 and *Salmonella* in U.S. Abattoirs That Process Fewer than 1000 Head of Cattle per Day. *J. Food Prot.* **2009**, *72*, 1272–1278. [CrossRef] [PubMed]

8. Zhao, S.; Blickenstaff, K.; Glenn, A.; Ayers, S.L.; Friedman, S.L.; Abbott, J.W.; McDermott, P.F. Lactam Resistance in Salmonella Strains Isolated from Retail Meats in the United States by the National Antimicrobial Resistance Monitoring System between 2002 and 2006. *Appl. Environ. Microbiol.* **2009**, *75*, 7624–7630. [CrossRef]

9. Centers for Disease Control and Prevention (US). *Antibiotic Resistance Threats in the United States*; Centers for Disease Control and Prevention (US): Atlanta, GA, USA, 2013.

10. Ao, T.T.; Feasey, N.A.; Gordon, M.A.; Keddy, K.H.; Angulo, F.J.; Crump, J.A. Global Burden of Invasive Nontyphoidal *Salmonella* Disease, 2010 (1). *Emerg. Infect. Dis.* **2015**, *21*, 941. [CrossRef]

11. World Health Organization. Global Burden of Food Safety. Available online: Http://Www.Who.Int/ Foodsafety/Areas_work/Foodborne-Diseases/Ferg/En/ (accessed on 1 January 2019).

12. Grimont, P.A.; Weill, F.-X. Antigenic Formulae of the *Salmonella* Serovars. *WHO Collab. Cent. Ref. Res. Salmonella* **2007**, *9*, 1–166.

13. Kearney, K. Food consumption trends and drivers. *Philos. Trans. R. Soc. B* **2010**, *365*, 2793–2807. [CrossRef] [PubMed]

14. Cohen, M.L. Changing patterns of infectious disease. *Nature* **2000**, *406*, 762–767. [CrossRef] [PubMed]

15. Thiennimitr, P.; Winter, S.E.; Winter, M.G.; Xavier, M.N.; Tolstikov, V.; Huseby, D.L.; Sterzenbach, T.; Tsolis, R.M.; Roth, J.R.; Bäumler, A.J. Intestinal inflammation allows *Salmonella* to use ethanolamine to compete with the microbiota. *Proc. Natl. Acad. Sci. USA* **2011**, *108*, 17480–17485. [CrossRef] [PubMed]

16. Winter, S.E.; Thiennimitr, P.; Winter, M.G.; Butler, B.P.; Huseby, D.L.; Crawford, R.W.; Russell, J.M.; Bevins, C.L.; Adams, L.G.; Tsolis, R.M.; et al. Gut inflammation provides a respiratory electron acceptor for *Salmonella*. *Nature* **2010**, *467*, 426–429. [CrossRef] [PubMed]

17. Stecher, B.; Robbiani, R.; Walker, A.W.; Westendorf, A.M.; Barthel, M.; Kremer, M.; Chaffron, S.; Macpherson, A.J.; Buer, J.; Parkhill, J.; et al. *Salmonella enterica* serovar *typhimurium* exploits inflammation to compete with the intestinal microbiota. *PLoS Biol.* **2007**, *5*, 2177–2189. [CrossRef] [PubMed]

18. Calder, P.C.; Kew, S. The immune system: A target for functional foods? *Br. J. Nutr.* **2002**, *88*, S165–S177. [CrossRef]

19. Harrison, L.M.; Balan, K.V.; Babu, U.S. Dietary fatty acids and immune response to food-borne bacterial infections. *Nutrients* **2013**, *5*, 1801–1822. [CrossRef] [PubMed]

20. Taylor, A.K.; Cao, W.; Vora, K.P.; Cruz, J.D.L.; Shieh, W.J.; Zaki, S.R.; Katz, J.M.; Sambhara, S.; Gangappa, S. Protein energy malnutrition decreases immunity and increases susceptibility to influenza infection in mice. *J. Infect. Dis.* **2012**, *207*, 501–510. [CrossRef] [PubMed]

21. Hekmatdoost, A.; Wu, X.; Morampudi, V.; Innis, S.M.; Jacobson, K. Dietary oils modify the host immune response and colonic tissue damage following *Citrobacter rodentium* infection in mice. *Am. J. Physiol.* **2013**, *304*, G917–G928. [CrossRef]

22. Chandra, R.K. Nutrition, immunity and infection: From basic knowledge of dietary manipulation of immune responses to practical application of ameliorating suffering and improving survival. *Proc. Natl. Acad. Sci. USA* **1996**, *93*, 14304–14307. [CrossRef]

23. He, R.R.; Wang, M.; Wang, C.Z.; Chen, B.T.; Lu, C.N.; Yao, X.S.; Chen, J.X.; Kurihara, H. Protective effect of apple polyphenols against stress-provoked influenza viral infection in restraint mice. *J. Agric. Food Chem.* **2011**, *59*, 3730–3737. [CrossRef]

24. Daglia, M. Polyphenols as antimicrobial agents. *Curr. Opin. Biotechnol.* **2012**, *23*, 174–181. [CrossRef] [PubMed]

25. Roberts, C.L.; Keita, Å.V.; Parsons, B.N.; Prorok-Hamon, M.; Knight, P.; Winstanley, C.; Niamh, O.; Söderholm, J.D.; Rhodes, J.M.; Campbell, B.J. Soluble plantain fibre blocks adhesion and M-cell translocation of intestinal pathogens. *J. Nutr. Biochem.* **2013**, *24*, 97–103. [CrossRef] [PubMed]

26. Long, K.Z.; Santos, J.I.; Rosado, J.L.; Estrada-Garcia, T.; Haas, M.; Al Mamun, A.; DuPont, H.L.; Nanthakumar, N.N. Vitamin A supplementation modifies the association between mucosal innate and adaptive immune responses and resolution of enteric pathogen infections. *Am. J. Clin. Nutr.* **2011**, *93*, 578–585. [CrossRef] [PubMed]

27. Dhaliwal, W.; Shawa, T.; Khanam, M.; Jagatiya, P.; Simuyandi, M.; Ndulo, N.; Bevins, C.L.; Sanderson, I.R.; Kelly, P. Intestinal antimicrobial gene expression: Impact of micronutrients in malnourished adults during a randomized trial. *J. Infect. Dis.* **2010**, *202*, 971–978. [CrossRef] [PubMed]

28. Hung, C.C.; Garner, C.D.; Slauch, J.M.; Dwyer, Z.W.; Lawhon, S.D.; Frye, J.G.; McClelland, M.; Ahmer, B.M.; Altier, C. The intestinal fatty acid propionate inhibits *Salmonella* invasion through the post-translational control of HilD. *Mol. Microbiol.* **2013**, *87*, 1045–1060. [CrossRef] [PubMed]

29. Agerberth, B.; Bergman, P.; Gudmundsson, G.H. Helping the host: Induction of antimicrobial peptides as a novel therapeutic strategy against infections. In *Antimicrobial Peptides and Innate Immunity*; Springer: Basel, Switzerland, 2013; Volume 14, pp. 359–375.

30. Ulluwishewa, D.; Anderson, R.C.; McNabb, W.C.; Moughan, P.J.; Wells, J.M.; Roy, N.C. Regulation of tight junction permeability by intestinal bacteria and dietary components. *J. Nutr.* **2011**, *141*, 769–776. [CrossRef] [PubMed]

31. Ramalingam, A.; Wang, X.; Gabello, M.; Valenzano, M.C.; Soler, A.P.; Ko, A.; Morin, P.J.; Mullin, J.M. Dietary methionine restriction improves colon tight junction barrier function and alters claudin expression pattern. *Am. J. Physiol. Cell Physiol.* **2010**, *299*, C1028–C1035. [CrossRef] [PubMed]

32. Kau, A.L.; Ahern, P.P.; Griffin, N.W.; Goodman, A.L.; Gordon, J.I. Human nutrition, the gut microbiome and the immune system. *Nature* **2011**, *474*, 327–336. [CrossRef] [PubMed]

33. Chai, S.J.; White, P.L.; Lathrop, S.L.; Solghan, S.M.; Medus, C.; McGlinchey, B.M.; Tobin-D'Angelo, M.; Marcus, R.; Mahon, B.E. *Salmonella enterica* Serotype *Enteritidis*: Increasing Incidence of Domestically Acquired Infections. *Clin. Infect. Dis.* **2012**, *54*, S488–S497. [CrossRef] [PubMed]

34. Arguello, H.; Alvarez-Ordonez, A.; Carvajal, A.; Rubio, P.; Prieto, M. Role of Slaughtering in *Salmonella* Spreading and Control in Pork Production. *J. Food Prot.* **2013**, *76*, 899–911. [CrossRef] [PubMed]

35. Metcalf, E.S.; Almond, G.W.; Routh, P.A.; Horton, J.R.; Dillman, R.C.; Orndorff, P.E. Experimental *Salmonella typhi* infection in the domestic pig, Sus scrofa domestica. *Microb. Pathog.* **2000**, *29*, 121–126. [CrossRef] [PubMed]

36. Santos, R.L.; Zhang, S.; Tsolis, R.M.; Kingsley, R.A.; Adams, L.G.; Bäumler, A.J. Animal models of *Salmonella* infections: Enteritis versus typhoid fever. *Microbes Infect.* **2001**, *3*, 1335–1344. [CrossRef]

37. Tsolis, R.M.; Kingsley, R.A.; Townsend, S.M.; Ficht, T.A.; Adams, L.G.; Bäumler, A.J. Of mice, calves, and men - Comparison of the mouse typhoid model with other *Salmonella* infections. *Adv. Exp. Med. Biol.* **1999**, *473*, 261–274. [PubMed]

38. Oussalah, M.; Caillet, S.; Saucier, L.; Lacroix, M. Inhibitory effects of selected plant essential oils on the growth of four pathogenic bacteria: *E. coli* O157: H7, *Salmonella* Typhimurium, *Staphylococcus aureus* and *Listeria monocytogenes*. *Food Control* **2007**, *18*, 414–420. [CrossRef]

39. Vikram, A.; Jesudhasan, P.R.; Jayaprakasha, G.K.; Pillai, S.D.; Jayaraman, A.; Patil, B.S. Citrus flavonoid represses *Salmonella* pathogenicity island 1 and motility in *S. typhimurium* LT2. *Int. J. Food Microbiol.* **2011**, *145*, 28–36. [CrossRef] [PubMed]

40. Ebani, V.V.; Nardoni, S.; Bertelloni, F.; Tosi, G.; Massi, P.; Pistelli, L.; Mancianti, F. In Vitro Antimicrobial Activity of Essential Oils against *Salmonella* enterica Serotypes Enteritidis and Typhimurium Strains Isolated from Poultry. *Molecules* **2019**, *24*, 900. [CrossRef]

41. Chen, C.Y.; Tsen, H.Y.; Lin, C.L.; Lin, C.K.; Chuang, L.T.; Chen, C.S.; Chiang, Y.C. Enhancement of the immune response against *Salmonella* infection of mice by heat-killed multispecies combinations of lactic acid bacteria. *J. Med. Microbiol.* **2013**, *62*, 1657–1664. [CrossRef]

42. Forbes, S.J.; Eschmann, M.; Mantis, N.J. Inhibition of *Salmonella enterica* serovar *Typhimurium* motility and entry into epithelial cells by a protective antilipopolysaccharide monoclonal immunoglobulin a antibody. *Infect. Immun.* **2008**, *76*, 4137–4144. [CrossRef]

43. Hansen-Wester, I.; Hensel, M. *Salmonella* pathogenicity islands encoding type III secretion systems. *Microbes Infect.* **2001**, *3*, 549–559. [CrossRef]

44. Bovee-Oudenhoven, I.M.J.; ten Bruggencate, S.J.M.; Lettink-Wissink, M.L.G.; van der Meer, R. Dietary fructo-oligosaccharides and lactulose inhibit intestinal colonisation but stimulate translocation of *Salmonella* in rats. *Gut* **2003**, *52*, 1572–1578. [CrossRef] [PubMed]

45. Hitchins, A.D.; Wells, P.; McDonough, F.E.; Wong, N.P. Amelioration of the adverse effect of a gastrointestinal challenge with *Salmonella enteritidis* on weanling rats by a yogurt diet. *Am. J. Clin. Nutr.* **1985**, *41*, 92–100. [CrossRef]

46. Kim, S.P.; Moon, E.; Nam, S.H.; Friedman, M. Composition of Herba Pogostemonis Water Extract and Protection of Infected Mice against *Salmonella* typhimurium-Induced Liver Damage and Mortality by Stimulation of Innate Immune Cells. *J. Agric. Food Chem.* **2012**, *60*, 12122–12130. [CrossRef] [PubMed]

47. Gabriel, P.O.; Aribisala, J.O.; Oladunmoye, M.K.; Arogunjo, A.O.; Ajayi-Moses, O.B. Therapeutic Effect of Goya Extra Virgin Olive Oil in Albino Rat Orogastricallly Dosed with Salmonella typhi. *South Asian J. Res. Microbiol.* **2019**, *3*, 1–9.

48. Kelly, P.; Shawa, T.; Mwanamakondo, S.; Soko, R.; Smith, G.; Barclay, G.R.; Sanderson, I.R. Gastric and intestinal barrier impairment in tropical enteropathy and HIV: Limited impact of micronutrient supplementation during a randomised controlled trial. *BMC Gastroenterol.* **2010**, *10*, 72. [CrossRef]

49. Tennant, S.M.; Hartland, E.L.; Phumoonna, T.; Lyras, D.; Rood, J.I.; Robins-Browne, R.M.; van Driel, I.R. Influence of gastric acid on susceptibility to infection with ingested bacterial pathogens. *Infect. Immun.* **2008**, *76*, 639–645. [CrossRef]

50. Lucas, J.S.; Cochrane, S.A.; Warner, J.O.; Hourihane, J.O. The effect of digestion and pH on the allergenicity of kiwifruit proteins. *Pediatr. Allergy Immunol.: Off. Publ. Eur. Soc. Pediatr. Allergy Immunol.* **2008**, *19*, 392–398. [CrossRef]

51. Henderson, L.M.; Brewer, G.J.; Dressman, J.B.; Swidan, S.Z.; DuRoss, D.J.; Adair, C.H.; Barnett, J.L.; Berardi, R.R. Effect of intragastric pH on the absorption of oral zinc acetate and zinc oxide in young healthy volunteers. *JPEN J. Parenter. Enter. Nutr.* **1995**, *19*, 393–397. [CrossRef]

52. Kararli, T.T. Comparison of the Gastrointestinal Anatomy, Physiology, and Biochemistry of Humans and Commonly Used Laboratory-Animals. *Biopharm. Drug Dispos.* **1995**, *16*, 351–380. [CrossRef]

53. Guilloteau, P.; Zabielski, R.; Hammon, H.M.; Metges, C.C. Nutritional programming of gastrointestinal tract development. Is the pig a good model for man? *Nutr. Res. Rev.* **2010**, *23*, 4–22. [CrossRef]

54. Michiels, J.; Missotten, J.; Rasschaert, G.; Dierick, N.; Heyndrickx, M.; De Smet, S. Effect of organic acids on *Salmonella* colonization and shedding in weaned piglets in a seeder model. *J. Food Prot.* **2012**, *75*, 1974–1983. [CrossRef]

55. Rajtak, U.; Boland, F.; Leonard, N.; Bolton, D.; Fanning, S. Roles of diet and the acid tolerance response in survival of common *Salmonella* serotypes in feces of finishing pigs. *Appl. Environ. Microbiol.* **2012**, *78*, 110–119. [CrossRef] [PubMed]

56. Boyen, F.; Haesebrouck, F.; Vanparys, A.; Volf, J.; Mahu, M.; Van Immerseel, F.; Rychlik, I.; Dewulf, J.; Ducatelle, R.; Pasmans, F. Coated fatty acids alter virulence properties of *Salmonella typhimurium* and decrease intestinal colonization of pigs. *Vet. Microbiol.* **2008**, *132*, 319–327. [CrossRef]

57. Chen, Y.; Chen, D.; Tian, G.; He, J.; Mao, X.; Mao, Q.; Yu, B. Dietary arginine supplementation alleviates immune challenge induced by *Salmonella enterica* serovar *Choleraesuis bacterin* potentially through the Toll-like receptor 4-myeloid differentiation factor 88 signalling pathway in weaned piglets. *Br. J. Nutr.* **2012**, *108*, 1069–1076. [CrossRef] [PubMed]

58. Omonijo, F.A.; Ni, L.; Gong, J.; Wang, Q.; Lahaye, L.; Yang, C. Essential oils as alternatives to antibiotics in swine production. *Anim. Nutr.* **2018**, *4*, 126–136. [CrossRef] [PubMed]

59. Boyen, F.; Haesebrouck, F.; Maes, D.; Van Immerseel, F.; Ducatelle, R.; Pasmans, F. Non-typhoidal *Salmonella* infections in pigs: A closer look at epidemiology, pathogenesis and control. *Vet. Microbiol.* **2008**, *130*, 1–19. [CrossRef] [PubMed]

60. Eberhard, M.; Hennig, U.; Kuhla, S.; Brunner, R.M.; Kleessen, B.; Metges, C.C. Effect of inulin supplementation on selected gastric, duodenal, and caecal microbiota and short chain fatty acid pattern in growing piglets. *Arch. Anim. Nutr.* **2007**, *61*, 235–246. [CrossRef] [PubMed]

61. Scharek, L.; Tedin, K. The porcine immune system–differences compared to man and mouse and possible consequences for infections by *Salmonella* serovars. *Berl. Und Munch. Tierarztl. Wochenschr.* **2007**, *120*, 347–354.

62. Higginson, E.E.; Simon, R.; Tennant, S.M. Animal models for salmonellosis: Applications in vaccine research. *Clin. Vaccine Immunol.* **2016**, *23*, 746–756. [CrossRef] [PubMed]

63. Costa, L.F.; Paixao, T.A.; Tsolis, R.M.; Baumler, A.J.; Santos, R.L. Salmonellosis in cattle: Advantages of being an experimental model. *Res. Vet. Sci.* **2012**, *93*, 1–6. [CrossRef] [PubMed]

64. Hill, T.M.; VandeHaar, M.J.; Sordillo, L.M.; Catherman, D.R.; Bateman Ii, H.G.; Schlotterbeck, R.L. Fatty acid intake alters growth and immunity in milk-fed calves. *J. Dairy Sci.* **2011**, *94*, 3936–3948. [CrossRef] [PubMed]

65. Hofmann, R.R. Evolutionary Steps of Ecophysiological Adaptation and Diversification of Ruminants - a Comparative View of Their Digestive-System. *Oecologia* **1989**, *78*, 443–457. [CrossRef] [PubMed]

66. Li, R.W.; Connor, E.E.; Li, C.J.; Baldwin, R.L.; Sparks, M.E. Characterization of the rumen microbiota of pre-ruminant calves using metagenomic tools. *Environ. Microbiol.* **2012**, *14*, 129–139. [CrossRef] [PubMed]

67. Sanchez-Vargas, F.M.; Abu-El-Haija, M.A.; Gomez-Duarte, O.G. *Salmonella* infections: An update on epidemiology, management, and prevention. *Travel Med. Infect. Dis.* **2011**, *9*, 263–277. [CrossRef] [PubMed]

68. Connor, B.A.; Schwartz, E. Typhoid and paratyphoid fever in travellers. *Lancet Infect. Dis* **2005**, *5*, 623–628. [CrossRef]

69. Hohmann, E.L. Nontyphoidal salmonellosis. *Clin. Infect. Dis.* **2001**, *32*, 263–269. [PubMed]

70. Centers for Disease Control and Prevention. National Outbreak Reporting System (NORS), 2018. Available online: https://www.cdc.gov/nors/index.html (accessed on 1 January 2019).

71. Glenn, L.M.; Lindsey, R.L.; Folster, J.P.; Pecic, G.; Boerlin, P.; Gilmour, M.W. Antimicrobial resistance genes in multidrug-resistant *Salmonella enterica* isolated from animals, retail meats, and humans in the United States and Canada. *Microb. Drug Resist.* **2013**, *19*, 175–184. [CrossRef] [PubMed]

72. Di Cagno, R.; De Angelis, M.; De Pasquale, I.; Ndagijimana, M.; Vernocchi, P.; Ricciuti, P.; Gagliardi, F.; Laghi, L.; Crecchio, C.; Guerzoni, M.E.; et al. Duodenal and faecal microbiota of celiac children: Molecular, phenotype and metabolome characterization. *BMC Microbiol.* **2011**, *11*, 219. [CrossRef]

73. Lara-Villoslada, F.; Sierra, S.; Boza, J.; Xaus, J.; Olivares, M. Beneficial effects of consumption of a dairy product containing two probiotic strains, *Lactobacillus coryniformis* CECT5711 and *Lactobacillus gasseri* CECT5714 in healthy children. *Nutr. Hosp.* **2007**, *22*, 496–502.

74. Rabbani, G.H.; Teka, T.; Zaman, B.; Majid, N.; Khatun, M.; Fuchs, G.J. Clinical studies in persistent diarrhea: Clinical studies in persistent diarrhea: Dietary management with green banana or pectin in Bangladeshi children. *Gastroenterology* **2001**, *121*, 554–560. [CrossRef]

75. Kingamkono, R.; Sjogren, E.; Svanberg, U. Enteropathogenic bacteria in faecal swabs of young children fed on lactic acid-fermented cereal gruels. *Epidemiol. Infect.* **1999**, *122*, 23–32. [CrossRef] [PubMed]

76. Szabó, I.; Wieler, L.H.; Tedin, K.; Scharek-Tedin, L.; Taras, D.; Hensel, A.; Appel, B.; Nöckler, K. Influence of a probiotic strain of Enterococcus faecium on Salmonella enterica serovar Typhimurium DT104 infection in a porcine animal infection model. *Appl. Environ. Microbiol.* **2009**, *75*, 2621–2628.

77. Pasetti, M.F.; Levine, M.M.; Sztein, M.B. Animal models paving the way for clinical trials of attenuated Salmonella enterica serovar Typhi live oral vaccines and live vectors. *Vaccine* **2003**, *21*, 401–418. [CrossRef]

78. Hofmann, A.F.; Hagey, R.L. Bile acids: Chemistry, pathochemistry, biology, pathobiology, and therapeutics. *Cell Mol. Life Sci.* **2008**, *65*, 2461–2483. [CrossRef]

79. Prieto, A.I.; Ramos-Morales, F.; Casadesus, J. Bile-induced DNA damage in *Salmonella enterica*. *Genetics* **2004**, *168*, 1787–1794. [CrossRef]

80. Merritt, M.E.; Donaldson, J.R. Effect of bile salts on the DNA and membrane integrity of enteric bacteria. *J. Med. Microbiol.* **2009**, *58*, 1533–1541. [CrossRef]

81. Begley, M.; Gahan, C.G.M.; Hill, C. The interaction between bacteria and bile. *FEMS Microbiol. Rev.* **2005**, *29*, 625–651. [CrossRef]

82. Antunes, L.C.M.; Andersen, S.K.; Menendez, A.; Arena, E.T.; Han, J.; Ferreira, R.B.; Borchers, C.H.; Finlay, B.B. Metabolomics Reveals Phospholipids as Important Nutrient Sources during *Salmonella* Growth in Bile in Vitro and in Vivo. *J. Bacteriol.* **2011**, *193*, 4719–4725. [CrossRef]

83. Kollanoor-Johny, A.; Mattson, T.; Baskaran, S.A.; Amalaradjou, M.A.R.; Hoagland, T.A.; Darre, M.J.; Khan, M.I.; Schreiber, D.T.; Donoghue, A.M.; Venkitanarayanan, K. Caprylic acid reduces *Salmonella enteritidis* populations in various segments of digestive tract and internal organs of 3- and 6-week-old broiler chickens, therapeutically. *Poult. Sci.* **2012**, *91*, 1686–1694. [CrossRef]

84. Inagaki, T.; Moschetta, A.; Lee, Y.K.; Peng, L.; Zhao, G.; Downes, M.; Ruth, T.Y.; Shelton, J.M.; Richardson, J.A.; Repa, J.J.; et al. Regulation of antibacterial defense in the small intestine by the nuclear bile acid receptor. *Proc. Natl. Acad. Sci. USA* **2006**, *103*, 3920–3925. [CrossRef]

85. Xu, Q.; Xue, C.; Zhang, Y.; Liu, Y.; Wang, J.; Yu, X.; Zhang, X.; Zhang, R.; Yang, X.; Guo, C. Medium-Chain Fatty Acids Enhanced the Excretion of Fecal Cholesterol and Cholic Acid in C57BL/6J Mice Fed a Cholesterol-Rich Diet. *Biosci. Biotechnol. Biochem.* **2013**, *77*, 1390–1396. [CrossRef] [PubMed]

86. Kato, T.; Hayashi, H.; Sugiyama, Y. Short- and medium-chain fatty acids enhance the cell surface expression and transport capacity of the bile salt export pump (BSEP/ABCB11). *Biochim. Biophys. Acta* **2010**, *1801*, 1005–1012. [CrossRef] [PubMed]

87. Costarelli, V.; Sanders, T.A. Acute effects of dietary fat composition on postprandial plasma bile acid and cholecystokinin concentrations in healthy premenopausal women. *Br. J. Nutr.* **2001**, *86*, 471–477. [CrossRef]

88. Yang, Q.; Lan, T.; Chen, Y.; Dawson, P.A. Dietary fish oil increases fat absorption and fecal bile acid content without altering bile acid synthesis in 20-d-old weanling rats following massive ileocecal resection. *Pediatr. Res.* **2012**, *72*, 38–42. [CrossRef] [PubMed]

89. Prouty, A.M.; Brodsky, I.E.; Manos, J.; Belas, R.; Falkow, S.; Gunn, J.S. Transcriptional regulation of *Salmonella enterica* serovar *typhimurium* genes by bile. *FEMS Immunol. Med. Microbiol.* **2004**, *41*, 177–185. [CrossRef] [PubMed]

90. Prouty, A.M.; Gunn, J.S. *Salmonella enterica* serovar *typhimurium* invasion is repressed in the presence of bile. *Infect. Immun.* **2000**, *68*, 6763–6769. [CrossRef]

91. Ye, W.; Li, Y.; Zhou, Z.; Wang, X.; Yao, J.; Liu, J.; Wang, C. Synthesis and antibacterial activity of new long-chain-alkyl bile acid-based amphiphiles. *Bioorg. Chem.* **2013**, *51*, 1–7. [CrossRef] [PubMed]

92. Antunes, L.C.; Wang, M.; Andersen, S.K.; Ferreira, R.B.; Kappelhoff, R.; Han, J.; Borchers, C.H.; Finlay, B.B. Repression of *Salmonella enterica* phoP expression by small molecules from physiological bile. *J. Bacteriol.* **2012**, *194*, 2286–2296. [CrossRef]

93. Matsumoto, K.; Yokoyama, S.; Gato, N. Bile acid-binding activity of young persimmon (Diospyros kaki) fruit and its hypolipidemic effect in mice. *Phytother. Res.* **2010**, *24*, 205–210. [CrossRef]

94. Andersson, K.E.; Immerstrand, T.; Swärd, K.; Bergenståhl, B.; Lindholm, M.W.; Öste, R.; Hellstrand, P. Effects of oats on plasma cholesterol and lipoproteins in C57BL/6 mice are substrain specific. *Br. J. Nutr.* **2010**, *103*, 513–521. [CrossRef]

95. Chen, H.L.; Lin, Y.M.; Wang, Y.C. Comparative effects of cellulose and soluble fibers (pectin, konjac glucomannan, inulin) on fecal water toxicity toward Caco-2 cells, fecal bacteria enzymes, bile acid, and short-chain fatty acids. *J. Agric. Food Chem.* **2010**, *58*, 10277–10281. [CrossRef]

96. Gunness, P.; Gidley, M.J. Mechanisms underlying the cholesterol-lowering properties of soluble dietary fibre polysaccharides. *Food Funct.* **2010**, *1*, 149–155. [CrossRef] [PubMed]

97. Hofmann, A.F.; Eckmann, L. How bile acids confer gut mucosal protection against bacteria. *Proc. Natl. Acad. Sci. USA* **2006**, *103*, 4333–4334. [CrossRef]

98. Zarepour, M.; Bhullar, K.; Montero, M.; Ma, C.; Huang, T.; Velcich, A.; Xia, L.; Vallance, B.A. The mucin Muc2 limits pathogen burdens and epithelial barrier dysfunction during *Salmonella enterica* serovar *Typhimurium* Colitis. *Infect. Immun.* **2013**, *81*, 3672–3683. [CrossRef] [PubMed]

99. Hansson, G.C. Role of mucus layers in gut infection and inflammation. *Curr. Opin. Microbiol.* **2012**, *15*, 57–62. [CrossRef] [PubMed]

100. Kim, Y.S.; Ho, S.B. Intestinal goblet cells and mucins in health and disease: Recent insights and progress. *Curr. Gastroenterol. Rep.* **2010**, *12*, 319–330. [CrossRef]

101. Kim, J.; Khan, W. Goblet Cells and Mucins: Role in Innate Defense in Enteric Infections. *Pathogens* **2013**, *2*, 55–70. [CrossRef] [PubMed]

102. Willemsen, L.E.; Koetsier, M.A.; van Deventer, S.J.; van Tol, E.A. Short chain fatty acids stimulate epithelial mucin 2 expression through differential effects on prostaglandin E (1) and E (2) production by intestinal myofibroblasts. *Gut* **2003**, *52*, 1442–1447. [CrossRef]

103. Hino, S.; Takemura, N.; Sonoyama, K.; Morita, A.; Kawagishi, H.; Aoe, S.; Morita, T. Small intestinal goblet cell proliferation induced by ingestion of soluble and insoluble dietary fiber is characterized by an increase in sialylated mucins in rats. *J. Nutr.* **2012**, *142*, 1429–1436. [CrossRef]

104. Kleessen, B.; Blaut, M. Modulation of gut mucosal biofilms. *Br. J. Nutr.* **2005**, *93*, S35–S40. [CrossRef]

105. Morita, T.; Tanabe, H.; Takahashi, K.; Sugiyama, K. Ingestion of resistant starch protects endotoxin influx from the intestinal tract and reduces D-galactosamine-induced liver injury in rats. *J. Gastroenterol. Hepatol.* **2004**, *19*, 303–313. [CrossRef] [PubMed]

106. Kosińska, A.; Andlauer, W. Modulation of Tight Junction Integrity by Food Components. *Food Res. Int.* **2013**, *54*, 951–960.

107. Liu, J.H.; Xu, T.T.; Liu, Y.J.; Zhu, W.Y.; Mao, S.Y. A high-grain diet causes massive disruption of ruminal epithelial tight junctions in goats. *Am. J. Psychol. Regul. Integr. Comp. Physiol.* **2013**, *305*, R232–R241. [CrossRef] [PubMed]

108. Nofrarias, M.; Martinez-Puig, D.; Pujols, J.; Majo, N.; Perez, J.F. Long-term intake of resistant starch improves colonic mucosal integrity and reduces gut apoptosis and blood immune cells. *Nutrition* **2007**, *23*, 861–870. [CrossRef] [PubMed]

109. Park, E.J.; Thomson, A.B.; Clandinin, M.T. Protection of intestinal occludin tight junction protein by dietary gangliosides in lipopolysaccharide-induced acute inflammation. *J. Pediatr. Gastroenterol. Nutr.* **2010**, *50*, 321–328. [CrossRef]

110. Suzuki, T.; Hara, H. Quercetin Enhances Intestinal Barrier Function through the Assembly of Zonnula Occludens-2, Occludin, and Claudin-1 and the Expression of Claudin-4 in Caco-2 Cells. *J. Nutr.* **2009**, *139*, 965–974. [CrossRef] [PubMed]

111. Das, S.; Devaraj, S.N. Protective role of Hemidesmus indicus R. Br. root extract against *Salmonella typhimurium*-induced cytotoxicity in Int 407 cell line. *Phytother. Res.* **2007**, *21*, 1209–1216. [CrossRef] [PubMed]

112. Willing, B.P.; Vacharaksa, A.; Croxen, M.; Thanachayanont, T.; Finlay, B.B. Altering Host Resistance to Infections through Microbial Transplantation. *PLoS ONE* **2011**, *6*, e26988. [CrossRef] [PubMed]

113. Croswell, A.; Amir, E.; Teggatz, P.; Barman, M.; Salzman, N.H. Prolonged impact of antibiotics on intestinal microbial ecology and susceptibility to enteric *Salmonella* infection. *Infect. Immun.* **2009**, *77*, 2741–2753. [CrossRef]

114. Kamada, N.; Chen, G.Y.; Inohara, N.; Nunez, G. Control of pathogens and pathobionts by the gut microbiota. *Nat. Immunol.* **2013**, *14*, 685–690. [CrossRef]

115. Stelter, C.; Käppeli, R.; König, C.; Krah, A.; Hardt, W.D.; Stecher, B.; Bumann, D. *Salmonella*-induced mucosal lectin RegIIIbeta kills competing gut microbiota. *PLoS ONE* **2011**, *6*, e20749. [CrossRef]

116. Gill, N.; Ferreira, R.B.; Antunes, L.C.M.; Willing, B.P.; Sekirov, I.; Al-Zahrani, F.; Hartmann, M.; Finlay, B.B. Neutrophil elastase alters the murine gut microbiota resulting in enhanced *Salmonella* colonization. *PLoS ONE* **2012**, *7*, e49646. [CrossRef]

117. Castillo, N.A.; de Moreno de LeBlanc, A.; Galdeano, C.M.; Perdigón, G. Probiotics: An alternative strategy for combating salmonellosis: Immune mechanisms involved. *Food Res. Int.* **2012**, *45*, 831–841. [CrossRef]

118. Burkholder, K.M.; Bhunia, A.K. *Salmonella enterica* serovar *typhimurium* adhesion and cytotoxicity during epithelial cell stress is reduced by *Lactobacillus rhamnosus* GG. *Gut Pathog.* **2009**, *1*, 14. [CrossRef]

119. Castillo, N.A.; Perdigon, G.; de Moreno de LeBlanc, A. Oral administration of a probiotic *Lactobacillus* modulates cytokine production and TLR expression improving the immune response against *Salmonella enterica* serovar *typhimurium* infection in mice. *BMC Microbiol.* **2011**, *11*, 177. [CrossRef] [PubMed]

120. Bayoumi, M.A.; Griffiths, M.W. Probiotics down-regulate genes in *Salmonella enterica* serovar *typhimurium* pathogenicity islands 1 and 2. *J. Food Prot.* **2010**, *73*, 452–460. [CrossRef]

121. McNulty, N.P.; Wu, M.; Erickson, A.R.; Pan, C.; Erickson, B.; Martens, E.C.; Pudlo, N.A.; Muegge, B.D.; Henrissat, B.; Hettich, R.L.; et al. Effects of Diet on Resource Utilization by a Model Human Gut Microbiota Containing Bacteroides cellulosilyticus WH2, a Symbiont with an Extensive Glycobiome. *PLoS Biol.* **2013**, *11*, e1001637. [CrossRef] [PubMed]

122. Turnbaugh, P.J.; Ridaura, V.K.; Faith, J.J.; Rey, F.E.; Knight, R.; Gordon, J.I. The Effect of Diet on the Human Gut Microbiome: A Metagenomic Analysis in Humanized Gnotobiotic Mice. *Sci. Transl. Med.* **2009**, *1*, 6–14. [CrossRef] [PubMed]

123. Toward, R.; Montandon, S.; Walton, G.; Gibson, G.R. Effect of prebiotics on the human gut microbiota of elderly persons. *Gut Microbes* **2012**, *3*, 57–60. [CrossRef]

124. Laparra, J.M.; Sanz, Y. Interactions of gut microbiota with functional food components and nutraceuticals. *Pharmacol. Res.: Off. J. Ital. Pharmacol. Soc.* **2010**, *61*, 219–225. [CrossRef]

125. Yuan, Q.; Walker, W.A. Innate immunity of the gut: Mucosal defense in health and disease. *J. Pediatr. Gastroenterol. Nutr.* **2004**, *38*, 463–473. [CrossRef] [PubMed]

126. Kawai, T.; Akira, S. The role of pattern-recognition receptors in innate immunity: Update on Toll-like receptors. *Nat. Immunol.* **2010**, *11*, 373–384. [CrossRef] [PubMed]

127. Abreu, M.T. Toll-like receptor signalling in the intestinal epithelium: How bacterial recognition shapes intestinal function. *Nat. Rev. Immunol.* **2010**, *10*, 131–143. [CrossRef] [PubMed]

128. Gordon, S. Pattern recognition receptors: Doubling up for the innate immune response. *Cell* **2002**, *111*, 927–930. [CrossRef]

129. Macpherson, A.J.; Mccoy, K.D.; Johansen, F.E.; Brandtzaeg, P. The immune geography of IgA induction and function. *Mucosal Immunol.* **2008**, *1*, 11–22. [CrossRef] [PubMed]

130. Fagarasan, S.; Honjo, T. Intestinal IgA synthesis: Regulation of front-line body defences. *Nat. Rev. Immunol.* **2003**, *3*, 63–72. [CrossRef] [PubMed]

131. Nanton, M.R.; Way, S.S.; Shlomchik, M.J.; McSorley, S.J. Cutting edge: B cells are essential for protective immunity against *Salmonella* independent of antibody secretion. *J. Immunol.* **2012**, *189*, 5503–5507. [CrossRef] [PubMed]

132. Cross, M.L. Microbes versus microbes: Immune signals generated by probiotic *lactobacilli* and their role in protection against microbial pathogens. *Fems Immunol. Med. Microbiol.* **2002**, *34*, 245–253. [CrossRef] [PubMed]

133. Jouanguy, E.; DÖffinger, R.; Dupuis, S.; Pallier, A.; Altare, F.; Casanova, J.L. IL-12 and IFN-gamma in host defense against mycobacteria and *salmonella* in mice and men. *Curr. Opin. Immunol.* **1999**, *11*, 346–351. [CrossRef]

134. Bao, S.; Beagley, K.W.; France, M.P.; Shen, J.; Husband, A.J. Interferon-gamma plays a critical role in intestinal immunity against *Salmonella typhimurium* infection. *Immunology* **2000**, *99*, 464–472. [CrossRef]

135. Arnold, J.W.; Niesel, D.W.; Annable, C.R.; Hess, C.B.; Asuncion, M.; Cho, Y.J.; Peterson, J.W.; Klimpel, G.R. Tumor necrosis factor-alpha mediates the early pathology in *Salmonella* infection of the gastrointestinal tract. *Microb. Pathog.* **1993**, *14*, 217–227. [CrossRef] [PubMed]

136. Kirby, A.C.; Yrlid, U.; Wick, M.J. The innate immune response differs in primary and secondary *Salmonella* infection. *J. Immunol.* **2002**, *169*, 4450–4459. [CrossRef] [PubMed]

137. Mastroeni, P.; Harrison, J.A.; Chabalgoity, J.A.; Hormaeche, C.E. Effect of interleukin 12 neutralization on host resistance and gamma interferon production in mouse typhoid. *Infect. Immun.* **1996**, *64*, 189–196. [PubMed]

138. Lapaque, N.; Walzer, T.; Meresse, S.; Vivier, E.; Trowsdale, J. Interactions between Human NK Cells and Macrophages in Response to *Salmonella* Infection. *J. Immunol.* **2009**, *182*, 4339–4348. [CrossRef]

139. Mittrucker, H.W.; Kaufmann, S.H.E. Immune response to infection with *Salmonella typhimurium* in mice. *J. Leukoc. Biol.* **2000**, *67*, 457–463. [CrossRef] [PubMed]

140. Michetti, P.; Mahan, M.J.; Slauch, J.M.; Mekalanos, J.J.; Neutra, M.R. Monoclonal Secretory Immunoglobulin-a Protects Mice against Oral Challenge with the Invasive Pathogen *Salmonella-typhimurium*. *Infect. Immun.* **1992**, *60*, 1786–1792. [PubMed]

141. Schley, P.D.; Field, C.J. The immune-enhancing effects of dietary fibres and prebiotics. *Br. J. Nutr.* **2002**, *87*, 221–230. [CrossRef] [PubMed]

142. Galdeano, C.M.; Núñez, I.N.; de Moreno de LeBlanc, A.; Carmuega, E.; Weill, R. Perdigón Impact of a probiotic fermented milk in the gut ecosystem and in the systemic immunity using a non-severe protein-energy-malnutrition model in mice. *BMC Gastroenterol.* **2011**, *11*, 64. [CrossRef] [PubMed]

143. Guerrant, R.L.; Oria, R.B.; Moore, S.R.; Oria, M.O.B.; Lima, A.A.M. Malnutrition as an enteric infectious disease with long-term effects on child development. *Nutr. Rev.* **2008**, *66*, 487–505. [CrossRef] [PubMed]

144. Ghoneum, M.; Matsuura, M. Augmentation of macrophage phagocytosis by modified arabinoxylan rice bran (MGN-3/biobran). *Int. J. Immunopathol. Pharm.* **2004**, *17*, 283–292. [CrossRef] [PubMed]

145. Wang, J.; Niu, X.; Du, X.; Smith, D.; Meydani, S.N.; Wu, D. Dietary supplementation with white button mushrooms augments the protective immune response to *salmonella* vaccine in mice. *J. Nutr.* **2014**, *144*, 98–105. [CrossRef] [PubMed]

![microorganisms logo] *microorganisms*

MDPI

Review

Microfluidic-Based Approaches for Foodborne Pathogen Detection

Xihong Zhao [1],*, Mei Li [1] and Yao Liu [2],*

[1] Research Center for Environmental Ecology and Engineering, Key Laboratory for Green Chemical Process of Ministry of Education, Key Laboratory for Hubei Novel Reactor & Green Chemical Technology, School of Environmental Ecology and Biological Engineering, Wuhan Institute of Technology, Wuhan 430205, China

[2] School of Pharmacy and Food Science, Zhuhai College of Jilin University, Zhuhai 519041, China

* Corresponding author: xhzhao2006@gmail.com (X.Z.); yauldliu@163.com (Y.L.)

Received: 20 August 2019; Accepted: 16 September 2019; Published: 23 September 2019

Abstract: Food safety is of obvious importance, but there are frequent problems caused by foodborne pathogens that threaten the safety and health of human beings worldwide. Although the most classic method for detecting bacteria is the plate counting method, it takes almost three to seven days to get the bacterial results for the detection. Additionally, there are many existing technologies for accurate determination of pathogens, such as polymerase chain reaction (PCR), enzyme linked immunosorbent assay (ELISA), or loop-mediated isothermal amplification (LAMP), but they are not suitable for timely and rapid on-site detection due to time-consuming pretreatment, complex operations and false positive results. Therefore, an urgent goal remains to determine how to quickly and effectively prevent and control the occurrence of foodborne diseases that are harmful to humans. As an alternative, microfluidic devices with miniaturization, portability and low cost have been introduced for pathogen detection. In particular, the use of microfluidic technologies is a promising direction of research for this purpose. Herein, this article systematically reviews the use of microfluidic technology for the rapid and sensitive detection of foodborne pathogens. First, microfluidic technology is introduced, including the basic concepts, background, and the pros and cons of different starting materials for specific applications. Next, the applications and problems of microfluidics for the detection of pathogens are discussed. The current status and different applications of microfluidic-based technologies to distinguish and identify foodborne pathogens are described in detail. Finally, future trends of microfluidics in food safety are discussed to provide the necessary foundation for future research efforts.

Keywords: foodborne pathogens; microfluidic chip; rapid detection; food safety; biosensors

1. Introduction

With the rapid development of the economy and the continuous improvement of living conditions, people today are paying more and more attention to health issues. At the same time, whether food is safe or not is also closely related to people's health, therefore, it is also very important to ensure the safety of food. Unfortunately, people sometimes unconsciously eat some foods that are harmful to the body in their daily lives, for example, food contaminated by pathogens [1,2]. If people eat food containing foodborne pathogens such as *Staphylococcus aureus*, *Salmonella*, and *Escherichia coli* O157:H7, they may suffer vomiting or even death, triggering consumer panic [3]. According to the statistics of Parisi et al. a quarter of the world's people are at higher risk of foodborne illnesses due to the current inefficient detection technology of bacteria, the imperfect food supervision system and high-speed economic development [4]. Overall, new strategies should be applied to improve food safety.

Foodborne illnesses are caused by pathogens or their toxins when they are contained in food or water. Pathogens causing foodborne illnesses include bacteria, viruses, fungi, and parasites [5]. For example, people infected by pathogenic *Escherichia coli* (*E. coli*) often experience severe diarrhea, and there are nearly 1.7 billion cases of diarrhoea every year in the world. More seriously, approximately 760,000 children under the age of five die each year from diarrhoeal diseases [6–8]. Most diseases are attributed to the common foodborne pathogens that include *Listeria monocytogenes*, *E. coli* O157:H7, *Staphylococcus aureus*, *Salmonella enterica*, *Bacillus cereus*, *Campylobacter jejunum*, and *Clostridium perfringens* [9]. Therefore, the effective detection of these pathogens is important.

At present, there are many methods to identify and detect pathogens, such as direct smear microscopy, nucleic acid hybridization, gene chip, polymerase chain reaction (PCR), gas chromatography and high performance liquid chromatography [10]. However, the most classic method is the plate cultivation method. However, this method requires three to seven days for bacterial culture, making it inappropriate for the rapid on-site detection of pathogens [11]. Additionally, PCR is also sometimes prone to false positive results due to DNA contamination [12,13]. Thus, to assess food safety, it is necessary to develop a rapid and simple method with high sensitivity, good reproducibility, and good on-site interpretation ability [14].

For this reason, microfluidics with the advantages of portability, miniaturization and automation have been widely introduced to detect different substances in the fields of chemistry, biomedicine, optics and information science, such as dyes, bacteria or heavy metals [15,16]. Microfluidics are typically made of silicon, glass, quartz or thermoplastic materials. Then, micro-processing techniques are used to integrate micro-valves, micro-pumps, micro-mixers, micro-electrodes onto a micro/nanoscale chip to form a network-like system that can achieve pretreatment, mixing, reaction, separation or detection of the sample, which is not possible in traditional laboratories [17]. Microfluidics have several different types of basic mixer structures, as shown in Figure 1. For example, a microfluidic fluorescence quantitative PCR system with pneumatic valve and a tree structure was developed by using 3D printing technology. Due to its good temperature uniformity and thermal conductivity of PCR-based microfluidics, the rapid detection of hepatitis B virus nucleic acid in blood samples was realized in 50 min [18]. At present, the miniaturization, integration and automation of these devices combined with multiple processes have made microfluidic chips popular options for use in a wide range of fields, and the following are the applications and research status of microfluidics for bacterial detection.

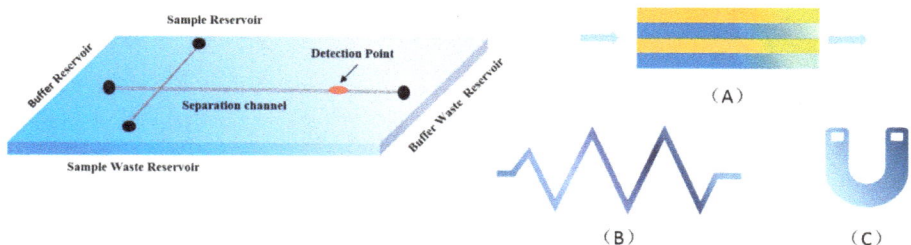

Figure 1. The structure of common microfluidic chip channel and variable styles of passive mixers. (**A**) Lamination; (**B**) Zigzag channels; (**C**) Serpentine.

In general, traditional microbial culture techniques require the use of tubes, culture dishes, multiwall plates and flasks, which makes the detection of bacteria more complicated. However, Wang et al. only combined a nano-dielectrophoretic enrichment-based microfluidic platform with surfaced-enhanced Raman scattering (SERS) to successfully and automatically monitor *Escherichia coli* O157:H7 in drinking water (the detection limited to single cell level) [19]. Wan et al. also developed a digital microfluidic system based on loop-mediated isothermal amplification (LAMP) for the detection of pathogen nucleic acids. In this experiment, only 1 μL of LAMP reaction sample that belong to purified *Trypanosoma cruzi* DNA was required, which reduced a 10-fold of reagent consumption compared to conventional LAMP. If the sample of LAMP is unknown, it also can be finished in 40 min with a detection limit of 10 copies/reverse. Moreover, the system can be thermally adjusted in real time, which is possible for the miniaturized, portable and on-site application in detecting bacteria in the future [20]. However, in order to reduce the costs and improve the portability of detection, high-performance materials such as paper-based microfluidic chips have been applied. Jokerst et al. designed a paper-based assay device for detecting *E. coli* O157:H7, *Salmonella typhimurium* and *Listeria monocytogenes*. The paper that was used for the preparation of the microfluidic system was wax printing on filter paper, which was achieved by measuring the color change of the response of the enzyme associated with the pathogen of interest to the chromogenic substrate. When combined with an enrichment procedure, the method allowed for an enrichment time of 12 h or less and was able to detect the bacteria in meat at a detection limit of 10 colony-forming units/cm^2 [21].

As described above, microfluidic devices are simple, automated, and portable miniaturized systems that can perform functions more efficiently and conveniently than the common techniques such as PCR and LAMP [22,23]. Although there are some reviews on the application of microfluidic chip technology in food safety, their overall systematisms and integrity are not enough. This review not only describes the latest developments in integrated-microfluidic systems for detecting foodborne pathogens, but also discusses the most promising strategies to address current challenges for the faster and more accurate detection of foodborne pathogens by microfluidic chips.

2. Microfluidic Chips

Microfluidic chips refer to the science and technology of systems that process or use very small volumes of liquids in channels with the dimensions of tens to hundreds of micrometres [24,25]. Microfluidics also are described as lab-on-a-chip (LOC) or miniaturized total analysis systems (μ-TAS), which integrate the sample preparation, reaction, separation, detection, and other basic operating units onto a centimeter-scale chip with a network of microchannels [26]. Microfluidics is an interdisciplinary field, including aspects of physics, chemistry, engineering, and biotechnology [24,27]. Due to the characteristics of electro-hydrodynamics with small size parameters and short detection times, electrodynamics, and thermal capillary phenomena, microfluidic devices have been developed to address specific scientific problems that are not able to be easily solved by traditional techniques [28,29].

Manz et al. first proposed the concept of a micro total analysis system (μ-TAS) [30]. In 1992, micro-electro-mechanical machining technology was used to etch micro-pipes on flat glass to prepare a chip capillary electrophoresis device, and the device realized the separation of fluorescently labeled amino acids and pioneered microfluidic chip technology [31]. In 1995, Woolley and Mathies successfully performed DNA sequencing using their own electrophoresis chip system, reading 150 bases in 540s with an accuracy rate of 97% [32]. Subsequently, Woolley et al. integrated PCR and capillary electrophoresis on a microfluidic chip, facilitating genetic analysis [33]. In 1998, Brahmasandra et al. (1998) used photolithography technology to fabricate a microfluidic chip that included a liquid sampler, a mixer, a positioning system, a temperature-controlled reaction chamber, an electrophoresis separation system, and a fluorescence detector system for DNA analysis [34]. In 2000, Anderson et al. developed a highly integrated chip that can be used to process a series of complex processes for multiple samples, and this device was applied for extracting concentrated nucleic acids from a liquid sample for microcrystalline chemical amplification, enzymatic reaction, hybridization, mixing, and measurement, allowing more than 60 consecutive operations of a dozen reactants [35].

Microfluidic devices are mainly operated by manipulating fluids in microfabricated channel and chamber structures. Additionally, microfluidics can be combined with diverse detection techniques including PCR, LAMP, mass spectroscopy, or fluorescence spectroscopy, for on-chip or after-chip detection of analytes [36–39]. Microfluidic chips are made of silicon, glass, quartz, organic polymer, and composite materials by micromachining technology. Figure 2 shows the preparation process of polydimethylsiloxane (PDMS) microfluidics. Recently, paper-based microfluidic chips with low cost, portability and easy operation have been developed in the food industry [40]. The selection of a certain material for a device is important for its functions. The different materials exploited for the fabrication of microfluidic chips and the advantages and disadvantages of these materials are listed in Table 1.

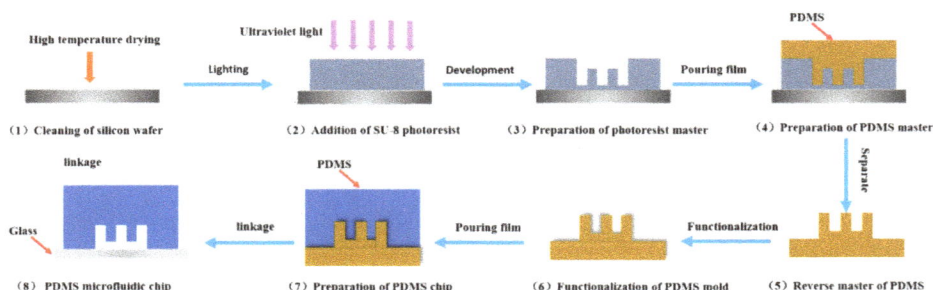

Figure 2. The preparation process of a polydimethylsiloxane (PDMS) microfluidic chip by the molding method.

Table 1. The application of a microfluidic system made of different materials.

Material Type	Classification	Representative	Methods of Preparation	Advantages	Disadvantages	Application	References
organic material	——	glass/quartz	photolithography and etching techniques	cheap and easy to obtain, reusable, good light transmission and electroosmosis, good electrical insulation and corrosion resistance	complex manufacturing process, time-consuming and high cost, fragile	gas chromatography and capillary electrophoresis (CE) and electrochemical detection, organic synthesis and droplet formation, PCR	[41,42]
	silicon material	silicon/silicon dioxide	etching techniques	mature process, good thermal stability and inertness.	high cost of materials, opaque, brittle, poor electrical insulation, and low adhesion coefficient	organic synthesis and droplet formation, PCR and CE	[43,44]
	elastomers	polydimethylsiloxane (PDMS)	molding and soft lithography	Low cost and easy to use, non-toxic and transparent, excellent chemical inertness and light transmission	Incompatibility of organic solvents and poor pressure resistance, low thermal conductivity and immature processing technology	protein crystallization and bioculture, PCR	[45,46]
Polymer materials	thermosets	SU-8 photoresist and polyimide	photopolymerization and casting	High resistance of temperature and most solvents, transparent and reusable	high cost of materials	CE, organic synthesis and droplet formation, PCR	[47,48]
	thermoplastics	poly (methyl methacrylate (PMMA) polystyrene (PS) and polycarbonate (PC)	hot embossing and laser ablation	good electrical insulation and light transmission, low cost and easy to use, simple preparation and high precision	Non-breathable, high-cost preparation equipment and rough process	CE and PCR, droplet formation	[49,50]
	perfluoropolymers	perfluoroalkoxy (PFA) and fluorinated ethylene propylene	photolithography	Good inertness and antifouling properties, transparent and soft	poor adhesion	environmental monitoring and food analysis	[51]

Table 1. *Cont.*

Material Type	Classification	Representative	Methods of Preparation	Advantages	Disadvantages	Application	References
Special materials	hydrogels	polyvinyl alcohol (PVA)	photopolymerization, casting	high permeability and controllable aperture, allowing small molecules or even biological particles to diffuse, and biocompatible	difficult to store	3D bioculture	[52]
	ceramics	polysiloxane	soft lithography and laser ablation	high resistance of temperature and pressure	poor light transmission, fragile	suitable for applications under harsh conditions	[53]
	paper	analysis filter paper	photolithography and printing	high permeability and low cost, portable and easy to use	easy to damage and disposable	bioculture	[54]

Compared to traditional methods such as PCR, enzyme-linked immunosorbent or DNA probes, microfluidic devices allow for a flexible combination of multiple operating units and overall controllability, so some steps such as sample pretreatment, mixing or reaction can be integrated into a single chip. Additionally, because the channel structure in the chip is micron-scale or even nanoscale, it has a high specific surface area, a high diffusion coefficient, and fast heat transfer, effectively accelerating the reaction in the channels and greatly shortening the overall analysis time [55,56]. For example, Zhang et al. developed a novel microfluidic liquid phase nucleic acid purification chip that can selectively separate DNA or RNA from 5000 μL to single cell bacterial cells. The sample volume is only 1 μL or 125 nL, which can be directly quantified by a chip in approximately 30 min. Thus, these small devices also require much lower amounts of reagents and samples, which greatly reduces the cost of detection and enables fast and low-cost detection [57]. Nevertheless, everything in the world has two sides. Without exception, microfluidic technology has its own disadvantages. For example, there is no skillful and mature technique for preparing a good microfluidic system and there is a lack of good and perfect preparation materials. Overall, the advantages of microfluidic technology confer promising potential for high-efficiency screening, environmental monitoring, clinical monitoring, on-site analysis, and DNA sequencing applications.

3. Sample Preparation in Microfluidics

3.1. For Single Component

For a single component, there is no special and complicated separation and purification treatment of the sample needed. However, how to improve the sensitivity, speed and accuracy of the detection component is particularly important. At present, the molecular technologies, such as PCR, first need to extract the DNA of the bacteria, and also add the required reagents by labour, which is extremely time-consuming and troublesome. In addition, if the concentration of the analyte does not reach a measurable level, it is also necessary to concentrate the bacterial DNA concentration multiple times [58]. Therefore, other methods, such as optical analysis, fluorescence detection or electrochemical analysis, can avoid the pretreatment of samples and achieve automated, simple and rapid bacterial detection.

PCR is routinely applied to detect some components in food. Therefore, Tachibana et al. developed a new PCR-based microfluidic technique for the successful detection of 0.031 μg/μL of *E. coli* O157:H7 genomic DNA, which was completed in 18 min and provided a new platform for a rapid, simple and low-cost detection assay for this pathogen. However, there is still a need to further improve the bacterial pre-enrichment and DNA purification steps to lower the detection limit of *E. coli* O157:H7 in the integrated PCR system (10^3 CFU/mL) [59]. Zhang et al. used magnetic silica beads and a special coaxial channel to optimize the detection of *E. coli* O157:H7. This special channel allows the improved separation and capture of the lysed DNA of *E. coli* O157:H7 using magnetic materials. With this modified system, *E. coli* was successfully detected by microfluidic PCR with a detection limit of only 12 CFU/mL [57].

To lower the cost and the difficulty of sample preparation, and increase the portability of testing, high-performance materials, such as paper-based microfluidic chips, are being developed to detect such pathogenic bacteria. Wang et al. proposed a paper-based impedance immunosensor for detecting *E. coli* O157:H7. Gold nanoparticles grew on the working electrode and anti-*E. coli* O157:H7 antibody immobilized on the paper electrode was used to capture the target bacteria, which changed the resistance of the reaction in different environments and successfully detected *E. coli* O157:H7 from ground beef (LOD of 1.5×10^4 CFU/mL) and cucumber (LOD of 1.5×10^3 CFU/mL) [60]. Moreover, there are also numerous introductions and research on paper-based microfluidics. Cate et al. reviewed the preparation, principles, and application of paper-based microfluidic chips [61]. Liu et al. also described recent developments, trends, and challenges of paper-based microfluidic chips for food safety applications [62]. Therefore, the detection process for pathogens is mainly to optimize the detection technology.

3.2. Complex Components in Food Matrix

To date, most techniques that are used to determine some experimental samples are relatively simple just for a single component. However, for practical use, complex samples and variability in environmental conditions may result in reduced sensitivity and specificity of microfluidic technology, so a device should be designed to analyze more complex samples, such as soil, sewage, or food samples [63]. Due to the physical and chemical properties of each component to be tested in the sample may not be much different than those of single components (such as the detection of different bacteria), it may be difficult to achieve simultaneous detection of multiple components. Additionally, the variety and content of other substances in complex food matrices may interfere with the detection and reduce the accuracy of the assay. For example, if the aim is to detect *E. coli* in food samples, there is definitely more than one kind of bacteria in this concentrated sample. Thus, in order to eliminate these interferences, some specific bio-recognition molecules can be integrated into the detection system. It is by increasing the concentration and specificity of the sample that makes it easier to detect the components from complex mixtures.

3.2.1. Special Materials and Sampling Methods

The main cause of low reproducibility or the inability of microfluidic devices to detect analytes in complex food substrates is due to the concentration of analytes below the detection limit. Thus, the separation and enrichment of targets from a food matrix are needed to increase the efficiency for detecting analytes. The sample concentration can be improved using a variety of techniques such, as magnetic beads or filter membranes. Among them, the magnetic beads generally have superior paramagnetism, such as Fe_3O_4, which is able to separate from the sample to be tested with the help of a magnetic field and a rich surface-active group. The filtration membranes are usually made of a variety of ultra-high-performance polymers, which have acid and alkali resistance or oxidation resistance to achieve the separation and purification of the samples. Furthermore, since the reaction is performed in a microfluidic system, different injection methods may improve the concentration of the target [18,64].

Immunomagnetic separation (IMS) can be used to concentrate bacterial cells present at lower concentrations, but it is just suited to a small volume sample (e.g., 1 mL), which is far smaller than the large volume of enrichment culture (e.g., 250 mL). To address this issue, Ganesh et al. integrated IMS of bacterial cells into microfluidic devices for the preconcentration of 50 mL volume samples. PCR was then applied for the qualitative and quantitative detection of the *E. coli* O157:H7 in less than two hours. This platform decreased both the required sample volume and the overall time of the reaction [65]. Oh et al. combined loop-mediated isothermal amplification (LAMP) with a disk-shaped centrifugal microfluidic device to successfully detect four foodborne pathogens (*Escherichia coli* O157:H7, *Salmonella typhimurium*, *Vibrio parahaemolyticus* and *Listeria monocytogenes*) in contaminated milk samples with bacteria. The use of Eriochrome Black T (EBT) in the system allowed the colorimetric detection of the LAMP reaction, and this process enabled a fully automated detection of bacteria with a detection limit of 10 bacterial cell level in 65 min [66]. However, the colorimetric measurement of this platform is identified by the naked eye, which may cause some errors in the interpretation of the experimental results. For this reason, Sayad et al. utilized calcein as an indicator and combined it with LAMP for a genotypic analysis of eight strains of the foodborne pathogens *E. coli* O157:H7, *Salmonella* and *Vibrio cholerae*, for a total of 24 pathogenic bacteria being detected. The result of the colorimetric method was analyzed and transmitted to a smartphone using a developed electronic system that interfaced with bluetooth wireless technology in 60 min. This system avoids artificial subjective errors and achieves a fully automated, quick and on-site test [67].

The above experiments use special materials to detect bacteria, but filter membranes can also be used to increase the concentration of the target. Li et al. used a poly sulfone hollow-fiber membrane module to separate and concentrate bacterial cells from chicken homogenates in cross-flow microfiltration. This special microfluidic system can effectively recover 70% of the analytes in the mixture in 30-45 min, greatly improving the concentration of analytes and decreasing experimental time (approximately 6 h in the industry) [68]. However, special microfluidic injection channels can also be used. For example, Shu et al. integrated multiple PCR steps into microfluidics by preparing special continuous-flow channels. With this special device, the genes of *S. enterica*, *L. monocytogenes*, *E. coli* O157:H7, and *S. aureus* could be simultaneously amplified and detected from banana, milk, and sausage samples. The whole experiment required only 19 min, with a detection limit as low as 10^2 copies/μL [69].

3.2.2. Bio-Recognition Molecules

Even if the concentration of analytes can be increased, the detection of the target in the presence of some similar components is challenging. Therefore, some biomarkers capable of specifically recognizing the analyte are required to achieve the rapid and accurate detection of the target. As shown in Figure 3, there is the high specific interaction between some surface antigen biomarkers and recognition molecules.

Antibodies are one of the most common bio-recognition molecules. Savas et al. used a biosensor-conjugated antibody on gold nanoparticles to successfully detect *Salmonella* from human stool samples. The fully automated microfluidic electrochemical sensor allowed *Salmonella*, as low as 1 CFU/mL, to be sensitively and specifically detected in mixed samples by a specific reaction between the specific antibody and the antigen on the surface of the bacteria [70]. As an alternative to antibodies, aptamers are single-stranded nucleic acid molecules that are stable, easy to synthesize, and cheaper than antibodies. Aptamers can also specifically bind target molecules and can be modified with various fluorescent dyes or other labels. Wu et al. first separated and concentrated analytes from a mixed solution using the property of Fe_3O_4 magnetic nanoparticles. Next, according to the specificity of the aptamer of different bacteria, color-changing upconverting nanoparticles conjugated with different aptamers were used as a signal probe to detect three corresponding pathogenic bacteria. The color change of the multi-color upconverting nanoparticle composite indicated whether the bacteria existed in the mixture to achieve simultaneous, sensitive and selective detection [71].

Lectins can also be used as a bio-recognition molecule. Kang et al. studied different sizes of magnetic nanoparticles coated with lectins for the capture of pathogenic bacteria from mixed solutions. The result showed that magnetic nanoparticles with a radius of 250 nm were the most effective method for separating and detecting *S. aureus* in a mixed solution (10^2 CFU/ mL) [72]. Another study used concanavalin A (ConA), a mannose/glucose-binding lectin that can be used to recognize lipopolysaccharides exposed to bacterial surfaces. Dao et al. combined ConA-functionalized microfluidic chips with LAMP to capture and enrich *Salmonella typhimurium* in urine samples (10 mL).

Through this integrated system, the label-free, fast and real-time detection of *Salmonella typhimurium* with a concentration as low as 5 CFU/mL was completed in 100 min [73].

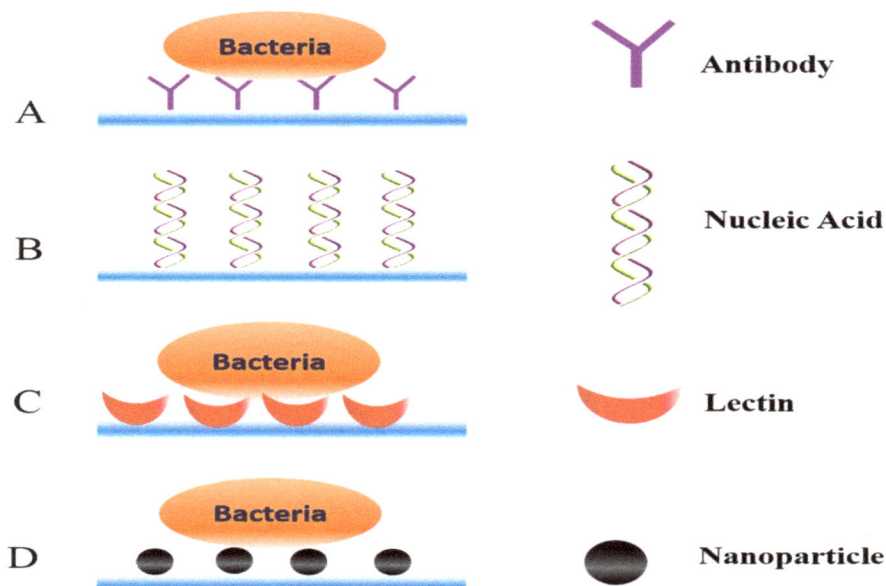

Figure 3. Schematic view of different bio-recognition elements in microfluidics. (**A**) Antibody; (**B**) Nucleic Acid; (**C**) Lectin; (**D**) Nanoparticle.

4. Application of Microfluidic Combined with Different Technologies

Currently, traditional technologies, such as PCR, ELISA and LAMP, are accurate and effective, but they may be costly and complicated [74,75]. Furthermore, for food or other complex environmental samples, the acquisition of the analytes may be difficult or it may be challenging to completely integrate the separation and detection processes in a single microfluidic chip [76]. In particular, if the physical and chemical properties of each component to be tested in the sample are similar, it may be difficult to simultaneously distinguish and detect various substances. The future work should aim to decrease pre-processing or to combine pre-processing steps with detection for the analysis of foodborne pathogens. The successful application mainly depends on high efficiency, high speed, and the automation of microfluidic technology, combined with different technologies, such as electrochemical biosensors, optical biosensors, immunoassays and nucleic acid-based methods [77].

4.1. Biosensor-Based Microfluidics for the Detection of Foodborne Pathogens

Biosensors are developed based on knowledge from the disciplines of biology, chemistry, physics, medicine, and electronic technology. A biosensor is sensitive to biological substances and can convert signals, such as the concentration and activity of analytes, into electrical signals for rapid detection [78–80]. Safavieh et al. used a microfluidic electrochemical biosensor that combined with LAMP for the detection and quantification of *E. coli*. There is no need of probe immobilization, and bacterial detection can be done in a single chamber without DNA extraction and purification steps. This experiment can detect and quantify bacteria to 24 CFU/mL and 8.6 fg/μL of DNA within 60 min [81]. This shows the use of biosensors in microfluidic chips may provide integrated systems with improved sensitivity and rapid and on-line detection. Biosensors include both the optical biosensors and electrochemical biosensors [82].

4.1.1. Microfluidic Chips with Optical Detection

Surface Plasmon Resonance (SPR) Biosensors

Surface plasmon resonance (SPR) is a high-sensitivity and real-time spectral analysis technique that measures the change in the refractive index of a surface material on a metal film. The application of SPR for the detection of analytes is shown in Figure 4. The advantage of SPR is that the object tested is label-free, and the method is easy and quick, allowing dynamic and real-time monitoring of the reaction [83,84]. At present, SPR has been used to detect pathogenic bacteria, allergens, and toxins [85].

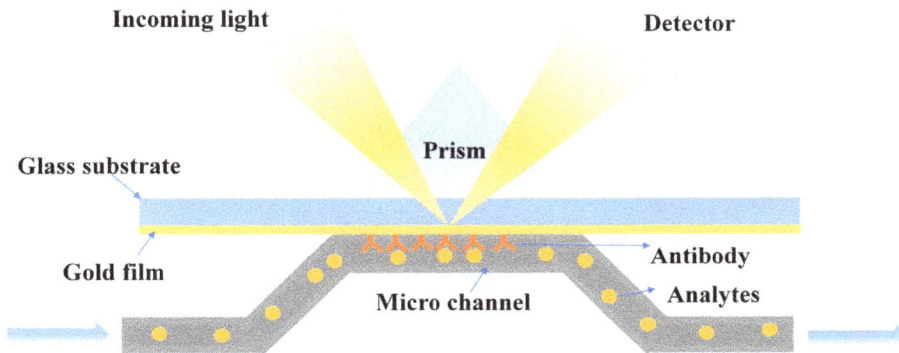

Figure 4. The principle of surface plasmon resonance detection.

Zordan et al. designed a hybrid microfluidic biochip for the detection of pathogens using SPR combined with fluorescence imaging. An array of gold spots was included in the microfluidic system to specifically capture the specific pathogens. A closed polydimethylsiloxane (PDMS) microfluidic flow chamber was used to transport and magnetically concentrate the sample to be tested. SPR and fluorescence were then used for the successful detection of *E. coli* O157:H7 [86]. The Zordan's group also developed a biosensor array chip to specifically detect the presence of different pathogens. In this design, the PDMS microfluidic system allowed SPR and fluorescence imaging for simultaneous, rapid, label-free, real-time and multiple detection of foodborne pathogens. Furthermore, the functionalized magnetic particles were applied to a hybrid microfluidic biochip [87]. Tokel et al. prepared a portable, low cost and multiplexed microfluidic system that used SPR to detect and quantify *E. coli* and *S. aureus*. As a result, 100 μL of *E. coli* or *S. aureus* in phosphate buffered saline and peritoneal dialysis solution at a concentration of 10^5 to 3.2×10^7 CFU/mL can be reliably and specifically detected within 20 min [88].

Optical Fibre Biosensors

Fiber-optic biosensors can selectively interact with a specific biosensor (i.e., antigen-antibody or enzymes), resulting in the production of biological or chemical information that can be converted into a transmitted light signal captured by the optical fiber, with varying light intensities, light amplitude, or phases [89]. An ideal sensor has good selectivity and high sensitivity for bacterial pathogens, pesticides, and toxins [90,91]. However, the spectra generated by the complexes or products formed in the experiments are similar, so the fibers are unable to be easily distinguished and detected. Therefore, the indicators or labels, such as enzymes, fluorescent substances, acid-based indicators, and lanthanide complexes are often used. Instead, optical fiber biosensors are mostly used in conjunction with various spectroscopy techniques such as absorption, fluorescence, or surface enhanced Raman spectroscopy (SERS) to improve sensitivity.

The Raman signal from molecules located near a nano-structured metallic surface and excited by visible light can be strongly enhanced, a process known as surface enhanced Raman scattering (SERS).

SERS is widely used in the detection of foodborne pathogens. Li et al. invented a microfluidic chip with an integrated nanoporous gold disk array, a highly effective SERS substrate. The integrated system has an order of magnitude of a larger surface area than its projected disk area, corresponding to a great improvement of the Raman signal. Rhodamine was used to test the performance of the microfluidic device, showing excellent and rapid detection [92]. Mungroo et al. developed a microfluidic device with silver nanoparticles to improve the detection of pathogenic bacteria. The data analysis included homometric, principle component, and linear discriminant analyses. This platform allowed the detection and discrimination of multiple major foodborne pathogens: *E. coli* O157:H7, *Salmonella*, *S. enteritidis*, *P. aeruginosa*, *L. monocytogenes*, and *L. innocua* [93].

Gilli et al. designed a disposable plastic sensing device that utilized a total internal reflection fluorescence optical. There is no interference caused by non-specific binding or noise, and the microfluidic chip is connected with automated and sensitive customized software to realize the multiplex detection of the different targets [94].

4.1.2. Microfluidic Chip with Electrochemical Detection

Electrochemical biosensors use electrodes as conversion elements and immobilize bio-sensitive substances including antigens, antibodies, or enzymes onto the electrode to detect target molecules by specific bio-recognition and antigen interaction [95]. The above reactions can be transformed into electrical signals, such as capacitance, current, potential, or conductivity, to achieve the qualitative or quantitative detection of analytes, resulting in powerful tools for the detection of biological samples [96–98].

Tan et al. developed a stable PDMS microfluidic device with an impedance immunosensor by grafting modified silane and an antibody on nanoporous membranes for the specific measurement of *E. coli* O157:H7 and *S. aureus*. The difference between these bacteria was expressed by monitoring the amplitude change of the impedance spectrum before and after the bacteria captured by complimentary antibodies on the nanoporous alumina membrane, which achieved a rapid and sensitive bacterial assay of 10^2 CFU/mL in 2 h [46]. Chen et al. developed a fast, sensitive and complex microfluidic device that integrated electrochemical impedance analysis and urease catalysis to measure *Listeria*. The bacteria cells, the modified magnetic nanoparticles (MNPs) with anti-*Listeria* monoclonal antibodies, anti-*Listeria* polyclonal antibodies, and the urease modified gold nanoparticles (AuNPs), were incubated in an integrated microfluidic chip with active mixing to form MNP-*Listeria*-AuNP-urease sandwich complexes. Through this platform, *Listeria* can be detected as low as 1.6×10^2 CFU/mL in one hour [99].

Overall, the use of online, automated, and sensitive microfluidic impedance biosensors for bacterial separation and detection is promising. To improve the effectiveness of these systems, Liu et al. integrated dielectrophoresis and electrochemical impedance into microfluidics for in-situ impedance detection of bacteria. The dielectrophoresis technique was applied to enrich trace bacteria. The microarray electrode microfluidic chips can specifically detect bacteria from microsystems. The detection limit of *E. coli* O157:H7 in this device was 5×10^4 CFU/mL in 6 min [100]. This integrated microfluidic analysis microsystem is the first step for the rapid real-time in situ detection of bacteria.

The above devices are impedance-based for the detection of foodborne pathogens. In addition, there are voltametry-based microfluidics. Safavieh et al. used LAMP in a microfluidic system for the quantitative detection of *E. coli* O157:H7 and *S. aureus* using the linear sweep voltametry method. The foodborne pathogens with a detection limit of 48 CFU/mL were detected in 35 min. Unlike other electrochemical techniques, this method does not require a complex probe immobilization process, and bacterial detection can be performed in the chamber structure without the need for DNA extraction and purification steps [81].

4.2. Immunoassay-Based Microfluidics for the Detection of Foodborne Pathogens

The immunological methods offer high specificity, high sensitivity, and high analytical capacity based on the specific reaction between the antigen and antibody to form a complex, as shown in

Figure 5. The immunological approaches have been used to detect bacteria, viruses, fungi, various toxins, parasites, proteins, hormones, other physiologically active substances, drug residues, and antibiotics [101]. The determination of pathogenic bacteria by immunological methods alone is prone to cross-contamination risks and negative results, and requires trained personnel. However, when combined with microfluidic technology and immunoassays, specific antigen-antibody reactions can enhance the specificity and sensitivity of microfluidic analysis. Additionally, the use of microfluidics is rapid, has low-consumption, and automated compared to traditional immunology techniques, such as ELISA, lateral flow assays (LFAs), or radioimmunoassays (RIAs), which may require long detection times, expensive reagents, or complicated procedures [102].

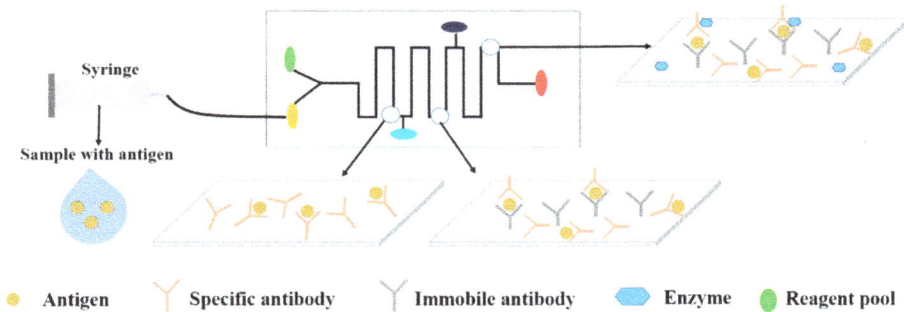

Figure 5. The basic principles of immunoassay-based microfluidics for the detection of foodborne pathogens.

4.2.1. Enzyme-Linked Immunosorbent Assay (ELISA)

In ELISA, a known antigen or antibody on the surface of a solid phase carrier (polystyrene microplate) is bound in an enzyme-labeled antigen-antibody reaction, and any free components in the liquid phase are washed away. This method has been applied to the effective and specific detection of pathogenic bacteria [103,104].

Thaitrong et al. designed a microfluidic sandwich ELISA for the rapid determination of plant pathogens. The microfluidic concentrator was fabricated using a microchannel, and the all reactions were in a microfluidic channel with the help of capillary force to drive the flow of the reactants [104]. Compared to traditional methods, this microfluidic system is faster, more portable, energy-efficient, and protected against sample contamination, providing a new approach for the detection of pathogens [92]. In a similar device by Wu et al. analytes were concentrated by mixing iron particles with PDMS to form an electromagnetically-driven microdevice that could be controlled by the application of a magnetic field [105], as described by Yanagisawa and Dutta [106].

4.2.2. Immunomagnetic Fluorescence Assay (IMS)

The IMS assay is also based on the reaction between the antigen and antibody. When the IMS assay combined with microfluidic-based technology, the performance of the IMS assay can be more highly specific and sensitive, fast, and convenient. Zhang et al. connected an optical fiber spectrometer with a microfluidic device to achieve the rapid and sensitive detection of avian influenza virus. The integrated device allowed the immunomagnetic capture, concentration, and fluorescence detection of foodborne pathogens [107]. Similarly, Kanayeva et al. combined magnetic nanoparticles, a microfluidic chip, and an interdigitated microelectrode to integrate an impedance immunosensor for the efficient separation and sensitive detection of *L. monocytogenes* [108].

There are several other microfluidic based immunological methods, such as lateral flow assays (LFAs) [109,110] and RIAs [111]. Although the combination of immunology and microfluidics has greatly improved its performance, further improvement is possible. For example, non-specific binding is a problem and increased objectivity is required for result interpretation, as it can lead to wrong results and affect later experiments [112].

4.3. Nucleic Acid-Based Microfluidics for the Detection of Foodborne Pathogens

Nucleic acid-based methods can be used to detect a certain sequence of DNA or RNA from pathogens, using capture and detector probes (short DNA or RNA sequences). These methods can provide more specific and accurate results than the above methods. The integration of nucleic acid-based detection technology in microfluidic devices has been widely applied in various fields due to the small-volume sample requirement, fast detection time, and simple sample processing, especially for the detection of foodborne pathogens [56,113]. Nucleic acid-based detection methods include PCR, LAMP, and recombinase polymerase amplification (RPA).

4.3.1. Polymerase Chain Reaction (PCR)

Ganesh et al. designed an integrated microfluidic PCR system consisting of two main components: A preconcentration chamber for the immunomagnetic separation of bacterial and a PCR chamber for DNA amplification. Further, *E. coli* O157:H7 with the detection limit of 10^3 CFU/mL was successfully detected by the integrated system [65]. Zhang and Wang developed an integrated microfluidic platform with silica superparamagnetic particle-based solid phase extraction for cell lysis, DNA binding, washing, elution, and PCR on a single platform [114]. The preparation, principle, and usage of specific PCR-based microfluidic chips have been described [58,59].

4.3.2. Multiplex PCR

Zhang et al. reported a flow-based multiplex PCR microfluidic system capable of high-throughput and rapid DNA amplification to detect foodborne pathogens. The system consisted of four reaction channels to simultaneously detect *L. monocytogenes*, *E. coli* O 157:H7, and *S. enterica* from food samples. Multiplex PCR with a special injection device of oscillatory-flow used only 5µL of the sample and the reaction can be completed in 13 min, being one sixth of the time required for conventional PCR (70 min) [115]. Similarly, Shu et al. prepared a segmented continuous-flow multiplex PCR on a special spiral channel microfluidic device that consists of a disposable polytetrafluoroethylene capillary microchannel coiled on three isothermal blocks. The microfluidic device rapidly identified a variety of foodborne pathogens, including *S. enterica*, *L. monocytogenes*, *E. coli* O157:H7 and *S. aureus*. After optimizing the parameters, their genomic DNA of four bacteria were amplified simultaneously at 19 min with a minimum detection limit of 10^2 copies/µL [69].

4.3.3. Loop-Mediated Isothermal Amplification (LAMP)

Compared with enzyme-linked immunosorbent assays, LAMP is a rapid and specific method for nucleic acid amplification. LAMP does not require thermal denaturation, temperature cycling, electrophoresis, or ultraviolet detection, and it shows better sensitivity, specificity, cost and detection range than PCR. Additionally, LAMP does not require a complex temperature gradient regulation for high-throughput rapid detection.

Tourlousse et al. developed a cheap, portable, easy-to-use, single use polymeric microfluidic chip for the quantitative detection of different pathogens by isothermal nucleic acid amplification. The microfluidic chips were able to rapidly and quantitatively detect bacteria DNA of 10–100 genomes/µL in 20 min [116]. Uddin et al. prepared a rapid, automatic and novel microfluidic compact disk platform combined with LAMP and a color sensor for the sensitive detection of different DNA concentrations for *Salmonella*. Furthermore, a disk platform can achieve a simultaneous detection of multiple sets of samples [117]. For simultaneous detection of complex samples, Sun et al. described an eight-chamber microfluidic chip that takes advantage of magnetic bead-based sample preparation and LAMP for the rapid quantitative detection of *Salmonella* in food samples. The system can measure *Salmonella* at concentrations of 50 cells per test within 40 min for rapid on-site screening of foodborne pathogens [118].

However, nucleic acid-based microfluidics for the detection of foodborne pathogens include RPA, a nucleic acid detection technology that allows for single-molecule nucleic acid detection at room temperature within 15 min. This technology is truly portable and fast, with acid-based detection for analytes and low requirements for hardware equipment [119,120]. Other methods for pathogen detection include nuclear acid sequence-based amplification (NASBA) and nuclear acid sequence-based amplification (HAD) [121–123].

5. Challenges and Opportunities

Microfluidics integrates the functions of a full laboratory into a single device, including sampling, dilution, reagent addition, reaction, separation, and detection. The potential applications of microfluidics in the food industry include the detection of foodborne pathogens, but also the detection of pesticide residues, heavy metals, or food additives. Microfluidic devices require lower consumption of reagents, and provide faster screening with shorter reaction times and lower costs. Thus, microfluidic technology provides promising approaches to solve key and complex problems in food safety.

However, the application of microfluidics based on different technologies for the detection of foodborne pathogens is still in its infancy. Although some special materials and bio-recognition molecules can be used to improve the detection of targets in actual samples, there may be some non-specific binding that can influence the results of the experiment. The severity of this problem, based on the composition of the food samples and the variation of the sample pretreatment processes, is described by Li. et al. [124]. However, current analytical systems are relatively immature, so the detection of pathogenic bacteria is not yet precise [72]. Furthermore, a major challenge to be overcome is that existing microfluidic systems are complex or expensive to easily integrate into a functional system, and such ease of integration is required for convenient use in food safety. However, there are great expectations for further innovation and development of highly efficient microfluidic technologies for measuring pathogenic bacteria.

6. Conclusions

Food safety is closely related to human health, therefore, powerful, sensitive and effective tools are needed to ensure food safety, such as the detection of foodborne pathogens. The high selectivity, sensitivity, and efficiency of microfluidic technology can employ these devices to replace some traditional labor-intensive and slow-culture methods for the detection of pathogens in foods. However, capturing effective pathogens from complex food samples for high-throughput multiplex analysis remains inefficient. Consequently, when some traditional methods are combined with microfluidic technologies that can be more effective, the preconcentration and sample preparation steps are typically improved and simplified. This article described the incorporation of rapid detection techniques such as SPR, ELISA, PCR, and LAMP in microfluidic devices for improving the detection efficiency of foodborne pathogens.

Nevertheless, the technologies for preparing microfluidic devices and integrating microfluidics with other detection technologies are imperfect and are in the initial stages of industrialization. Therefore, further exploration and research is needed to expand the application of microfluidic technology in different industries. The authors are confident that microfluidics will be more broadly applied in multiple fields, once these problems are addressed in future studies.

Funding: This work has been supported by the National Natural Science Foundation of China (31501582), Hubei Provincial Natural Science Foundation of China (2018CFB514), Graduate Innovative Fund of Wuhan Institute of Technology (CX2018163 and CX2018167) and Innovation Cultivation Project of School of Pharmacy and Food Science, Zhuhai College of Jilin University(2018YSCP01).

Conflicts of Interest: The authors declare no conflict of interest.

References

1. Chapman, B.; Gunter, C. Local Food Systems Food Safety Concerns. *Microbiol. Spectr.* **2018**, *6*, 34–39. [CrossRef] [PubMed]

2. Zhao, X.; Li, M.; Xu, Z. Detection of Foodborne Pathogens by Surface Enhanced Raman Spectroscopy. *Front. Microbiol.* **2018**, *9*, 1236. [CrossRef] [PubMed]

3. Jones, T.F.; Yackley, J. Foodborne Disease Outbreaks in the United States: A Historical Overview. *Foodborne Pathog. Dis.* **2018**, *15*, 11–15. [CrossRef] [PubMed]

4. Parisi, A.; Crump, J.A.; Glass, K.; Howden, B.P.; Furuya-Kanamori, L.; Vilkins, S.; Gray, D.J.; Kirk, M.D. Health Outcomes from Multidrug-Resistant Salmonella Infections in High-Income Countries: A Systematic Review and Meta-Analysis. *Foodborne Pathog. Dis.* **2018**, *15*, 428–436. [CrossRef] [PubMed]

5. Zhao, X.; Lin, C.W.; Wang, J.; Oh, D.H. Advances in rapid detection methods for foodborne pathogens. *J. Microbiol. Biotechnol.* **2014**, *24*, 297. [CrossRef] [PubMed]

6. Liu, J.Y.; Zhou, R.; Li, L.; Peters, B.M.; Li, B.; Lin, C.W.; Peters, B.M.; Chuang, T.L.; Chen, D.Q.; Zhao, X.H.; et al. Viable but non-culturable state and toxin gene expression of enterohemorrhagic *Escherichia coli* O157 under cryopreservation. *Res. Microbiol.* **2017**, *168*, 188–193. [CrossRef] [PubMed]

7. Torgerson, P.R.; Devleesschauwer, B.; Praet, N.; Speybroeck, N.; Willingham, A.L.; Kasuga, F.; Rokni, M.B.; Zhou, X.N.; Fèvre, E.M.; Sripa, B.; et al. World Health Organization Estimates of the Global and Regional Disease Burden of 11 Foodborne Parasitic Diseases, 2010: A Data Synthesis. *PLoS Med.* **2015**, *12*, e1001920. [CrossRef]

8. Yang, S.C.; Lin, C.H.; Aljuffali, I.A.; Fang, J.Y. Current pathogenic Escherichia coli foodborne outbreak cases and therapy development. *Arch. Microbiol.* **2017**, *199*, 811–825. [CrossRef] [PubMed]

9. Zhao, X.H.; Wei, C.J.; Zhong, J.L.; Jin, S.W. Research advance in rapid detection of foodborne Staphylococcus aureus. *Biotechnol. Biotechnol. Equip.* **2016**, *30*, 827–833. [CrossRef]

10. Umesha, S.; Manukumar, H.M. Advanced Molecular Diagnostic Techniques for Detection of Food-borne Pathogens: Current Applications and Future Challenges. *Crit. Rev. Food Sci. Nutr.* **2018**, *58*, 84–104. [CrossRef]

11. Wei, C.J.; Zhong, J.L.; Hu, T.; Zhao, X.H. Simultaneous detection of Escherichia coli O157:H7, Staphylococcus aureus and Salmonella by multiplex PCR in milk. *3 Biotech* **2018**, *8*, 76. [CrossRef] [PubMed]

12. Zhao, X.; Zhao, F.; Wang, J.; Zhong, N. Biofilm formation and control strategies of foodborne pathogens: Food safety perspectives. *RSC Adv.* **2017**, *7*, 36670–36683. [CrossRef]

13. Zhong, J.; Zhao, X. Isothermal amplification technologies for the detection of foodborne pathogens. *Food Anal. Methods* **2018**, *11*, 1543–1560. [CrossRef]

14. Zhao, X.; Zhong, J.; Wei, C.; Lin, C.W.; Ding, T. Current perspectives on viable but non-culturable state in foodborne pathogens. *Front. Microbiol.* **2017**, *8*, 580. [CrossRef] [PubMed]

15. Holmes, D.; Gawad, S. The application of microfluidics in biology. *Methods Mol. Biol.* **2010**, *583*, 55–80. [PubMed]

16. Yujie, L.I.; Huo, Y.; Di, L.I.; Tang, X.; Shi, F.; Wang, C. Technology, application and development of microfluidics. *J. Hebei Univ. Sci. Technol.* **2014**, *35*, 11.

17. Wen, N.; Zhao, Z.; Fan, B.; Chen, D.; Men, D.; Wang, J.; Chen, J. Development of Droplet Microfluidics Enabling High-Throughput Single-Cell Analysis. *Molecules* **2016**, *21*, 881. [CrossRef] [PubMed]

18. Wang, K.-K.; Ke, Y.; Jun, Z.; Can-Can, Z.; Ling, Z.; Yong, L. Rapid Detection of Hepatitis B Virus Nucleic Acid Based on Microfluidic Chip Using Fluorescence Quantitative PCR. *J. Anal. Sci.* **2018**, *34*, 11–15.

19. Wang, C.; Madiyar, F.; Yu, C.; Li, J. Detection of extremely low concentration waterborne pathogen using a multiplexing self-referencing SERS microfluidic biosensor. *J. Biol. Eng.* **2017**, *11*, 9. [CrossRef]

20. Wan, L.; Chen, T.; Gao, J.; Dong, C.; Wong, A.H.; Jia, Y.; Mak, P.; Deng, C.X.; Martins, R.P. A digital microfluidic system for loop-mediated isothermal amplification and sequence specific pathogen detection. *Sci. Rep.* **2017**, *7*, 14586. [CrossRef]

21. Jokerst, J.C.; Adkins, J.A.; Bisha, B.; Mentele, M.M.; Goodridge, L.D.; Henry, C.S. Development of a Paper-Based Analytical Device for Colorimetric Detection of Select Foodborne Pathogens. *Anal. Chem.* **2012**, *84*, 2900–2907. [CrossRef] [PubMed]

22. Long, H.; Bao, L.J.; Habeeb, A.A.; Lu, P.X. Effects of doping concentration on the surface plasmonic resonances and optical nonlinearities in AGZO nano-triangle arrays. *Opt. Quantum Electron.* **2017**, *49*, 345. [CrossRef]

23. Sun, Y.; Liu, D.M.; Lu, P.; Sun, Q.Z.; Yang, W.; Wang, S.; Liu, L.; Zhang, J.S. Dual-Parameters Optical Fiber Sensor with Enhanced Resolution Using Twisted MMF Based on SMS Structure. *IEEE Sens. J.* **2017**, *17*, 3045–3051. [CrossRef]

24. Squires, T.M.; Quake, S.R. Microfluidics: Fluid physics at the nanoliter scale. *Rev. Mod. Phys.* **2005**, *77*, 977–1026. [CrossRef]

25. Whitesides, G.M. The origins and the future of microfluidics. *Nature* **2006**, *442*, 368–373. [CrossRef] [PubMed]

26. Weng, X.; Neethirajan, S. Paper-based microfluidic aptasensor for food safety. *J. Food Saf.* **2017**, *38*, e12412. [CrossRef]

27. Xu, J.; Kawano, H.; Liu, W.W.; Hanada, Y.; Lu, P.X.; Miyawaki, A.; Midorikawa, K.; Sugioka, K. Controllable alignment of elongated microorganisms in 3D microspace using electrofluidic devices manufactured by hybrid femtosecond laser microfabrication. *Microsyst. Nanoeng.* **2017**, *3*, 16078. [CrossRef] [PubMed]

28. Huang, Z.Q.; He, X.Q.; Liew, K.M. A sensitive interval of imperfect interface parameters based on the analysis of general solution for anisotropic matrix containing an elliptic inhomogeneity. *Int. J. Solids Struct.* **2015**, *73–74*, 67–77. [CrossRef]

29. Hou, M.; Wang, Y.; Liu, S.; Guo, J.; Li, Z.; Lu, P. Sensitivity-Enhanced Pressure Sensor with Hollow-Core Photonic Crystal Fiber. *J. Lightwave Technol.* **2014**, *32*, 4637–4641.

30. Manz, A.; Graber, N.; Widmer, H.M. Miniaturized total chemical analysis systems: A novel concept for chemical sensing. *Sens. Actuators B Chem.* **1990**, *1*, 244–248. [CrossRef]

31. Manz, A.; Harrison, D.J.; Verpoorte, E.M.J.; Fettinger, J.C.; Paulus, A.; Lüdi, H.; Widmer, H.M. Planar chips technology for miniaturization and integration of separation techniques into monitoring systems: Capillary electrophoresis on a chip. *J. Chromatogr. A* **1992**, *593*, 253–258. [CrossRef]

32. Woolley, A.T.; Mathies, R.A. Ultra-high-speed DNA sequencing using capillary electrophoresis chips. *Anal. Chem.* **1995**, *67*, 3676–3680. [CrossRef] [PubMed]

33. Woolley, A.T.; Hadley, D.; Landre, P.; Demello, A.J.; Mathies, R.A.; Northrup, M.A. Functional integration of PCR amplification and capillary electrophoresis in a microfabricated DNA analysis device. *Anal. Chem.* **1996**, *68*, 4081–4086. [CrossRef] [PubMed]

34. Brahmasandra, S.N.; Johnson, B.N.; Webster, J.R.; Burke, D.T.; Mastrangelo, C.H.; Burns, M.A. On-chip DNA band detection in microfabricated separation systems. *Proc. Spie—Int. Soc. Opt. Eng.* **1998**, *3515*, 242–251.

35. Anderson, J.R.; Chiu, D.T.; Jackman, R.J.; Cherniavskaya, O.; Mcdonald, J.C.; Wu, H.; Whitesides, S.H.; Whitesides, G.M. Fabrication of topologically complex three-dimensional microfluidic systems in PDMS by rapid prototyping. *Anal. Chem.* **2000**, *72*, 3158–3164. [CrossRef] [PubMed]

36. Ghaemmaghami, A.M.; Hancock, M.J.; Harrington, H.; Kaji, H.; Khademhosseini, A. Biomimetic tissues on a chip for drug discovery. *Drug Discov. Today* **2012**, *17*, 173–181. [CrossRef]

37. Marre, S.; Jensen, K.F. Synthesis of micro and nanostructures in microfluidic systems. *Chem. Soc. Rev.* **2010**, *39*, 1183–1202. [CrossRef] [PubMed]

38. Mu, X.; Zheng, W.; Sun, J.; Zhang, W.; Jiang, X. Microfluidics for Manipulating Cells. *Small* **2013**, *9*, 9–21. [CrossRef]

39. Zhao, D.; Liu, W.W.; Ke, S.L.; Liu, Q.J. Large lateral shift in complex dielectric multilayers with nearly parity-time symmetry. *Opt. Quantum Electron.* **2018**, *50*, 323. [CrossRef]

40. Chuang, T.L.; Chang, C.C.; Chu-Su, Y.; Wei, S.C.; Zhao, X.H.; Hsueh, P.R.; Lin, C.W. Disposable surface plasmon resonance aptasensor with membranebased sample handling design for quantitative interferon-gamma detection. *Lab Chip* **2014**, *14*, 2968–2977. [CrossRef]

41. Atsushi, K.; Akiko, I.; Tamotsu, Y.; Yoshiaki, U.; Eiichi, T.; Yuzuru, T. Highly sensitive elemental analysis for Cd and Pb by liquid electrode plasma atomic emission spectrometry with quartz glass chip and sample flow. *Anal. Chem.* **2011**, *83*, 9424–9430.

42. Francisca, A.; Neha, S.; Mohammad-Ali, S.; Dongfei, L.; Bárbara, H.B.; Makila, E.M.; Jarno, J.S.; Jouni, T.H.; Pedro, L.G.; Bruno, S.; et al. Microfluidic Assembly of a Multifunctional Tailorable Composite System Designed for Site Specific Combined Oral Delivery of Peptide Drugs. *Acs Nano* **2015**, *9*, 8291–8302.

43. Xuan, T.V.; Stockmann, R.; Wolfrum, B.; Offenhäusser, A.; Ingebrandt, S. Fabrication and application of a microfluidic-embedded silicon nanowire biosensor chip. *Phys. Status Solidi* **2010**, *207*, 850–857.

44. Zhang, H.; Dongfei, L.; Mohammad-Ali, S.; Ermei, M.K.; Bárbara, H.B.; Jarno, S.; Hirvonen, J.; Santos, H.A. Fabrication of a multifunctional nano-in-micro drug delivery platform by microfluidic templated encapsulation of porous silicon in polymer matrix. *Adv. Mater.* **2014**, *26*, 4497–4503. [CrossRef] [PubMed]

45. Crabtree, H.J.; Morrissey, Y.C.; Taylor, B.J.; Liang, T.; Johnstone, R.W.; Stickel, A.J.; Manage, P.; Atrazhev, A.; Backhouse, C.J.; Pilarski, L.M. Inhibition of on-chip PCR using PDMS–glass hybrid microfluidic chips. *Microfluid. Nanofluid.* **2012**, *13*, 383–398. [CrossRef]

46. Tan, F.; Leung, P.H.M.; Liu, Z.B.; Zhang, Y.; Xiao, L.; Ye, W.; Zhang, X.; Yi, L.; Yang, M. A PDMS microfluidic impedance immunosensor for E. coli O157:H7 and Staphylococcus aureus detection via antibody-immobilized nanoporous membrane. *Sens. Actuators B Chem.* **2011**, *159*, 328–335. [CrossRef]

47. Al-Shehri, S.; Palitsin, V.; Webb, R.P.; Grime, G.W. Fabrication of three-dimensional SU-8 microchannels by proton beam writing for microfluidics applications: Fluid flow characterisation. *Nucl. Instrum. Methods Phys. Res. B* **2015**, *348*, 223–228. [CrossRef]

48. Dy, A.J.; Cosmanescu, A.; Sluka, J.; Glazier, J.A.; Stupack, D.; Amarie, D. Fabricating microfluidic valve master molds in SU-8 photoresist. *J. Micromech. Microeng.* **2014**, *24*, 057001. [CrossRef]

49. Floquet, C.F.A.; Sieben, V.J.; Milani, A.; Joly, E.P.; Ogilvie, I.R.G.; Morgan, H.; Mowlem, M.C. Nanomolar detection with high sensitivity microfluidic absorption cells manufactured in tinted PMMA for chemical analysis. *Talanta* **2011**, *84*, 235–239. [CrossRef]

50. Wu, N.; Zhu, Y.G.; Brown, S.; Oakeshott, J.; Peat, T.; Surjadi, R.; Easton, C.; Leech, P.W.; Sexton, B.A. A PMMA microfluidic droplet platform for in Vitro protein expression using crude E. Coli S30 extract. *Lab Chip* **2009**, *9*, 3391–3398. [CrossRef]

51. Stojkovič, G.; Krivec, M.; Vesel, A.; Marinšek, M.; Žnidaršič-Plazl, P. Surface cell immobilization within perfluoroalkoxy microchannels. *Appl. Surf. Sci.* **2014**, *320*, 810–817. [CrossRef]

52. Detlev, B.; Alfred, D.; Frank, K.; Martin, L. Poly (vinyl alcohol)-coated microfluidic devices for high-performance microchip electrophoresis. *Electrophoresis* **2015**, *23*, 3567–3573.

53. Huang, K.W.; Wu, Y.C.; Lee, J.A.; Chiou, P.Y. Microfluidic integrated optoelectronic tweezers for single-cell preparation and analysis. *Lab Chip* **2013**, *13*, 3721–3727. [CrossRef] [PubMed]

54. Hong, B.; Xue, P.; Wu, Y.; Bao, J.; Chuah, Y.J.; Kang, Y. A concentration gradient generator on a paper-based microfluidic chip coupled with cell culture microarray for high-throughput drug screening. *Biomed. Microdevices* **2016**, *18*, 21. [CrossRef] [PubMed]

55. Li, W.L.; Wu, A.; Li, Z.C.; Zhang, G.; Yu, W.Y. A new calibration method between an optical sensor and a rotating platform in turbine blade inspection. *Meas. Sci. Technol.* **2017**, *28*, 035009. [CrossRef]

56. Zeng, D.; Chen, Z.; Jiang, Y.; Xue, F.; Li, B. Advances and Challenges in Viability Detection of Foodborne Pathogens. *Front. Microbiol.* **2016**, *7*, 1833. [CrossRef]

57. Zhang, R.; Hai-Qing, G.; Xudong, Z.; Chaoping, L.; Chunchau, S. A microfluidic liquid phase nucleic acid purification chip to selectively isolate DNA or RNA from low copy/single bacterial cells in minute sample volume followed by direct on-chip quantitative PCR assay. *Anal. Chem.* **2013**, *85*, 1484–1491. [CrossRef]

58. Wu, J.; Kodzius, R.; Cao, W.; Wen, W. Extraction, amplification and detection of DNA in microfluidic chip-based assays. *Microchim. Acta* **2014**, *181*, 1611–1631. [CrossRef]

59. Tachibana, H.; Saito, M.; Shibuya, S.; Tsuji, K.; Miyagawa, N.; Yamanaka, K.; Tamiya, E. On-chip quantitative detection of pathogen genes by autonomous microfluidic PCR platform. *Biosens. Bioelectron.* **2015**, *74*, 725–730. [CrossRef]

60. Wang, Y.; Jianfeng, P.; Zunzhong, Y.; Jian, W.; Yibin, Y. Impedimetric immunosensor based on gold nanoparticles modified graphene paper for label-free detection of Escherichia coli O157:H7. *Biosens. Bioelectron.* **2013**, *4*, 492–498. [CrossRef]

61. Cate, D.M.; Adkins, J.A.; Mettakoonpitak, J.; Henry, C.S.; Chem, A. Recent Developments in Paper-Based Microfluidic Devices. *Anal. Chem.* **2015**, *87*, 19–41. [CrossRef] [PubMed]

62. Liu, C.C.; Wang, Y.N.; Fu, L.M.; Chen, K.L. Microfluidic paper-based chip platform for benzoic acid detection in food. *Food Chem.* **2018**, *249*, 162–167. [CrossRef] [PubMed]

63. Wang, M.F.; Juan, H.U.; Zheng, G.; Zhao, G.H. Application of microfluidic chip in food safety analysis. *Sci. Technol. Food Ind.* **2011**, *32*, 401–404.

64. Zheng, Y.; Mao, S.; Liu, S.; Wong, S.H.; Wang, Y.W. Normalized Relative RBC-Based Minimum Risk Bayesian Decision Approach for Fault Diagnosis of Industrial Process. *IEEE Trans. Ind. Electron.* **2016**, *63*, 7723–7732. [CrossRef]

65. Ganesh, I.; Tran, B.M.; Kim, Y.; Kim, J.; Cheng, H.; Lee, N.Y.; Park, S. An integrated microfluidic PCR system with immunomagnetic nanoparticles for the detection of bacterial pathogens. *Biomed. Microdevices* **2016**, *18*, 116. [CrossRef] [PubMed]

66. Oh, S.J.; Park, B.H.; Choi, G.; Seo, J.H.; Jung, J.H.; Choi, J.S.; Kim, D.H.; Seo, T.S. Fully automated and colorimetric foodborne pathogen detection on an integrated centrifugal microfluidic device. *Lab Chip* **2016**, *16*, 1917–1926. [CrossRef]

67. Sayad, A.; Ibrahim, F.; Mukim, S.U.; Cho, J.; Madou, M.; Thong, K.L. A microdevice for rapid, monoplex and colorimetric detection of foodborne pathogens using a centrifugal microfluidic platform. *Biosens. Bioelectron.* **2017**, *100*, 96–104. [CrossRef]

68. Li, X.; Ximenes, E.; Amalaradjou, M.A.; Vibbert, H.B.; Foster, K.; Jones, J.; Liu, X.Y.; Bhunia, A.K.; Ladisch, M.R. Rapid sample processing for detection of food-borne pathogens via cross-flow microfiltration. *Appl. Environ. Microbiol.* **2013**, *79*, 7048–7054. [CrossRef]

69. Shu, B.; Zhang, C.; Xing, D. Segmented continuous-flow multiplex polymerase chain reaction microfluidics for high-throughput and rapid foodborne pathogen detection. *Anal. Chim. Acta* **2014**, *826*, 51–60. [CrossRef]

70. Savas, S.; Ersoy, A.; Gulmez, Y.; Kilic, S.; Levent, B.; Altintas, Z. Nanoparticle Enhanced Antibody and DNA Biosensors for Sensitive Detection of *Salmonella*. *Materials* **2018**, *11*, 1541. [CrossRef]

71. Wu, S.; Duan, N.; Shi, Z.; Fang, C.; Wang, Z. Simultaneous aptasensor for multiplex pathogenic bacteria detection based on multicolor upconversion nanoparticles labels. *Anal. Chem.* **2014**, *86*, 3100–3107. [CrossRef] [PubMed]

72. Kang, J.H.; Um, E.; Diaz, A.; Driscoll, H.; Rodas, M.J.; Domansky, K.; Rodas, M.J.; Watters, A.L.; Super, M.; Stone, H.A.; et al. Optimization of Pathogen Capture in Flowing Fluids with Magnetic Nanoparticles. *Small* **2015**, *11*, 5657–5666. [CrossRef] [PubMed]

73. Dao, T.N.T.; Yoon, J.; Jin, C.E.; Koo, B.; Han, K.; Yong, S.; Lee, T.Y. Rapid and Sensitive Detection of Salmonella based on Microfluidic Enrichment with a Label-free Nanobiosensing Platform. *Sens. Actuators B Chem.* **2017**, *262*, 588–594. [CrossRef]

74. Deshmukh, R.A.; Joshi, K.; Bhand, S.; Roy, U. Recent developments in detection and enumeration of waterborne bacteria: A retrospective minireview. *Microbiologyopen* **2016**, *5*, 901–922. [CrossRef] [PubMed]

75. Law, J.W.; Ab Mutalib, N.S.; Chan, K.G.; Lee, L.H. Rapid methods for the detection of foodborne bacterial pathogens: Principles, applications, advantages and limitations. *Front. Microbiol.* **2015**, *5*, 770. [CrossRef] [PubMed]

76. Ríos, Á.; Zougagh, M. Modern qualitative analysis by miniaturized and microfluidic systems. *Trends Anal. Chem.* **2015**, *69*, 105–113. [CrossRef]

77. Kant, K.; Shahbazi, M.A.; Dave, V.P.; Ngo, T.A.; Chidambara, V.A.; Linh, Q.T.; Dang, D.B.; Anders, W. Microfluidic devices for sample preparation and rapid detection of foodborne pathogens. *Biotechnol. Adv.* **2018**, *36*, 1003–1024. [CrossRef]

78. Chi, L.W.; Olivo, M. Surface Plasmon Resonance Imaging Sensors: A Review. *Plasmonics* **2014**, *9*, 809–824.

79. Lee, H.; Xu, L.; Koh, D.; Nyayapathi, N.; Oh, K.W. Various on-chip sensors with microfluidics for biological applications. *Sensors* **2014**, *14*, 17008–17036. [CrossRef]

80. Liu, S.H.; Tian, J.; Liu, N.L.; Lu, P.X. Temperature Insensitive Liquid Level Sensor Based on Antiresonant Reflecting Guidance in Silica Tube. *J. Lightwave Technol.* **2016**, *34*, 5239–5243. [CrossRef]

81. Safavieh, M.; Ahmed, M.U.; Tolba, M.; Zourob, M. Microfluidic electrochemical assay for rapid detection and quantification of Escherichia coli. *Biosens. Bioelectron.* **2012**, *31*, 523–528. [CrossRef] [PubMed]

82. Narsaiah, K.; Jha, S.N.; Bhardwaj, R.; Sharma, R.; Kumar, R. Optical biosensors for food quality and safety assurance-a review. *J. Food Sci. Technol.* **2012**, *49*, 383–406. [CrossRef] [PubMed]

83. Li, Y.; Liu, X.; Lin, Z. Recent developments and applications of surface plasmon resonance biosensors for the detection of mycotoxins in foodstuffs. *Food Chem.* **2012**, *132*, 1549–1554. [CrossRef] [PubMed]

84. Wang, D.S.; Fan, S.K. Microfluidic Surface Plasmon Resonance Sensors: From Principles to Point-of-Care Applications. *Sensors* **2016**, *16*, 1175. [CrossRef] [PubMed]

85. Pennacchio, A.; Ruggiero, G.; Staiano, M.; Piccialli, G.; Oliviero, G.; Lewkowicz, A.; Synak, A.; Bojarski, P.; D'Auriaa, S. A surface plasmon resonance based biochip for the detection of patulin toxin. *Opt. Mater.* **2014**, *36*, 1670–1675. [CrossRef]

86. Zordan, M.D.; Grafton, M.M.G.; Acharya, G.; Reece, L.M.; Cooper, C.L.; Aronson, A.I.; Park, K.; Leary, J.F. Detection of pathogenic E. coli O157:H7 by a hybrid microfluidic SPR and molecular imaging cytometry device. *Cytom. Part A* **2010**, *75*, 155–162.

87. Zordan, M.D.; Grafton, M.M.G.; Acharya, G.; Reece, L.M.; Aronson, A.I.; Park, K.; Leary, J.F. A microfluidic-based hybrid SPR/molecular imaging biosensor for the multiplexed detection of foodborne pathogens. *Proc. Spie—Int. Soc. Opt. Eng.* **2009**, *7167*, 1–10.

88. Tokel, O.; Yildiz, U.H.; Inci, F.; Durmus, N.G.; Ekiz, O.O.; Turker, B.; Cetin, C.; Rao, S.; Sridhar, K.; Natarajan, N.; et al. Portable Microfluidic Integrated Plasmonic Platform for Pathogen Detection. *Sci. Rep.* **2015**, *5*, 9152. [CrossRef] [PubMed]

89. Reig, B.; Bardinal, V.; Camps, T.; Doucet, J.B. A miniaturized VCSEL-based system for optical sensing in a microfluidic channel. *Sensors* **2012**, *23*, 1–4.

90. Ohk, S.H.; Koo, O.K.; Sen, T.; Yamamoto, C.M.; Bhunia, A.K. Antibody–aptamer functionalized fibre-optic biosensor for specific detection of Listeria monocytogenes from food. *J. Appl. Microbiol.* **2010**, *109*, 808–817. [CrossRef] [PubMed]

91. Zhou, W.; Zhang, W.; Wang, Z.; Liu, T.; Zhang, Y. Progress on fiber-optic evanescent wave biosensor technique in food safety detection. *J. Food Saf. Qual.* **2014**, *5*, 3971–3974.

92. Li, M.; Zhao, F.; Zeng, J.; Qi, J.; Lu, J.; Shih, W.C. Microfluidic surface-enhanced Raman scattering sensor with monolithically integrated nanoporous gold disk arrays for rapid and label-free biomolecular detection. *J. Biomed. Opt.* **2014**, *19*, 111611. [CrossRef] [PubMed]

93. Mungroo, N.A.; Oliveira, G.; Neethirajan, S. SERS based point-of-care detection of food-borne pathogens. *Microchim. Acta* **2016**, *183*, 697–707. [CrossRef]

94. Gilli, E. Optical biosensor system with integrated microfluidic sample preparation and TIRF based detection. *Proc. Spie—Int. Soc. Opt. Eng.* **2013**, *8774*, 140–144.

95. Setterington, E.B.; Alocilja, E.C. Electrochemical Biosensor for Rapid and Sensitive Detection of Magnetically Extracted Bacterial Pathogens. *Biosensors* **2012**, *2*, 15–31. [CrossRef]

96. Campuzano, S.; Yanez-Sedeno, P.; Pingarron, J.M. Molecular Biosensors for Electrochemical Detection of Infectious Pathogens in Liquid Biopsies: Current Trends and Challenges. *Sensors* **2017**, *17*, 2533. [CrossRef]

97. Dong, S.; Zhou, J.; Hui, D.; Pang, X.; Wang, Q.; Zhang, S.; Wang, L. Interaction between edge dislocations and amorphous interphase in carbon nanotubes reinforced metal matrix nanocomposites incorporating interface effect. *Int. J. Solids Struct.* **2014**, *51*, 1149–1163. [CrossRef]

98. Ligaj, M.; Tichoniuk, M.; Gwiazdowska, D.; Filipiak, M. Electrochemical DNA biosensor for the detection of pathogenic bacteria Aeromonas hydrophila. *Electrochim. Acta* **2014**, *128*, 67–74. [CrossRef]

99. Chen, Q.; Wang, D.; Cai, G.; Xiong, Y.; Li, Y.; Wang, M.; Hou, H.; Lin, J. Fast and sensitive detection of foodborne pathogen using electrochemical impedance analysis, urease catalysis and microfluidics. *Biosens. Bioelectron.* **2016**, *86*, 770–776. [CrossRef]

100. Liu, H.T.; Wen, Z.Y.; Xu, Y.; Shang, Z.G.; Peng, J.L.; Tian, P. An integrated microfluidic analysis microsystems with bacterial capture enrichment and in-situ impedance detection. *Mod. Phys. Lett. B* **2017**, *31*, 196–199. [CrossRef]

101. Wang, X.; Niessner, R.; Tang, D.; Knopp, D. Nanoparticle-based immunosensors and immunoassays for aflatoxins. *Anal. Chim. Acta* **2016**, *912*, 10–23. [CrossRef] [PubMed]

102. Zhu, L.; Jing, H.; Cao, X.; Huang, K.; Luo, Y.; Xu, W. Development of a double-antibody sandwich ELISA for rapid detection of Bacillus Cereus in food. *Sci. Rep.* **2016**, *6*, 16092. [CrossRef] [PubMed]

103. Rasooly, A.; Bruck, H.A.; Kostov, Y. An ELISA Lab-on-a-Chip (ELISA-LOC). *Humana Press* **2013**, *949*, 451–471.

104. Thaitrong, N.; Charlermroj, R.; Himananto, O.; Seepiban, C.; Karoonuthaisiri, N. Implementation of microfluidic sandwich ELISA for superior detection of plant pathogens. *PLoS ONE* **2013**, *8*, e83231. [CrossRef] [PubMed]

105. Wu, J.H.; Ma, Y.D.; Chung, Y.D.; Lee, G.B. An integrated microfluidic system for dual aptamer assay utilizing magnetic-composite-membranes. *IEEE Int. Conf. Nano/Micro Eng. Mol. Syst.* **2017**, *4*, 438–441.

106. Yanagisawa, N.; Dutta, D. Enhancement in the Sensitivity of Microfluidic Enzyme-Linked Immunosorbent Assays through Analyte Preconcentration. *Anal. Chem.* **2012**, *84*, 7029. [CrossRef] [PubMed]

107. Zhang, R.Q.; Liu, S.L.; Zhao, W.; Zhang, W.P.; Yu, X.; Li, Y.; Li, A.J.; Pang, D.W.; Zhang, Z.L. A Simple Point-of-Care Microfluidic Immunomagnetic Fluorescence Assay for Pathogens. *Anal. Chem.* **2013**, *85*, 2645–2651. [CrossRef]

108. Kanayeva, D.A.; Wang, R.; Rhoads, D.; Erf, G.F.; Slavik, M.F.; Tung, S.; Li, Y. Efficient separation and sensitive detection of Listeria monocytogenes using an impedance immunosensor based on magnetic nanoparticles, a microfluidic chip, and an interdigitated microelectrode. *J. Food Prot.* **2012**, *75*, 1951–1959. [CrossRef]

109. Dector, A.; Galindo-De-La-Rosa, J.; Amaya-Cruz, D.M.; Ortíz-Verdín, A.; Guerra-Balcázar, M.; Olivares-Ramírez, J.M.; Arriaga, L.G.; Ledesma-García, J. Towards autonomous lateral flow assays: Paper-based microfluidic fuel cell inside an HIV-test using a blood sample as fuel. *Int. J. Hydrog. Energy* **2017**, *42*, 29–32. [CrossRef]

110. Hsieh, H.; Dantzler, J.; Weigl, B. Analytical Tools to Improve Optimization Procedures for Lateral Flow Assays. *Diagnostics* **2017**, *7*, 29. [CrossRef]

111. Doller, C.; Jakubik, J. Direct solid-phase radioimmunoassay for the detection of Aujeszky's disease antibodies. *Zent. Bakteriol. A* **1980**, *247*, 1–7.

112. Yao, L.; Wang, L.; Huang, F.; Cai, G.; Xi, X.; Lin, J. A microfluidic impedance biosensor based on immunomagnetic separation and urease catalysis for continuous-flow detection of E. coli O157:H7. *Sens. Actuators B Chem.* **2018**, *259*, 2657. [CrossRef]

113. Mangal, M.; Bansal, S.; Sharma, S.K.; Gupta, R.K. Molecular Detection of Food Borne Pathogens: A Rapid and Accurate Answer to Food Safety. *Crit. Rev. Food Sci. Nutr.* **2016**, *56*, 1568–1584. [CrossRef] [PubMed]

114. Zhang, Y.; Wang, T.H. An automated all-in-one microfludic device for parallel solid phase DNA extraction and droplet-in-oil PCR analysis. *IEEE Int. Conf. Micro Electro Mech. Syst.* **2010**, *9*, 971–974.

115. Zhang, C.; Wang, H.; Xing, D. Multichannel oscillatory-flow multiplex PCR microfluidics for high-throughput and fast detection of foodborne bacterial pathogens. *Biomed. Microdevices* **2011**, *13*, 885. [CrossRef]

116. Tourlousse, D.M.; Ahmad, F.; Stedtfeld, R.D.; Seyrig, G.; Tiedje, J.M.; Hashsham, S.A. A polymer microfluidic chip for quantitative detection of multiple water- and foodborne pathogens using real-time fluorogenic loop-mediated isothermal amplification. *Biomed. Microdevices* **2012**, *14*, 769–778. [CrossRef] [PubMed]

117. Uddin, S.M.; Ibrahim, F.; Sayad, A.A.; Thiha, A.; Pei, K.X.; Mohktar, M.S.; Hashim, U.; Cho, J.M.; Thong, K.L. A portable automatic endpoint detection system for amplicons of loop mediated isothermal amplification on microfluidic compact disk platform. *Sensors* **2015**, *15*, 5376–5389. [CrossRef]

118. Sun, Y.; Quyen, T.L.; Hung, T.Q.; Chin, W.H.; Wolff, A.; Bang, D.D. A lab-on-a-chip system with integrated sample preparation and loop-mediated isothermal amplification for rapid and quantitative detection of Salmonella spp. in food samples. *Lab Chip* **2015**, *15*, 1898–1904. [CrossRef]

119. Lutz, S.; Weber, P.; Focke, M.; Faltin, B.; Hoffmann, J.; Müller, C.; Mark, D.; Roth, G.; Munday, P.; Armes, N.; et al. Microfluidic lab-on-a-foil for nucleic acid analysis based on isothermal recombinase polymerase amplification (RPA). *Lab Chip* **2010**, *10*, 887–893. [CrossRef]

120. Tortajada-Genaro, L.A.; Santiago-Felipe, S.; Amasia, M.; Russom, A.; Maquieira, Á. Isothermal solid-phase recombinase polymerase amplification on microfluidic digital versatile discs (DVDs). *RSC Adv.* **2015**, *5*, 29987–29995. [CrossRef]

121. Mauk, M.G.; Liu, C.; Song, J.; Bau, H.H. Integrated Microfluidic Nucleic Acid Isolation, Isothermal Amplification, and Amplicon Quantification. *Microarrays* **2015**, *4*, 474–489. [CrossRef] [PubMed]

122. Pang, B.; Fu, K.; Liu, Y.; Ding, X.; Hu, J.; Wu, W.; Xu, K.; Song, X.L.; Wang, J.; Mu, Y.; et al. Development of a self-priming PDMS/paper hybrid microfluidic chip using mixed-dye-loaded loop-mediated isothermal amplification assay for multiplex foodborne pathogens detection. *Anal. Chim. Acta* **2018**, *1040*, 81–89. [CrossRef] [PubMed]

123. Zhong, J.L.; Zhao, X.H. Detection of viable but non-culturable Escherichia coli O157:H7 by PCR in combination with propidium monoazide. *3 Biotech* **2017**, *8*, 28. [CrossRef] [PubMed]

124. Li, M.F.; Li, L.M.; Liu, R.Y. Application of paper based microfluidic chip technology in food safety detection. *J. Food Saf. Qual.* **2018**, *38*, e12412.

microorganisms MDPI

Review

Role of Natural Volatiles and Essential Oils in Extending Shelf Life and Controlling Postharvest Microorganisms of Small Fruits

Toktam Taghavi *, Chyer Kim and Alireza Rahemi

Agricultural Research Station, Virginia State University, Petersburg, VA 23806, USA; ckim@vsu.edu (C.K.); a_rahemi@yahoo.com (A.R.)
* Correspondence: ttaghavi@vsu.edu

Received: 31 August 2018; Accepted: 3 October 2018; Published: 5 October 2018

Abstract: Small fruits are a multi-billion dollar industry in the US, and are economically important in many other countries. However, they are perishable and susceptible to physiological disorders and biological damage. Food safety and fruit quality are the major concerns of the food chain from farm to consumer, especially with increasing regulations in recent years. At present, the industry depends on pesticides and fungicides to control food spoilage organisms. However, due to consumer concerns and increasing demand for safer produce, efforts are being made to identify eco-friendly compounds that can extend the shelf life of small fruits. Most volatiles and essential oils produced by plants are safe for humans and the environment, and lots of research has been conducted to test the in vitro efficacy of single-compound volatiles or multi-compound essential oils on various microorganisms. However, there are not many reports on their in vivo (in storage) and in situ (in the field) applications. In this review, we discuss the efficacy, minimum inhibitory concentrations, and mechanisms of action of volatiles and essential oils that control microorganisms (bacteria and fungi) on small fruits such as strawberries, raspberries, blueberries, blackberries, and grapes under the three conditions.

Keywords: postharvest diseases; food borne pathogens; bacteria; fungi; food safety; plant extracts; small fruits; grape; strawberry; blueberry; raspberry; blackberry; essential oils

---

1. Introduction

There are two major concerns for the handling and consumption of small fruits. First is their short shelf life, and second is food safety. Fresh produce in particular is gaining increasing attention in the current food safety and regulatory climate. Centers for Disease Control and Prevention [1], has developed a comprehensive list of food sources of all foodborne illnesses in the United States (U.S.) and fresh produce accounted for nearly half of illnesses among all types of foods reported during 1998–2008.

Standard small fruit production practices discourage growers from washing fruits after harvest and during storage. This may contribute to the buildup of food-borne bacteria and may pose a food safety risk to consumers since small fruits are eaten fresh. Also, small fruits have a very short shelf life due to their susceptibility to spoilage pathogens, physiological disorders and mechanical injury [2]. This is partly due to their high nutrient content and water activity, and low pH which make them susceptible to fungal attack. Thus, losses during harvest, handling and marketing are sometimes as high as 50% [3]. Pathogens may also produce mycotoxins that make the fruit harmful to consumers [4].

Low temperature storage is not adequate to extend fruit shelf-life for prolonged storage or for delivery to distant markets. Therefore, controlled temperature combined with high CO_2 in modified atmospheres has been recommended to reduce spoilage [5]. Although small fruits can tolerate high CO_2 levels in storage, it causes off-flavors in the long term [6], and application of fungicides is still the main method to control postharvest diseases in small fruit [6–8]. However, under optimal conditions for disease development, fungicides are not effective and growers may try to overcome this challenge by increasing fungicide applications, which in turn leads to higher residue levels in produce [7]. In the long term, the population of beneficial organisms declines as residual fungicides lead to the development of resistant fungi [6,8,9]. High levels of pesticide residues on berries are of particular concern because they are consumed fresh shortly after harvest. This has placed strawberries at the top of the "Dirty Dozen" list [10] and raised concerns about human health and pronounced oncogenic risk [8,11]. Other concerns with excessive pesticide application are increased cost, handling hazards, and threats to the environment [12]. Therefore, many restrictions are imposed on pesticide application in many countries around the world [9,12].

Nowadays, consumers concerned about fungicide residues in small fruits demand safer alternatives to replace synthetic pesticides [13]. Therefore, new disease control technologies that are safe for humans and the environment are needed [12]. Biological control, such as the use of plant volatiles is an exciting alternative [8]. Plants synthesize an enormous number of volatiles (phenols or their by-products). Many of them are responsible for flavor, while some have useful medicinal compounds [12].

Natural plant products tend to have low mammalian toxicity, broad-spectrum antimicrobial activity, are less hazardous than synthetic compounds, and are generally more acceptable to the public [4,7,12,14–16]. At present, there are many research results on the efficacy of plant volatiles to control fungal and microbial growth in in vitro (laboratory) conditions [17,18], but very few discuss their efficacy to extend the shelf life of fresh produce.

In this review, we discuss the efficacy of volatiles (single compounds) and essential oils (EOs, group of compounds) produced by plants to control food-borne bacteria (*Salmonella* spp., *Escherichia coli*, and *Listeria*) and fruit pathogens (mainly *Botrytis* and *Colletotrichum*) on extending the shelf life of small fruits, such as strawberries (*Fragaria* × *ananassa* Duch), raspberries (*Rubus* spp.), blueberries (*Vaccinium* spp.) and grapes (*Vitis vinifera* L.). We also provide insight into their in vitro, in vivo (in storage) and in situ (in the field) applications, and their mechanisms of action. A summary of the papers reviewed is presented in Table 1.

Table 1. In vitro, in vivo (in storage), and in situ (in field) applications of volatiles (single compounds) and essential oils (EOs, multiple compounds) on diseases and plants, and parameters measured. The abbreviations are explained in the notes.

Volatile, EO, plant extract	Concentration	Disease/plant	Parameters measured	Reference
In vitro application				
(E)-hex-2-enal, hexanal, (E)-non-2-enal, nonanal, 2-carene, limonene and aldehydes upon wounding of tomato leaves	10% solution (w/w) of the test compound and diluted serially (w/w) to 1 and 0.1% solutions.	Alternaria alternata and Botrytis cinerea	Hyphal length	Hamilton-Kemp et al. [19]
Aldehydes hexanal and (E)-hex-2-enal; the alcohols hexan-1-ol, (Z)-hex-3-en-1-ol, and (E)-hex-2-en-1-ol; and the esters (Z)-hex-3-enyl acetate, (Z)-hex-2-enyl acetate, and hexyl acetate.	33.78 to 1351.35 µL L^{-1}	Colletotrichum acutatum	Mycelial growth, conidial germination, development of appressoria, MID95, ID50	Arroyo et al. [9]
Fifteen compounds from aldehydes, alcohols, ketones, an ester and a mixed alcohol and ketone moiety.	0, 0.02, 0.04, 0.1, 0.4 µL. mL^{-1}	B. cinerea	Growth of fungi, % of control	Vaughn et al. [20]
Hexanal in β-cyclodextrin complex	0, 1, 1.5, 2, 4, 5, 7, 10 µL	C. acutatum, A. alternata and B. cinerea	Radial growth of cultures in cm^2	Almenar et al. [14]
Extracts from 345 plants and 49 essential oils	10% plant extract solution, 50, 25, 12.5, 6.25, 3.13, 1.56, 0.78, and 0.39% EOs	B. cinerea	Reduction in spore germination	Wilson et al. [18]
Twenty six essential oils of ten plants (Chenopodium ambrosioides, Eucalyptus citriodora, Eupatorium cannabinum, Lawsonia inermis, Ocimum canum, O. gratissimum, O. sanctum, Prunus persica, Zingiber cassumunar and Zingiber officinale)	500 ppm, different for MIC and MFC	B. cinerea, and only three selected essential oils on 15 other fruit rotting pathogens	% mycelial inhibition, MIC	Tripathi et al. [4]
Oregano, thyme, dictamnus, marjoram (carvacrol), lavender (linalool, linalyl acetate), rosemary, sage (eucalyptol) and pennyroyal (cis-menthone) EOs	Not specified	B. cinerea, Fusarium sp., Clavibacter michiganensis	Radial growth on PDA	Daferera et al. [7]
EOs of two clonal types of Thymus vulgaris	50, 100, 200 ppm	B. cinerea and Rhizopus stolonifer	% inhibition of radial growth	Bhaskara Redy et al. [6]
Lemongrass (Cympopogon citratus)	25, 50, 100, 500 ppm	B. cinerea, Colletotrichum coccodes, C. herbarum, R. stolonifer, Aspergillus niger	Pathogen development, spore production, spore germination, germ tube length	Tzortzakis & Economakis [13]
Eighteen EOs	50-3000 µLL^{-1}	5 pathogens from 5 crops including B. cinerea from grape	Visual inspection, inhibition of mycelial growth (%)	Combrinck et al. [21]
Thyme (P-cymene, thymol, α-terpineol, carvacrol, Cinnamon bark (cinnameldehyde, cinnamyl acetate), Clove bud (eugenol, β-caryophyllene)	13 concentrations from 0.067 to 667 µL L^{-1} of media	C. acutatum	Mycelial growth, conidial germination, appressoria formation	Duduk et al. [2]

Table 1. *Cont.*

Volatile, EO, plant extract	Concentration	Disease/plant	Parameters measured	Reference
In vitro application				
Origanum vulgare L. essential oil	40, 20, 10, 5, 2.5, 1.25, 0.06 µL mL^{-1}	*R. stolonifer* and *A. niger*	Mycelial growth of the test fungi, spore germination and morphological changes	Santos et al. [22]
Essential oils of seven Moroccan Labiatae	0, 10, 50, 100, 150, 200 and 250 ppm	*B. cinerea*	Percentage of inhibition of radial growth vs control	Bouchra et al. [23]
In vivo (in storage) application				
(E)-hex-2-enal	4, 5, 10, 20, and 50 µL L^{-1}	*C. acutatum* inoculated strawberry fruit	Incidence of infected fruits, scale 0,1	Arroyo et al. [9]
Three groups of naturally occurring volatile compoundscontains 24 volatiles	2, 10, 100 µL/250 mL bottle	Strawberry, blackberry & grape	Lesion appearance and size, phytotoxicity	Archbold et al. [24]
Fifteen volatiles released by red raspberries and strawberries	0.4 µL mL^{-1}	*B. cinerea* on raspberry and strawberry	Rated for development of fungi and damage of volatile	Vaughn et al. [20]
Thymol, menthol, eugenol	200 mg L^{-1}	Strawberry	Sugar, acid, anthocyanin, TPC, ORAC, DPPH, HRS, SARS, flavonoids	Wang et al. [16]
Carvacrol, anethole, cinnamaldehyde, cinnamic acid, perillaldehyde, linalool, and p-cymene	200 mg L^{-1}	Blueberries	ORAC) and hydroxyl radical (•OH) scavenging, total anthocyanins, total phenolics capacity, sugars, organic acids, % of fruit showing fungal symptoms	Wang et al. [25]
Volatile substances emitted by 'Isabella' grapes	0, 300, 400, or 500 g of 'Isabella' grapes	*B. cinerea* in kiwifruit	# infected kiwifruit, # kiwifruit on which fungal fruiting bodies had appeared	Kulakiotu, et al. [8]
Eucalyptus and cinnamon EOs	50, 500 ppm	Strawberry, tomato	Degree of visual infection, weight loss, TSS, firmness, TA, TPC,	Tzortzakis [12]
Thyme, Cinnamon bark, Clove bud EO	13 conc. (0.067–667)	*C. acutatum*	# of diseased fruit or mycelial growth	Duduk et al. [2]
Essential oils from thyme (*T. vulgaris*), clove (*Syzygium aromaticum*), and massoialactone (bark of the tree *Cryptocarya massoia*)	(0, 0.033, 0.1, 0.33, 1.0 and 3.3%) for phytotoxicity, *B. cinerea* inoculated on heated lesions and treated with (0.033, 0.1, 0.33%) EOs	*B. cinerea* in grapes	Scaling the formation of necrosis on the underside of the leaves	Walter et al. [26]

Table 1. *Cont.*

Volatile, EO, plant extract	Concentration	Disease/plant	Parameters measured	Reference
In vivo (in storage) application				
Eugenol or thymol	75 or 150 µL/bag (vol. was not mentioned)	Grape	Ethylene, weight loss, color and firmness, TSS, TA, sensory analysis, decay, microorganism analysis, antioxidant activity, TPC, total anthocyanins, organic acids, and sugar contents	Valero et al. [27]
O. vulgare L. essential oil	40, 20, 10, 5, 2.5, 1.25, 0.06 µL mL^{-1}	Grapes	TSS, TA, weight loss, color, firmness, anthocyanin, and sensory characteristics of the fruits during storage	Santos et al. [22]
O. sanctum, P. persica and *Z. officinale*	200,100 and 100 ppm	*B. cinerea* in grapes	Initiations of rotting of the fruits	Tripathi et al. [4]
Thymus damensis and *T. carmanicus*	150, 300, 600, 1200 µL L^{-1}	*R. stolonifer, Penicillium digitatum, A. niger* and *B. cinerea* strawberry	Disease incidence (%)	Nabigol & Morshedi [28]
EOs of two *T. vulgaris* clones	50, 100, 200 ppm	*B. cinerea* and *R. stolonifer*	Decay of fruit	Bhaskara Reddy et al. [6]
Carvacrol, anethole, cinnamic acid, perillaldehyde, cinnamaldehyde, and linalool	200 mg L^{-1}	Raspberries	SOD, CAT, G-POD, AsA-POD, GR, GSH-POD, MDAR, DHAR, Protein content, TPC, Total anthocyanins, ORAC, HOSC, DPPH, flavonoids	Jin et al. [29]
Carvacrol, cinnamaldehyde	0.50%	*Escherichia coli* and *P. digitatum* on blueberries	Microbial populations, fruit firmness	Sun et al. [30]
Bergamot EO, on grape	2% *w/v*	Grape cv Muscatel	Microbial counts, weight loss, °Brix, total phenols, antioxidant activity, color and texture, respiration rate	Sánchez-González et al. [17]
In situ (field application)				
Thyme R oil in two years and massoialactone year 2	0.033 Thyme, 0.1 massoialactone	*Botrytis* in grape	*B. cinerea* sporulation on leaves, % of berries showing *B. cinerea* sporulation	Walter et al. [26]

Notes: The aforementioned abbreviations stand for PDA (Potato dextrose agar), MIC (Minimum inhibitory concentration), MID (minimum inhibitory doses), ID (inhibitory doses), ORAC (Oxygen radical absorbance capacity), TPC (Total phenolic content), HRS (Hydroxyl Radical Scavenging), SARS (Superoxide Anion Radical Scavenging), TSS (Total soluble solids), TA (Titratable acidity), SOD (Superoxide dismutas, EC 1.15.1.1), CAT (Catalase, EC 1.11.1.6), G-POD (Guaiacol peroxidase, EC 1.11.1.7), AsA-POD (Ascorbate peroxidase, EC 1.11.1.11), GR (Glutathione reductase, EC1.6.4.2), GSH-POD (Glutathione peroxidase, EC 1.11.1.9), MDAR (Monodehydroascorbate reductase, EC 1.6.5.4), DHAR (Dehydroascorbate reductase, EC 1.8.5.1), HOSC (Hydroxyl radical scavenging capacity (°OH; HOSC) assay), DPPH (2,2-Di (4-tert-octylphenyl) -1-picrylhydrazyl (DPPH) scavenging capacity assay).

2. Control of Food Born Bacteria by Volatiles and Essential Oils in Small Fruits

Most of the studies on the application of natural volatiles and essential oils (EOs) to control foodborne pathogens, such as *Salmonella Typhimurium, Salmonella enterica, E. coli,* and *Listeria monocytogenes,* etc., have been conducted in vitro in petri dishes [31]. However, very few reports discuss the efficacy of plant volatiles and essential oils to control foodborne bacteria on small fruits during storage. Sun et al. [30] studied chitosan coating mixed with six different essential oils in vitro to control *E. coli* and *Penicillium digitatum* in blueberries during storage. Carvacrol and cinnamaldehyde had high antimicrobial capacity and were selected for in vivo studies to control blueberry pathogens.

Sánchez-González et al. [17] tested biodegradable coatings with and without bergamot essential oil (from *Citrus × bergamia* Risso & Poit.) on table grapes (*V. vinifera* L.) during storage. They found that incorporation of bergamot essential oil improved the antimicrobial activity of the coatings and significantly reduced mold, yeast, and mesophile counts.

3. Control of Fungal Diseases in In Vitro Conditions

3.1. Effect of Volatiles on Fungal Diseases in In Vitro

Plant volatiles are mainly aldehydes, alcohols, acids and esters synthesized through the hydroperoxide lyase (HPL) and liposygenase (LOX) pathways. These products are involved in plant wounding responses and may have physiological roles (such as self-defense and antimicrobial properties) beside their role in aroma biosynthesis [9].

The effect of single compound terpene, terpenoids, and aldehyde vapors and their mixture from crushed tomato (*Solanum* sp. 'Mountain Pride') leaves were tested on conidiospores from *Alternaria alternata* and *Botrytis cinerea*. (E)-hex-2-enal and hexanal inhibited *A. alternata* hyphal growth and *B. cinerea* spore growth, while the terpene limonene and 2-carene did not affect hyphal growth [19]. However, the synergistic effect of other bioactive volatiles should not be excluded. The unsaturated aldehydes (E)-non-2-enal and (E)-hex-2-enal, were significantly more active than saturated hexanal, in inhibiting fungal growth [19].

Similarly, Arroyo et al. [9] examined the antifungal activity of eight aroma compounds of strawberry fruit against *Colletotrichum acutatum in vitro*. The aldehyde (E)-hex-2-enal significantly inhibited mycelial growth and spore germination (minimum inhibitory dose, MID of 33. 6 µL L^{-1}). Hexanal was only effective when applied in higher doses of up to 10 times [9]. The inhibitory concentrations of hexanal on pathogens reported by others are quite different. Almenar et al. [14] observed that hexanal completely inhibited the growth of *C. acutatum, B. cinerea,* and *A. alternata* at low concentrations of 1.1, 1.3 and 2.3 µL L^{-1} air respectively. Andersen et al. [32] reported that 0.00007 µL L^{-1} air of hexanal reduced *A. alternata* growth on agar by 50%. Song et al. [33] reported that 0.5 µL L^{-1} air inhibited *B. cinerea* growth on potato dextrose agar (PDA) media and Archbold et al. [24] showed 0.008 µL L^{-1} (2 µL/250 ml container) hexanal reducing *B. cinerea* in inoculated strawberries by 90%. In contrast to all these reports, Hamilton-Kemp et al. [19] observed a stimulatory effect of hexanal on *A. alternata* and *B. cinerea* mycelial growth at low levels.

In an experiment, among five antifungal volatiles (hexan-1-ol, benzaldehyde, 2-nonanone, (Z)-hex-3-en-1-ol and (E)-hex-2-enal) released by raspberry and strawberry fruits, (E)-hex-2-enal and benzaldehyde were the most effective against three fungal (*A. alternata, B. cinerea,* and *Colletotrichum gloeosporioides*) species in vitro [20].

To our knowledge, only one study reported the application of slow-release hexanal in postharvest fungal control in small fruits. To develop a controlled release mechanism, hexanal was incorporated into a polymer (β-cyclodextrin) and then evaluated in vitro against *B. cinerea, C. acutatum* and *A. alternata*. Hexanal was fungistatic on all three pathogens and fungicidal on *C. acutatum*. *C. acutatum* was more responsive to hexanal than the other two pathogens. The data suggested that a slow-release compound provided a more uniform volatile dose during storage and was more effective in reducing postharvest diseases [14].

3.2. Effect of Essential Oils on Fungal Diseases In Vitro

Essential oils (EO) are naturally occurring substances extracted from plant material and are a complex of volatiles produced in different plant parts. They constitute different plant secondary metabolites mainly from terpenes, and other aromatic compounds [34]. They have various functions in plants such as resistance against pest and diseases [13]. Essential oils with active ingredients can be used as an alternative management method to control microorganisms. They have shown antibacterial, antifungal, and antioxidant properties against some important food-borne and plant pathogens.

Essential oils have three main features. Since they are produced by plants, they are safer to the environment and humans than synthetic alternatives. Second, they have low risk of resistance development by pathogens because they are a mixture of oil components and pathogens lack the mechanism to develop resistance against a range of chemicals. Third, there is a broad range of natural compounds that have the potential to be used for controlling pathogens [7]. Essential oils are generally considered safe for treating food products, and sometimes even beneficial in extending their shelf life. However, phytotoxicity, off-flavors and off-odors have also been reported.

A large number of papers have been published on the biological activity of essential oils. The large variability among the data may be attributed to factors such as genetics, climate, geographical and seasonal conditions, distillation techniques and time of harvest [4]. Consequently, the chemical composition of essential oils needs to be standardized and reproducible [7].

Most of the reports on the inhibition of post-harvest fungal pathogens by essential oils focus on in vitro conditions [28]. The efficacy of several essential oils and their antimicrobial compounds were studied in extending the shelf life and inhibiting decay of strawberries. Only a limited number of literature investigated the effect of essential oils and/or plant extracts against *B. cinerea* or other small fruit pathogens [4,6,7,18].

Two studies [4,18] screened a large number of essential oils and plant extracts for their efficacy against *B. cinerea*. Wilson et al. [18] screened 345 plant extracts for their antifungal activity and thirteen were highly antifungal. The extracts from *Capsicum* and *Allium* genera were the most effective and inhibited *B. cinerea* spore germination after 48 and 24 hours, respectively. Among 49 essential oils tested, red thyme (*Thymus zygis* L.), palmarosa (*Cymbopogon martini* (Roxb.) Will. Watson), clove buds (*Syzygium aromaticum (L.) Merr. & L.M. Perry*), and cinnamon leaf (*Cinnamomum zeylanicum* Blume) greatly inhibited germination of *B. cinerea* spores at the lowest concentration tested (0.78% dilution; initial concentration was not provided). The essential oils with high antifungal activity and the highest frequency were: cineole, limonene, α-pinene, β-myrcene, camphor and β-pinene [18].

Tripathi et al. [4] tested essential oils of 26 plants against B. cinerea. Among them, ten plants (*Zingiber officinale* Roscoe, *Z. cassumunar* Roxb., *Prunus persica* (L.) Batsch, *Ocimum sanctum* L., *Ocimum gratissimum* L., *Ocimum canum* Sims, *Lawsonia inermis* L., *Eupatorium cannabinum* L., *Eucalyptus citriodora* Hook., and *Chenopodium ambrosioides* L.) showed 100% toxicity at 500 (mg L^{-1}). The essential oils were a mixture of nine major and sixteen minor compounds with possible synergistic effects between major and minor constituents [4].

In another study, growth of *B. cinerea* mycelium was completely inhibited by thyme, oregano, dictamnus and marjoram essential oils at 150, 200, 200 and 300 mgml^{-1}, respectively. The main component of oregano oil was thymol, and the main component of marjoram, thyme and dictamnus oils was carvacrol [7].

Bhaskara Reddy et al. [6] reported that, there is a significant difference between the antifungal activity of two clones of *Thymus vulgaris* (Laval-1 and Laval-2) against two strawberry storage pathogens (*Rhizopus stolonifer* and *B. cinerea*). Oil from Laval-2 clones showed higher antifungal activity in vitro and decay control action in vivo which is mainly due to its higher carvacrol, thymol and linalool contents. Lemongrass oil reduced germ tube length and spore germination in *R. stolonifer*, *Colletotrichum coccodes*, and *B. cinerea* [13].

Combrinck et al. [21] tested eighteen essential oils on five pathogens, including *B. cinerea* isolated from grapes. *B. cinerea* was the most susceptible to cinnamon, thyme, and eugenol at the lowest concentration tested (500 μL L^{-1}).

In the study conducted by Duduk et al. [2], clove bud, thyme and cinnamon bark essential oils were fungistatic on mycelial growth of *C. acutatum* (667 μL L^{-1} of medium) and their volatiles disabled appressoria formation (1.53 μL L^{-1} of air) and prevented conidia germination (76.5, 15.3 and 1.5 μL L^{-1} of air respectively).

Santos et al. [22] tested the efficacy of chitosan packages incorporated with *Origanum vulgare* L. essential oil on *Aspergillus niger* and *R. stolonifer* on grapes. Mycelial growth was inhibited by essential oil in in vitro conditions.

Bouchra et al. [23] evaluated essential oils of seven Labiatae plants grown in Morocco for their antifungal activity in in vitro conditions against *B. cinerea*. *Thymus glandulosus* Req. and *Origanum compactum* Benth., with thymol and carvacrol as main compounds, inhibited mycelium growth more effectively than others.

3.3. Mechanism of Action of Volatiles and Essential Oils on Pathogen Growth in In Vitro Conditions

Different concentrations of hexanal are effective against fungal growth. Caccioni et al. [35] believed that, the actual vapor pressure rather than the concentration of volatiles determines their effectiveness. Therefore, the effectiveness of hexanal depends on its vapor pressure (effective concentration), its initial amount and the tested fungus. In essential oils, the antifungal activity strongly depends on the proportion of the individual components [36].

Gardini et al. [37] posited that hexanal is not soluble in water and therefore in the culture medium. However, Almenar et al. [14] showed that PDA medium absorbed hexanal and prevented *A. alternata* and *C. acutatum* growth (3.0 and 3.3 μL respectively). Therefore, if media can absorb henxanal, we hypothesize that fresh produce can also absorb small amounts of volatiles and extend their shelf life by extruding it.

The saturation status of volatile compounds seems to also affect their efficacy with unsaturated compounds being more reactive than saturated ones. Hamilton-Kemp et al. [19] and Andersen et al. [32] demonstrated that (E)-hex-2-enal (unsaturated aldehyde) was more effective against *B. cinerea* and *A. alternata* hyphal growth than hexanal (saturated aldehyde). Arroyo et al. [9] showed that aldehydes ((E)-hex-2-enal and hexanal) which are less saturated than alcohols and esters, are more effective in fungal inhibition. In contrast, among the alcohols tested, (E)-hex-2-en-1-ol, and (Z)-hex-3-en-1-ol which are less saturated than hexan-1-ol, showed lower antifungal activity against *C. acutatum* mycelial growth [9]. The reason could be due to a different reaction of *C. acutatum* to volatiles compared to *B. cinerea* and *A. alternata* as was mentioned by Almenar et al. [14]. Arroyo et al. [9] studied the conidial cells of *C. acutatum* treated with (E)-hex-2-enal by transmission electron microscopy. They observed that cell components were highly disorganized and cell wall and plasma membrane were disrupted, resulting in lysis of organelles and cell death.

Almenar et al. [14] reported that *B. cinerea* and *A. alternata* reacted to hexanal similarly but different from *C. acutatum*. They suggested that the inhibition mechanisms are different, but no further explanation was provided. However, Hamilton-Kemp et al. [19] reported that *C. acutatum* and *B. cinerea* responded similarly meaning hyphal growth was inhibited by aldehyde but not terpenes. However, they suggested that more cautious interpretation is needed as variable culture conditions and organism strain may affect the results.

The difference in the affinity of these compounds with microbial membranes seems to affect their effectiveness and their toxicity [14]. Cell membrane permeability and its interaction with volatiles are key elements in the effectiveness of gaseous compounds [38]. Changes in the permeability of cell membranes, and degeneration of the ion gradients adversely affect vital cell processes and lead to cell death [16].

In this regard, Kulakiotu et al. [8] reported cell protoplast secretion without initial cell wall disruption in mycelial hyphae of *B. cinerea* exposed to volatiles of grape (*Vitis labrusca*, cv. Isabella) resulting in cell wall deformation. Hyphae from the control treatment was healthy, possessing conidiophores with abundant conidia [8]. The mechanism of action of volatiles inhibiting hyphal growth and their effect on membranes needs further investigation. Thymol and carvacrol also showed antimicrobial activity against natural grape yeasts in red wines via damage to the membrane, leakage of cytoplasmic content and finally inhibition of erogosterol biosynthesis [39].

The fact that different natural compounds have a different antifungal profile suggests that different modes of action exist. In human pathogens, several mechanisms of actions have been reported for antifungal activity of natural plant compounds. Sivakumar et al. [40] reported that chalcones (aromatic ketons found in a few plant species) are able to disrupt fungal cell wall formation and cell lysis by inhibiting synthesis of the cell wall polysaccharide 1,3-beta-D-glucan. They can also interrupt cell division by inhibiting the conversion of tubulin into microtubules [41].

Aldehydes, another group of volatiles, inhibit fungal cell division by reacting with, and inactivating sulfhydryl, the functional group involved in fungal cell division. Some aldehydes (such as cinnamaldehyde, citral and perillaldehyde) are good electron acceptors and form a charge transfer complex with electron donors present in fungal cells, thus interfering with fungal metabolism [42].

Volatiles with α,β-unsaturated cabonyl groups (such as enones and enals) often react with nucleophilus in fungi through the Michael reaction to create chemical modifications and interrupt fungal growth [43] of *C. acutatum*, *Colletotrichum fragariae*, *C. gloeosporioides*, *Fusarium oxysporum*, *B. cinerea*, and *Phomopsis* sp. Nevertheless, the accessibility of α,β-unsaturated carbonyl groups to bulky nucleophile biomolecules plays an important role in their antifungal activity [44].

It has been reported that mitochondrial electron-transfer respiration systems in mammals were inhibited by *C. ambrosioides* essential oil component such as carvacrol [36]. Another component of the EO (ascaridole) formed toxic radicals in the presence of Fe^{2+} and reduced hemin [36]. The phenol carvacrol has shown antibacterial properties due to its ability to distribute into membranes and to permeate in them causing a breakdown of ion gradients. Carvacrol also increases the passive permeability of the cell membrane and modulates certain Ca^{2+} permeable transient receptor potential channels, inhibits sarcoplasmic reticulum Ca^{2+} ATPase, and activates ryanodine receptors in skeletal muscle, thereby influencing intracellular calcium homeostasis (Monzote et al. [36] and references therein).

4. Control of Fungal Diseases in In Vivo (in Storage) Conditions

4.1. Effect of Volatiles on Fungal Diseases In Vivo

Arroyo et al. [9] reported complete inhibition of *C. acutatum* development after five days by (E)-hex-2-enal in inoculated strawberry fruit. This compound also effectively reduced *B. cinerea* disease symptoms in grapes, blackberry (*Rubus* spp.) and strawberry [24]. Vaughn et al. [20], evaluated fifteen volatiles from strawberries and red raspberries during ripening and found that five compounds (hexan-1-ol, benzaldehyde, 2-nonanone, (Z)-hex-3-en-1-ol and (E)-hex-2-enal) completely inhibited *A. alternata*, *B. cinerea*, and *C. gloeosporioides* on strawberry and raspberry fruits.

In another study, treatments with thymol, eugenol, and menthol reduced decay in strawberries, with thymol being the most effective at slowing berry decay compared to the other two compounds [16]. Similarly, p-cymene, linalool, carvacrol, anethole, and perillaldehyde effectively retarded blueberry mold formation, while cinnamic acid and cinnamaldehyde were less effective in suppressing blueberry mold growth [25]. It has also been reported that grape (cv. Isabella) volatiles reduce *B. cinerea* activity and inoculum density, thereby limiting incidence and infection in kiwifruits [8].

4.2. Effect of Essential Oils on Fungal Diseases In Vivo

Essential oils from a number of plants have shown activity against strawberry pathogens. For example, eucalyptus (*Eucalyptus globulus* L.) and cinnamon oil (*C. zeylanicum*, Blume) improved fruit quality and reduced fruit decay [12], while cinnamon bark EO reduced both *C. acutatum* and *B. cinerea* penetration, development and number of infected fruits at concentrations above 76.5 μL L^{-1} of air [2]. Thyme (*T. vulgaris* L.) EO had higher antifungal activity and suppressed pathogen development at concentrations above 15.3 μL L^{-1} of air. At a higher concentration (153 μL L^{-1} of air), thyme EO inhibited *C. acutatum* development on inoculated fruits [2]. Also, strawberry decay caused by *R. stolonifer* and *B. cinerea* was controlled by volatiles of *T. vulgaris* [6,28] with both reports showing differences between thyme genotypes in terms of chemistry and efficacy.

Essential oils have also been tested on grape diseases. In a laboratory study, Walter et al. [26] found that EOs from clove (*S. aromaticum (L.) Merr. & L.M. Perry*), thyme, and massoialactone (derived from the bark of the tree *Cryptocarya massoia* R.Br.) significantly reduced necrotic lesions on grape leaves caused by *B. cinerea* when treated with thyme, and massoialactone. In a separate study, Valero et al. [27] showed that thymol or eugenol (volatile from clove) reduced the loss of sensory quality and microbial spoilage of grape in modified atmosphere storage.

Chitosan packages infused with *O. vulgare* essential oil changed spore and mycelia morphology and inhibited growth of *A. niger*, and *R. stolonifer* spore germination on inoculated grapes in storage [22]. The oils of *Z. officinale*, *O. sanctum* and *P. persica* were also effective in controlling storage rot of grapes during in vivo trials, with treated grapes showing an improved shelf life for up to 6 days [4].

4.3. Effect of Volatiles and Essential Oils on Fruit Quality

Wang et al. [16] found that treatment with thymol and eugenol extended strawberry shelf life and increased fruit free radical scavenging capacity, thereby enhancing resistance to spoilage and deterioration. Essential oil treated fruits were found to have higher amounts of sugars, flavonoids, anthocyanins, organic acid and phenolic compounds. The increase in phenolic content may have led to increased oxygen radical absorbance capacity (ORAC) and decreased fruit spoilage in essential oil treated strawberries [16].

Carvacrol, anethole, and perillaldehyde also enhanced antioxidant activity and increased total anthocyanins, phenolics and hydroxyl radical (•OH) scavenging capacity in blueberry fruit tissues [25]. Chitosan coating with carvacrol and cinnamaldehyde was also found to maintain blueberry firmness and effectively extended blueberry shelf life [30].

Similarly, postharvest essential oil treatments (perillaldehyde, linalool, cinnamaldehyde, cinnamic acid, anethole, and carvacrol) enhanced antioxidant capacity in raspberries with perillaldehyde being the most effective [29].

Grape berries in a modified atmosphere package with eugenol or thymol had lower weight loss, lower changes in skin color, reduced ripening, and lower decay throughout storage [27]. Lower weight loss was also seen in gerbera (*Gerbera jamesonii* cv. Dune) flowers treated with thymol [45].

Sánchez-González et al. [17] tested chitosan-coated packages of grapes alone or in mixture with bergamot essential oil and found chitosan mixed with bergamot oil to be the most effective to control respiration rate, water loss and antimicrobial activity during storage. Chitosan packages infused with *O. vulgare* essential oil also preserved the quality, physical, physiochemical and sensory attributes of grapes in storage [22].

4.4. Mechanism of Fruit Resistance to Fungal Attack Treated by Volatiles and Essential Oils

Plant volatiles play a major role in self-defense. Pe´rez et al. [46] observed that a 25% decrease in (E)-hex-2-enal content during strawberry development and ripening coincides with activation of latent infection, and appearance of visible disease symptoms, leading to extensive fruit damage.

Six and nine-carbon volatiles are produced in plant tissues in response to wounding in LOX (liposygenase) and HPL (hydroperoxide lyase) pathways. In addition to their role in aroma biosynthesis, products of LOX and HPL pathways, have antimicrobial and antifungal activities and may have a role in plant-pathogen interactions. Some have also shown a stimulatory effect on some pathogenic fungi. In *Arabidopsis thaliana*, (E)-hex-2-enal activated several self-defense genes (i.e., lipoxygenase 2 and allene oxide synthase), lignified leaves and accumulated pathogenesis-related proteins (i.e., 3 transcripts, and camalexin), which reveals that volatile treatments stimulate a wound-repair mechanism, and will ultimately act as a physical barrier to the pathogen penetration [47]. Externally applied essential oils improved fruit resistance to fungal attack, mostly due to increased antiproliferative activity and free radical scavenging capacity in strawberries [16]. This has also been reported in kiwifruit (*Actinidia deliciosa* cv. 'Hayward') against *B. cinerea* with the application of grape (cv. Isabella) volatiles which induced mechanisms of resistance in the host [8].

Since plant volatiles inhibit fungal growth and are produced rapidly through the lipoxygenase pathway (in response to wounding), more studies are needed to further clarify their role in plant self-defense systems.

In addition to self-defense and induced resistance, essential oils affect pathogens by contact either in media or as volatiles in storage. Essential oils also lead to the secretion of cell protoplasts without previous cell wall disruption, as shown by the presence of chlamydospores and deformation of cell walls in *B. cinerea* hyphae treated with grape (cv. Isabella) volatiles [8].

The bioactivity of essential oils in the vapor phase is a very attractive characteristic that makes them suitable for use as fumigants in small fruits not suited to aqueous sanitation [16]. Easy application of essential oil vapors during storage makes them more attractive than dipping [48]. Essential oil application could also limit the spread of pathogens by suppressing spore production or reducing spore load on surfaces or in storage. However, the role of volatiles in spore germination and pathogen infectivity needs to be understood in order to develop new techniques for disease control [8].

4.5. Phytotoxicity, Off-Flavor and Off-Odor of Volatiles and Essential Oils on Fresh Produce

Extended exposure to volatiles has resulted in phytotoxicity on strawberry fruits and affected fruit quality [9]. The phytotoxicity of (E)-hex-2-enal has also been described on pears [49], beans [50], and sliced apples [51]. Volatiles 2-nonanone, hexan-1-ol and benzaldehyde slightly damaged strawberry fruit while (Z)-hex-3-en-1-ol and (E)-hex-2-enal caused extensive tissue necrosis [20]. Fallik et al. [52] reported strawberry weight loss and other side effects by (E)-hex-2-enal. Walter et al. [26] reported excessive water loss from detached grape branches (in the laboratory, but not in the field) exposed to thyme and massoialactone, and phytotoxicity on grape flowers in the field. Also, incorporation of bergamot essential oil in chitosan coating resulted in brown-colored grapes during storage [17].

In the authors' preliminary experiments with the application of thymol and carvacrol in air-tight containers, phytotoxicity was observed as increased mushiness and dullness of strawberry fruits after 24 hours (unpublished data). However, Bhaskara Reddy et al. [6] did not observe any visual phytotoxic symptoms after 4 days of fruit exposure to *T. vulgaris* essential oils. Further work is needed to identify and minimize the cause of adverse effects on fruit quality by volatiles [9].

Very few studies have reported on the sensory characteristics of fruits treated with essential oils. Neri et al. [47] reported off-odors in peach and nectarine fruits treated with (E)-hex-2-enal described as a "green" (leafy) and "butyric" aroma. However, off-odors decreased during ripening and nothing was perceived after ripening. No fruit off-odors were observed in peach and nectarine fruits treated with carvacrol or citral. Aloui et al. [53] reported complete absence of off-flavors and off-odors in dates treated with citrus essential oils. Serrano et al. [54] treated cherries with eugenol, thymol or menthol and reported no sensory effects. However, cherries treated with eucalyptol generated off-flavors. Prasad & Stadelbacher [55], reported that acetaldehyde treated strawberries did not show any off-flavor at a concentration of 1%, but showed off-flavor at 4%.

The reviewed articles suggest that most of the essential oils do not leave off-flavors or off-odors on intact fruits at the minimum inhibitory concentrations, especially when stored fruits finish their ripening process. However, there is variation between fruit species and cultivars and each commodity has to be tested individually. For example, Neri et al. [47], reported that stone fruits were more sensitive to (E)-hex-2-enal injury than pome fruits and showed off-flavor, probably because they absorb more (E)-hex-2-enal than pome fruits.

5. Control of Fungal Diseases in the Field

To our knowledge, only one research paper described the application of essential oils in the field. A single application of thyme oil or massoialactone application on grape bunches and leaves reduced *B. cinerea* in the field trial [26]. However, the efficacy of Thyme oil to control bunch rot needs further field evaluations, as floral tissue browning occurred [26].

6. Conclusions

In this review, the efficacy of essential oils (plant-based multi-compounds) and single compound volatiles was discussed in in vitro, in vivo (in storage) and in situ (in the field) conditions. Despite the importance of fresh fruits food safety, there are not many research articles on the effectiveness of essential oils to control bacteria populations on small fruits, which suggests the need for research in this area. Many research articles discussed the positive effects of volatiles and essential oils in in vitro conditions on media-grown fungal populations with a large variation in their efficacy and their minimum inhibitory concentrations. Most of these variations were related to the vapor pressure of volatiles.

The efficacy of multi-compound essential oils is even more variable due to genetic variation, environmental conditions, and synergistic effects of multiple compounds. Major compounds play an essential role in the antimicrobial activity of essential oils, however, minor compounds in the mixture may have synergistic effects. Future studies are needed to explain the role of single natural compounds or their combination in plant-fungal interactions.

(E)-hex-2-enal and hexanal were the most effective volatiles and thyme the most effective essential oil in controlling fungal diseases of small fruits both in vitro and in storage. Thyme oil has the potential as a postharvest treatment to extend shelf life of fresh fruits and vegetable and to replace commercial fungicides or controlled atmosphere storage. However, active packaging and improved formulation techniques are needed to prolong activity, reduce volatility and improve coverage.

Several researchers have mentioned the toxicity of volatiles and essential oils after extended exposure. Research is needed to study the effect of volatiles and essential oils on vegetative and reproductive phases of fungal development, fruit quality (firmness, sensory evaluations), cell wall structure and integrity and their role in plant self-defense and postharvest storage.

Acknowledgments: We would like to thank our colleague, Laban Rutto for the critical review of this manuscript. This article is a contribution of the Virginia State University (VSU), Agricultural Research Station (Journal Article Series Number 353).

References

1.   Centers for Disease Control and Prevention (CDC), National Outbreak Reporting System (NORS). Available online: https://wwwn.cdc.gov/norsdashboard/ (accessed on 4 June 2018).
2.   Duduk, N.; Markovic, T.; Vasic, M.; Duduk, B.; Vico, I.; Obradovic, A. Antifungal Activity of Three Essential Oils against *Colletotrichum acutatum*, the Causal Agent of Strawberry Anthracnose. *J. Essent. Oil Bear. Plant.* **2015**, *18*, 529–537. [CrossRef]
3.   Agrios, G.N. *Significance of plant diseases*; Plant pathology; Academic Press: London, UK, 2000; pp. 25–37.
4.   Tripathi, P.; Dubey, N.K.; Shukla, A.K. Use of some essential oils as post-harvest botanical fungicides in the management of grey mold of grapes caused by *Botrytis cinereal*. *World J. Microbiol. Biotechnol.* **2008**, *24*, 39–46. [CrossRef]

5. Perdones, A.; Sánchez-Gonzáleza, L.; Chiralt, A.; Vargas, M. Effect of chitosan–lemon essential oil coatings on storage-keeping quality of strawberry. *Postharvest Biol. Technol.* **2012**, *70*, 32–41. [CrossRef]

6. Bhaskara Reddy, M.V.; Angers, P.; Gosselin, A.; Arul, J. Characterization and use of essential oil from *Thymus vulgaris* against *Botrytis cinerea* and *Rhizopus stolonifer* in strawberry fruits. *Phytochemistry* **1998**, *47*, 1515–1520. [CrossRef]

7. Daferera, D.J.; Ziogas, B.N.; Polissiou, M.G. The effectiveness of plant essential oils on the growth of *Botrytis cinerea*, *Fusarium* sp., and *Clavibacter michiganeneis* subs. Michiganensis. *Crop Prot.* **2003**, *22*, 39–44. [CrossRef]

8. Kulakiotu, E.K.; Thanassoulopoulos, C.C.; Sfakiotakis, E.M. Postharvest biological control of *Botrytis cinerea* on kiwifruit by volatiles of 'Isabella' grapes. *Phytopathology* **2004**, *94*, 1280–1285. [CrossRef] [PubMed]

9. Arroyo, F.T.; Moreno, J.; Daza, P.; Boianova, L.; Romero, F. Antifungal activity of strawberry fruit volatile compounds against *Colletotrichum acutatum*. *J. Agric. Food Chem.* **2007**, *55*, 5701–5707. [CrossRef] [PubMed]

10. Environmental Working Group (EWG). Dirty Dozen list. Available online: https://www.ewg.org/foodnews/dirty_dozen_list.php (accessed on 9 March 2018).

11. Research Council; Board of Agriculture. *Regulating Pesticides in Food-The Delaney Paradox*; National Academy Press: Washington, DC, USA, 1987.

12. Tzortzakis, N.G. Maintaining postharvest quality of fresh produce with volatile compounds. *Innov. Food Sci. Emerg. Technol.* **2007**, *8*, 111–116. [CrossRef]

13. Tzortzakis, N.G.; Economakis, C.D. Antifungal activity of lemongrass (*Cympopogon citratus* L.) essential oil against key postharvest pathogens. *Innov. Food Sci. Emerg. Technol.* **2007**, *8*, 253–258. [CrossRef]

14. Almenar, E.; Auras, R.; Rubino, M.; Harte, B. A new technique to prevent the main post harvest diseases in berries during storage: Inclusion complexes β-cyclodextrin-hexanal. *Int. J. Food Microbiol.* **2007**, *118*, 164–172. [CrossRef] [PubMed]

15. Hamilton-Kemp, T.R.; Archbold, D.D.; Loughrin, J.H.; Andersen, R.A.; McCracken, C.T.; Collins, R.W. Stimulation and inhibition of fungal pathogens of plants by natural volatile phytochemicals and their analogs. *Curr. Top. Phytochem.* **2000**, *4*, 95–104.

16. Wang, C.Y.; Wang, S.Y.; Yin, J.J.; Parry, J.; Yu, L.L. Enhancing Antioxidant, Antiproliferation, and free radical scavenging activities in strawberries with essential oils. *J. Agric. Food Chem.* **2007**, *55*, 6527–6532. [CrossRef] [PubMed]

17. Sánchez-González, L.; Pastor, C.; Vargas, M.; Chiralt, A.; González-Martínez, C.; Cháfer, M. Effect of hydroxypropylmethylcellulose and chitosan coatings with and without bergamot essential oil on quality and safety of cold-stored grapes. *Postharvest Biol. Technol.* **2011**, *60*, 57–63. [CrossRef]

18. Wilson, C.L.; Solar, J.M.; El Ghaouth, A.; Wisniewski, M.E. Rapid evaluation of plant extracts and essential oils for antifungal activity against *Botrytis cinerea*. *Plant Dis.* **1997**, *81*, 204–210. [CrossRef]

19. Hamilton-Kemp, T.R.; Mccracken, C.T.; Loughrin, J.H.; Andersen, R.A.; Hildebrand, D.F. Effects of some natural volatile compounds on the pathogenic fungi *A. alternata* and *Botrytis cinereal*. *J. Chem. Ecol.* **1992**, *18*, 1083–1091. [CrossRef] [PubMed]

20. Vaughn, S.F.; Spencer, G.F.; Shasha, B.S. Volatile compounds from raspberry and strawberry fruit inhibit postharvest decay fungi. *J. Food Sci.* **1993**, *58*, 793–796. [CrossRef]

21. Combrinck, S.; Regniera, T.; Kamatoub, G.P.P. In vitro activity of eighteen essential oils and some major components against common postharvest fungal pathogens of fruit. *Ind. Crop. Prod.* **2011**, *33*, 344–349. [CrossRef]

22. Santos, N.S.T.; Aguiar, A.J.A.A.; de Oliveira, C.E.V.; de Sales, C.V.; Silva, S.M.; Silva, R.S.; Stamford, T.C.M.; de Souza, E.L. Efficacy of the application of a coating composed of chitosan and *Origanum vulgare* L. essential oil to control *Rhizopus stolonifer* and *Aspergillus niger* in grapes (*Vitis labrusca* L.). *Food Microbiol.* **2012**, *32*, 345–353. [CrossRef] [PubMed]

23. Bouchra, C.; Achouri, M.; Hassani, L.M.I.; Hmamouchi, M. Chemical composition and antifungal activity of essential oils of seven Moroccan Labiate against *Botrytis cinerea* Pers: Fr. *J. Ethnopharmacol.* **2003**, *89*, 165–169. [CrossRef]

24. Archbold, D.D.; Hamilton-Kemp, T.R.; Barth, M.M.; Langlois, B.E. Identifying Natural Volatile Compounds That Control Gray Mold (*Botrytis cinerea*) during Postharvest Storage of Strawberry, Blackberry, and Grape. *J. Agric. Food Chem.* **1997**, *45*, 4032–4037. [CrossRef]

25. Wang, C.Y.; Wang, S.Y.; Chen, C. Increasing antioxidant activity and reducing decay of blueberries by essential oils. *J. Agric. Food Chem.* **2008**, *56*, 3587–3592. [CrossRef] [PubMed]

26. Walter, M.; Jaspers, M.V.; Eade, K.; Frampton, C.M.; Stewart, A. Control of *Botrytis cinerea* in grape using thyme oil. *Australas. Plant Pathol.* **2001**, *30*, 21–25.

27. Valero, D.; Valverde, J.M.; Martı́nez-Romero, D.; Guille´n, F.; Castillo, S.; Serrano, M. The combination of modified atmosphere packaging with eugenol or thymol to maintain quality, safety and functional properties of table grapes. *Postharvest Biol. Technol.* **2006**, *41*, 317–327. [CrossRef]

28. Nabigol, A.; Morshedi, H. Evaluation of the antifungal activity of the Iranian thyme essential oils on the postharvest pathogens of strawberry fruits. *Afr. J. Biotechnol.* **2011**, *10*, 9864–9869.

29. Jin, P.; Wang, S.Y.; Gao, H.; Chen, H.; Zheng, Y.; Wang, C.Y. Effect of cultural system and essential oil treatment on antioxidant capacity in raspberries. *Food Chem.* **2012**, *132*, 399–405. [CrossRef] [PubMed]

30. Sun, X.; Narciso, J.; Wang, Z.; Ference, C.; Bai, J.; Zhou, K. Effects of Chitosan-Essential Oil Coatings on Safety and Quality of Fresh Blueberries. *J. Food Sci.* **2014**, *79*, 955–960. [CrossRef] [PubMed]

31. Calo, J.R.; Crandall, P.G.; Bryan, C.O.; Ricke, S. Essential oils as antimicrobials in food systems—A review. *Food Control.* **2015**, *54*. [CrossRef]

32. Andersen, R.A.; Hamilton-Kemp, T.R.; Hildebrand, D.F.; McCracken, C.T., Jr.; Collins, R.W.; Fleming, P.D. Structure-antifungal activity relationships among volatile C6 and C9 aliphatic aldehydes, ketones, and alcohols. *J. Agric. Food. Chem.* **1994**, *42*, 1563–1568. [CrossRef]

33. Song, J.; Leepipattanawit, R.; Deng, W.; Beaudry, R.M. Hexanal vapor is a natural metabolizable fungicide: Inhibition of fungal activity and enhancement of aroma biosynthesis in apple slices. *J. Am. Soc. Hortic. Sci.* **1996**, *121*, 937–942.

34. Bakkali, F.; Averbeck, S.; Averbeck, D.; Idaomar, M. Biological effects of essential oils—A review. *Food Chem. Taxicol.* **2008**, *46*, 446–475. [CrossRef] [PubMed]

35. Caccioni, D.R.L.; Gardini, F.; Lanciotti, R.; Guerzoni, M.E. Antifungal activity of natural volatile compounds in relation to their vapor pressure. *Sci. Aliments* **1997**, *17*, 21–34.

36. Monzote, L.; Stamberg, W.; Staniek, K.; Gille, L. Toxic effects of carvacrol, caryophyllene oxide, and ascaridole from essential oil of *Chenopodium ambrosioides* on mitochondria. *Toxicol. Appl. Pharmacol.* **2009**, *240*, 337–347. [CrossRef] [PubMed]

37. Gardini, F.; Lanciotti, R.; Caccioni, D.R.L.; Guerzoni, M.E. Antifungal activity of hexanal as dependent on its vapor pressure. *J. Agric. Food Chem.* **1997**, *33*, 50–55. [CrossRef]

38. Inouye, S.; Takizawa, T.; Yamaguchi, H. Antibacterial activity of essential oils and their major constituents against respiratory tract pathogens by gaseous contact. *J. Antimicrob. Chemother.* **2001**, *47*, 565–573. [CrossRef] [PubMed]

39. Chavan, P.S.; Tupe, S.G. Antifungal activity and mechanism of action of carvacrol and thymol against vineyard and wine spoilage yeasts. *Food Control* **2014**, *46*, 115–120. [CrossRef]

40. Sivakumar, P.M.; Kumar, T.M.; Doble, M. Antifungal activity, mechanism and QSAR studies on chalcones. *Chem. Biol. Drug Des.* **2009**, *74*, 68–79. [CrossRef] [PubMed]

41. Rozmer, Z.; Perjési, P. Naturally occurring chalcones and their biological activities. *Phytochem. Rev.* **2016**, *15*, 87–120. [CrossRef]

42. Kurita, N.; Miyaji, M.; Kurane, R.; Takahara, Y.; Ichimura, K. Antifungal activity and molecular orbital energies of aldehyde compounds from oils of higher plants. *Agric. Biol. Chem.* **1979**, *43*, 2365–2371.

43. Babu, K.S.; Li, X.C.; Jacob, M.R.; Zhang, Q.; Khan, S.; Ferreia, D.; Clark, A.M. Synthesis, antifungal activity, and structure-activity relationships of coruscanone A. analogs. *J. Med. Chem.* **2006**, *49*, 7877–7886. [CrossRef] [PubMed]

44. Wedge, D.E.; Galindo, J.C.G.; Macias, F.A. Fungicidal activity of natural and synthetic sesquiterpene lactone analogs. *Phytochemistry* **2000**, *53*, 747–757. [CrossRef]

45. Solgi, M.; Kafi, M.; Taghavi, T.S.; Naderi, R. Essential oils and silver nanoparticles (SNP) as novel agents to extend vase-life of gerbera (*Gerbera jamesonii* cv. 'Dune') flowers. *Postharvest Biol. Technol.* **2009**, *53*, 155–158. [CrossRef]

46. Pe´rez, A.G.; Sanz, C.; Olias, R.; Olias, J.M. Lipoxygenase and hydroperoxide lyase activities in ripening strawberry fruits. *J. Agric. Food Chem.* **1999**, *47*, 249–253. [CrossRef]

47. Neri, F.; Mari, M.; Brigati, S.; Bertolini, P. Fungicidal activity of plant volatile compounds for controlling *Monilinia laxa* in stone fruit. *Plant Dis.* **2007**, *91*, 30–35. [CrossRef]

48. Jobling, J. Essential oils: A new idea for postharvest disease control. *Good Fruit and Vegetables Magazine* **2000**, *11*, 50. Available online: http://www.postharvest.com.au/gfv_oils.pdf (accessed on 3 October 2018).

49. Neri, F.; Mari, M.; Brigati, S. Control de *Penicillium expansum* by plant volatile compounds. *Plant Pathol.* **2006**, *55*, 100–105. [CrossRef]

50. Croft, K.P.C.; Juttner, F.; Slusarenko, A.J. Volatile products of the lipoxygenase pathway evolved from *Phaseolus Vulgaris* (L.) leaves inoculated with *Pseudomonas syringae* pV phaseolicola. *Plant Physiol.* **1993**, *101*, 13–24. [CrossRef] [PubMed]

51. Corbo, M.R.; Lanciotti, R.; Gardini, F.; Sinigaglia, M.; Guerzoni, M.E. Effects of hexanal, trans-2-hexenal and storage temperature on shelf life of fresh sliced apples. *J. Agric. Food Chem.* **2000**, *48*, 2401–2408. [CrossRef] [PubMed]

52. Fallik, E.; Archbold, D.D.; Hamilton-Kemp, T.R.; Clements, A.M.; Collins, R.W.; Barth, M.M. (E)-2-Hexenal both stimulates and inhibits *Botrytis cinerea* growth in vitro and on strawberry fruit in vivo. *J. Am. Soc. Hortic. Sci.* **1998**, *123*, 875–881.

53. Aloui, H.; Khwaldia, K.; Licciardello, F.; Mazzaglia, A.; Muratore, G.; Hamdi, M.; Restucciad, C. Efficacy of the combined application of chitosan and Locust Bean Gum with different citrus essential oils to control postharvest spoilage caused by *Aspergillus flavus* in dates. *Int. J. Food Microbiol.* **2014**, *170*, 21–28. [CrossRef] [PubMed]

54. Serrano, M.; Martinez-Romero, D.; Castillo, S.; Guille, F.; Valero, D. The use of natural antifungal compounds improves the beneficial effect of MAP in sweet cherry storage. *Innov. Food Sci. Emerg. Technol.* **2005**, *6*, 115–123. [CrossRef]

55. Prasad, K.; Stadelbacher, G.J. Effect of acetaldehyde vapor on postharvest decay and market quality of fresh strawberries. *Phytopathology* **1974**, *64*, 948–951. [CrossRef]

MDPI

St. Alban-Anlage 66

4052 Basel

Switzerland

Tel. +41 61 683 77 34

Fax +41 61 302 89 18

www.mdpi.com

Microorganisms Editorial Office

E-mail: microorganisms@mdpi.com

www.mdpi.com/journal/microorganisms

www.ingramcontent.com/pod-product-compliance
Lightning Source LLC
Chambersburg PA
CBHW041216220326
41597CB00033BA/5990